CELESTIAL

ARCANA

Celestial Arcana

Precession, Tarot, and
the Secret Doctrine

Titus Salmon

Copyright © 2017 Titus Salmon
First edition hardback 2017

All rights reserved. No part of this work may be reproduced or utilized in any form or by any means, electronic or mechanical, including *xerography, photocopying, microfilm,* and *recording* or by any information storage system, without permission in writing from the publishers.

Published by
Mandrake of Oxford
P.O. Box 250
Oxford
OX11AP (UK)

Cover design and illustration: Titus Salmon
Interior design and illustrations: Titus Salmon

Contents

Origins and Attributions 1
The Tree of Life 13
The Domicile System and World Ages 29
The Enneagram, Zodiac, and Celestial Sphere 39
The Great Seal 58
The Porta Alchemica 70
I The Magician 78
II The High Priestess 84
III The Empress 102
IV The Emperor 110
V The Hierophant 117
VI The Lovers 124
VII The Chariot 129
VIII Strength 140
IX The Hermit 149
X Wheel of Fortune 155
XI Justice 164
XII The Hanged Man 175
XIII Death 183
XIV Temperance 196
XV The Devil 208
XVI The Tower 242
XVII The Star 251
XVIII The Moon 264
XIX The Sun 270
XX Judgment 280
XXI The World 289
0 The Fool 296
The Minor Arcana 302
A Proposed System 316

Conclusion	340
About the Author	345
Bibliography	346
Index	352

We have to recognize, in a word, that there is no canon of authority in the interpretation of Tarot symbolism. The field is open therefore: it is indeed so open that any one of my readers is free to produce an entirely new explanation, making no appeal to past speculations: but the adventure will be at his or her own risk and peril as to whether they can make it work and thus produce a harmony of interpretation throughout. The sentence to be pronounced on previous attempts is either that they do not work, because of their false analogies, or that the scheme of evolved significance is of no real consequence. There is an explanation of the Trumps Major which obtains throughout the whole series and belongs to the highest order of spiritual truth: it is not occult but mystical; it is not of public communication and belongs to its own Sanctuary. I can say only concerning it that some of the symbols have suffered a pregnant change. Here is the only answer to the question whether there is a deeper meaning in the Trumps Major than is found on their surface. And this leads up to my final point. If anyone feels drawn in these days to the consideration of Tarot symbolism they will do well to select the Trumps Major produced under my supervision by Miss Pamela Colman Smith. I am at liberty to mention these as I have no interest in their sale. If they seek to place upon each individually the highest meaning that may dawn upon them in a mood of reflection, then to combine the messages, modifying their formulation until the whole series moves together in harmony, the result may be something of living value to themselves and therefore true for them. - Arthur Waite.[1]

---

[1] Waite, Arthur Edward. "Great Symbols of the Tarot". 1926. *The Occult Review*. The "Trumps Major" referred to here comprise the Major Arcana, from the popular Rider-Waite Tarot. This deck was designed by Pamela Colman Smith, under the tutelage of Arthur Waite. Although it may be more appropriate to refer to this deck as the Smith-Waite or Waite-Smith Tarot, I have consistently used the original "Rider-Waite" appellation, as various derivative decks with subtle differences have arisen under various names, and I do not wish to unduly confuse matters.

Origins and Attributions

THE ORIGINS OF the Tarot remain a mystery some 600 years after the appearance of the first hand-painted decks in Western Europe.[2] Many theories have been advanced to account for the evolution of its form and iconography, but all remain unproven.[3]

Some suggest the Tarot emerged from the necessity of prehistoric humanity to record the duration of various celestial rhythms in order to survive.[4] Others suppose an Indian origin, where the cards represent different incarnations of Vishnu.[5] Still others have noted the similarity between images in ancient Mayan codices and the Major and Minor Arcana.[6] There is also the Egyptian origin theory,[7] which is perhaps one of the most influential in terms of the development of Western Tarot symbolism in the 18th century to the present. While each of these theories offers a unique perspective, they need not be mutually exclusive.

The late 18th century was the era immediately preceding the French "occult revival,"[8] during which Antoine Court de Gébelin wrote *Monde Primitif*. In the eighth volume, he suggests the origin of the Tarot to be connected with humanity's development of writing. The ancients' desire to preserve and transmit new discoveries led to the creation-by-necessity of ideograms, which they considered to have been invented by the Egyptian god Thoth. These symbols were nothing less than Thoth's original paintings of the gods: the first Tarot, or "Ta-Rosh," which de Gébelin translates as "Pictures of Mercury."[9]

This idea seems to have resonated with various influential esotericists in subsequent generations, the likes of which include Gérard Encausse

---

[2] Kaplan, Stuart. *The Encyclopedia of Tarot*. Volume 1. Page 12.
[3] Ibid. Page 345.
[4] Ibid. Page 12.
[5] Ibid. Page 18.
[6] Ibid. Page 19.
[7] Ibid. Page 12.
[8] Ibid. Page 22.
[9] De Gébelin. *Monde Primitif*. Volume 8. Book 1.

("Papus"), P. D. Ouspensky, Aleister Crowley, Paul Foster Case, and H. P. Blavatsky – who each seem to have implicitly or overtly accepted the theory of Egyptian origin.

Arthur Waite appears to have rejected this notion, and in his preface to A.E. Thierens' *General Book of the Tarot*, he criticizes the tendency for outright invention regarding the history and symbolism of the Tarot, notably taking to task Levi's various treatments of the subject as fantasies.

It is strange, then, that he would implicitly give the nod to Thierens' attribution scheme presented therein. But this may be an overgeneralization, for the only part of the book Waite specifically mentions in his preface is the section where Pisces is attributed to the Hanged Man, which he considered as being one of the more notable interpretations of that trump. His introductory remarks seem to admonish against excessive speculation, while obliquely calling attention to the possibilities inherent within an astrological/zodiacal system of trump attributions.

Thierens gives credit to Papus for having previously ventured into that sphere, but remarks that his correspondences are hopelessly wrong, lacking any logical basis whatsoever.[10] His solution is to match the first 12 trumps in sequence to the 12 astrological houses, attributing the remaining 10 to the Sun, Moon, and planets – inclusive of Uranus, Neptune, and Earth [Table 1].

Waite appends a footnote in his introduction to Thierens' work, in which he states that he would have expected more to have been done at that point with respect to an astrological and astronomical treatment of the Trumps Major vis-à-vis the Hebraic alphabetic structure,[11] but in the same introduction he appears to suggest that such a connection is fallacious.[12]

---

[10] Thierens, A. E. *General Book of the Tarot*. Page 38. <Web. Sacred-texts.com>: "Where Papus, in his book on *The Tarot of the Bohemians* ventures to indicate relationship between the symbols of the Greater Arcana and zodiacal and planetary principles, he is hopelessly wrong however. He does not insist on it nor does he appear to make any use at all of these relations or explain *how* they are to be found out. Logic is entirely absent in this particular side of his renderings."

[11] "Oswald Wirth has a short excursus on Astrology at the end of his work, in which he enumerates the zodiacal implicities allocated to the four elements, but no Tarot connection is suggested. It is rather curious that a study of the Sepher Yetzirah in conjunction with the Tree of Life and the triple marriage effected by Éliphas Lévi has not produced speculations long since on the astronomical and astrological correspondences of the Tarot Trumps." (Footnote from page 11 of Waite's Introduction to A. E. Thierens' *General Book of the Tarot*.)

[12] "De Gébelin was a man of learning at his own period and remained within the circle of facts, actual or supposed, as he saw and read them. His successor [Levi] was a man of extravagant mind, who contemplated past and future alike through a glass of vision, and so beheld all faërie unfold its images. The occult happenings of the past became in the process as much a matter of invention as his own notions. ... He took up the Tarot, and just as a cartomancist shuffles and deals and lays out its picture-symbols for the reading of things to come, so did he divine their past. He adopted the speculations of De Gébelin, and they dilated in his own mind. He dressed up the Trumps Major in Egyptian vestures and affirmed that he

ORIGINS AND ATTRIBUTIONS

Thierens' attribution scheme	
Trump	**Planet/Sign**
I Magician	♈ - Aries
II Priestess	♉ - Taurus
III Empress	♊ - Gemini
IV Emperor	♋ - Cancer
V Hierophant	♌ - Leo
VI Lovers	♍ - Virgo
VII Chariot	♎ - Libra
VIII Justice	♏ - Scorpio
IX Hermit	♐ - Sagittarius
X Fortune	♑ - Capricorn
XI Strength	♒ - Aquarius
XII Hanged Man	♓ - Pisces
XIII Death	♄ - Saturn
XIV Temperance	☿ - Mercury
XV Devil	♂ - Mars
XVI Tower	♅ - Uranus
XVII Star	♀ - Venus
XVIII Moon	☽ - Moon
XIX Sun	☉ - Sun
XX Judgment	♃ - Jupiter
XXI World	♆ - Neptune
0 Fool	⊕ - Earth

[Table 1: Thierens' Attribution Scheme as presented in *General Book of the Tarot*]

Fourteen years after publication of Thierens' *General Book of the Tarot*, Crowley published *The Book of Thoth* (1944), in which a completely different system of astrological correspondences was proposed. Instead of starting from trump I, matching that to the first zodiacal house, and working in sequential fashion through the rest of the cards, he draws upon the astrological characteristics appended to each letter of the Hebraic alphabet, using the resulting sequence as a template from which to organize the Major Arcana

---

had restored the Tarot in its primitive hieroglyphical form. ... The Tree of Life in Kabbalism has 22 Paths by which the Sephiroth or Numerations are connected one with another and late Kabbalism had married these Paths to the 22 Letters of the Hebrew Alphabet. But the Tarot Trumps Major are also 22, and Éliphas Lévi proclaimed another marriage, constituting a Trinity in unity of Cards and Paths and Letters. It has been the joy of all Occult hierophants and their believing disciples through the decades that followed." [From pages 7-8 of Waite's Introduction to A. E. Thierens' *General Book of the Tarot*. Ellipses and bracketed notes added.]

into three categories, corresponding to the 3 elemental "mother" letters, 7 planetary "double" letters, and 12 zodiacal "simple" letters.

There can be little doubt that Crowley derived his basic Major Arcana attribution scheme directly from the zodiacal assignments appended to the 12 "simple" letters of the Hebraic alphabet, as delineated within the *Sefer Yetzirah*. The sequence and attributions of the 12 zodiacal trumps in his deck correspond almost exactly to those of the Yetzirac system; however, the "Master Therion" introduces a double inversion involving four trumps [Table 2].

Hebraic Alphabet		Crowley's attribution scheme	attribution schemes of 5 versions of the *Sefer Yetzirah*				
			Gra	Short	Long	Saadia	Sheirat Yosef
Alef	א	0 Fool – Air - א	Air	Air	Air	Air	---
Beit	ב	I Magician - ☿ - ב	☽	♄	♄	♄	♄
Gimel	ג	II High Priestess - ☽ - ג	♂	♃	♃	♃	☉
Dalet	ד	III Empress - ♀ - ד	☉	♂	♂	♂	☽
Hei	ה	IV Emperor - ♈ - ה,צ	♈	♈	♈	♈	---
Vav	ו	V Hierophant - ♉ - ו	♉	♉	♉	♉	---
Zayin	ז	VI Lovers - ♊ - ז	♊	♊	♊	♊	---
Chet	ח	VII Chariot - ♋ - ח	♋	♋	♋	♋	---
Tet	ט	XI Strength - ♌ - ט	♌	♌	♌	♌	---
Yud	י	IX Hermit - ♍ - י	♍	♍	♍	♍	---
Kaf	כ,ך	X Fortune - ♃ - כ,ך	♀	☉	☉	☉	♂
Lamed	ל	VIII Justice - ♎ - ל	♎	♎	♎	♎	---
Mem	מ,ם	XII Hanged Man – Water - מ,ם	Water	Water	Water	Water	---
Nun	נ,ן	XIII Death - ♏ - נ,ן	♏	♏	♏	♏	---
Samech	ס	XIV Temperance - ♐ - ס	♐	♐	♐	♐	---
Ayin	ע	XV Devil - ♑ - ע	♑	♑	♑	♑	---
Pei	פ,ף	XVI Tower - ♂ - פ,ף	☿	♀	♀	♀	☿
Tzadik	צ,ץ	XVII Star - ♒ - ה	♒	♒	♒	♒	---
Kuf	ק	XVIII Moon - ♓ - ק	♓	♓	♓	♓	---
Reish	ר	XIX Sun - ☉ - ר	♄	☿	☿	☿	♃
Shin	ש	XX Judgment – Fire - ש	Fire	Fire	Fire	Fire	---
Tav	ת	XXI World - ♄ - ת	♃	☽	☽	☽	♀

[Table 2][13]

Here we see Crowley has inverted the alphabetical sequence of the Emperor and Star trumps, as well as the numerical sequence of Strength and Justice. The implicit reason for this is connected to the fact that the

---

[13] NOTE ABOUT TABLE 2: In its original form, the *Sefer Yetzirah* makes no mention of any sort of astro-planetary attribution scheme in relation to the alphabet; it is only with later versions of the text that such a device appears. The 5 attribution schemes of the *Sefer Yetzirah* have been tabulated from Aryeh Kaplan's commentary: *Sefer Yetzirah*. Revised edition. 1997. Redwheel, Weiser, LLC.

ORIGINS AND ATTRIBUTIONS

traditional numbering of the *Tarot de Marseille* inverts the order of Strength and Justice, relative to the rectified Rider-Waite numbering scheme.

Waite says the following regarding his decision to invert the sequence of Strength and Justice:

> For reasons which satisfy myself, this card has been interchanged with that of justice, which is usually numbered eight. As the variation carries nothing with it which will signify to the reader, there is no cause for explanation.[14]

Although he does not specifically say, the main reason seems quite obvious: the iconography of the Strength and Justice cards clearly relate to the constellations of Leo and Libra, respectively, and in order to fall naturally into the Yetzirac zodiacal attribution sequence, such inversion is unavoidable.

Crowley had apparently been thinking along similar lines as early as 1916. Paul Foster Case – his younger American contemporary – makes note of it in his series of articles published in *The Word* (a popular occult magazine, now out of print) from 1916-1917:

> The Hermetic Order of the Golden Dawn transposed the positions of these trumps, [Leo/Strength and Libra/Justice] for the principle figure in Strength is a lion, symbol of Leo, and the woman holding the sword and scales has been the emblem of Libra, as well as justice, for centuries. Aleister Crowley, once a member of the Golden Dawn, follows the same plan in his explanations of the Tarot. A.E. Waite has gone a step farther (in the right direction, I think), and has not only changed the positions of these cards, but has also reversed their numbers, making Strength VIII and Justice XI.[15]

He developed a similar system alongside both Crowley and Waite, accepting Waite's Strength/Justice inversion, but rejecting Crowley's Emperor/Star juxtaposition. His attribution scheme, as tabulated from his later work *The Tarot: A Key to the Wisdom of the Ages* (1947), is given in [Table 3].

Case suggests that his planetary attributions are supported by the Sefer Yetzirah, but this isn't quite true, since none of the official versions of the Kabbalistic document match his ascriptions (see [Table 2] and [Table 3]). Furthermore, the original versions of the Sefer Yetzirah never had any planetary designations. Apparently Case is using a version of the text that was edited by Golden Dawn co-founder S.L. Macgregor Mathers, who changed the planetary attributions from an earlier version.[16]

---

[14] Waite. *The Pictorial Key to the Tarot*. Part II: The Doctrine Behind the Veil. Section 1: The Tarot and Secret Tradition; VIII Strength, or Fortitude.
[15] Case. *The Secret Doctrine of the Tarot*. Chapter 1. Bracketed notes added.
[16] Hulse. *The Key of It All*.

Hebraic Alphabet		Case's attribution scheme as tabulated from *The Secret Doctrine of the Tarot*
Alef	א	0 Fool – Air - ♅
Beit	ב	I Magician - ☿
Gimel	ג	II High Priestess - ☽
Dalet	ד	III Empress - ♀
Hei	ה	IV Emperor - ♈
Vav	ו	V Hierophant - ♉
Zayin	ז	VI Lovers - ♊
Chet	ח	VII Chariot - ♋
Tet	ט	VIII Strength/Lust - ♌
Yud	י	IX Hermit - ♍
Kaf	כ,ך	X Fortune - ♃
Lamed	ל	XI Justice - ♎
Mem	מ,ם	XII Hanged Man – Water - ♆
Nun	נ,ן	XIII Death - ♏
Samech	ס	XIV Temperance - ♐
Ayin	ע	XV Devil - ♑
Pei	פ,ף	XVI Tower - ♂
Tzadik	צ,ץ	XVII Star - ♒
Kuf	ק	XVIII Moon - ♓
Reish	ר	XIX Sun - ☉
Shin	ש	XX Judgment – Fire - ♇
Tav	ת	XXI World - ♄

[Table 3]

For whatever reasons, Crowley did not accept Waite's Strength/Justice inversion. His solution was to introduce an inversion between the Yetzirac attributions of the Emperor and Star trumps, such that instead of *Hei*, the Emperor is ascribed *Tzadik*; and instead of *Tzadik*, the Star is ascribed *Hei*. According to Crowley:

> For "The Star" is referred to Aquarius in the Zodiac, and "The Emperor" to Aries. Now Aries and Aquarius are on each side of Pisces, just as Leo and Libra are on each side of Virgo; that is to say, the correction in the *Book of the Law* gives a perfect symmetry in the zodiacal attribution, just as if a loop were formed at one end of the ellipse to correspond exactly with the existing loop at the other end.
>
> These matters sound rather technical; in fact, they are; but the more one studies the Tarot, the more one perceives the admirable symmetry and perfection of the symbolism. Yet, even to the layman, it ought to be evident that balance and fitness are essential to any perfection, and the elucidation of

ORIGINS AND ATTRIBUTIONS

these two tangles in the last 150 years is undoubtedly a very remarkable phenomenon.[17]

In the first place, it should be evident "even to the layman" that there is no "double loop" in the zodiac [Figure 1]. Furthermore, Crowley does not follow through with his inversion of *Tzadik* and *Hei*, for as we see in [Table 4], although the Emperor is ascribed to *Tzadik*, the corresponding zodiacal attribution is given as Aquarius, when in fact it should – according to Crowley's logic – be Aries. Similarly, the Star should be ascribed to Aquarius, but he attributes Aries to it. This discrepancy is resolved in the actual card designs, but the fact remains that such an inversion is hardly the proof of inspiration from higher dimensional "Secret Chiefs" that Crowley suggests it to be.[18] A more parsimonious solution may have been to accept Waite's Strength/Justice inversion, thereby avoiding transpositions of both the Hebraic alphabetic sequence and the zodiac.

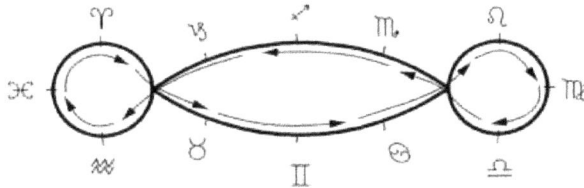

[Figure 1: Diagram of Crowley's "double inversion" in the Zodiac – Author's rendering, based on diagram in *The Book of Thoth*.]

According to Aryeh Kaplan, there are four "important" versions of the Sefer Yetzirah: the so-called Short, Long, Saadia, and Gra variations.[19] The Gra has historically garnered the most support in terms of authenticity as evidenced by consensus of Kabbalistic experts. It is this version that Kaplan focuses on in his commentary, although none of the versions referenced agree with Crowley's or Case's schemes, with regard to the 7 planetary trumps. The supposed reason for such a discrepancy is connected to the "esoteric blind."

Essentially, we are led to infer that of the 22 letters of the Hebraic alphabet, the seven doubles have been obfuscated with respect to their planetary attributions. By implementing the 0 (Fool) = 1 (*Alef*) formula

---

[17] Crowley, Aleister. *The Book of Thoth*. Page 10. York Beach, Maine: Weiser Books, 2000. Print. Originally published in 1944.

[18] From page 10 of *The Book of Thoth*: "Tzaddi is 'The Emperor'; and therefore the positions of XVII and IV must be counterchanged. This attribution is very satisfactory."

"Yes, but it is something a great deal more than satisfactory; it is, to clear thought, the most convincing evidence possible that the *Book of the Law* is a genuine message from the Secret Chiefs."

[19] Kaplan, Aryeh. *Sefer Yetzirah*. Page xxv of Introduction. Revised Edition, published 1997 by Redwheel/Weiser, LLC.

(which Case mentions as early as 1916, and Waite as early as 1923),[20] the Golden Dawn succeeded in massaging the zodiacal trumps into their "rightful" category and order. However, whereas the 7 planetary trumps had been relegated to their appropriate location in the generic sense, the specific ordering of those seven was deemed to be needful of a reorganization.

Hebraic Alphabet		Crowley's attribution scheme as tabulated from *The Book of Thoth*
Alef	א	0 Fool - Air
Beit	ב	I Magician - ☿
Gimel	ג	II High Priestess - ☾
Dalet	ד	III Empress - ♀
Tzadik	צ,ץ	IV Emperor - ♒
Vav	ו	V Hierophant - ♉
Zayin	ז	VI Lovers - ♊
Chet	ח	VII Chariot - ♋
Lamed	ל	XI Justice/Adjustment - ♎
Yud	י	IX Hermit - ♍
Kaf	כ,ך	X Fortune - ♃
Tet	ט	VIII Strength/Lust - ♌
Mem	מ,ם	XII Hanged Man - Water
Nun	נ,ן	XIII Death - ♏
Samech	ס	XIV Temperance - ♐
Ayin	ע	XV Devil - ♑
Pei	פ,ף	XVI Tower - ♂
Hei	ה	XVII Star - ♈
Kuf	ק	XVIII Moon - ♓
Reish	ר	XIX Sun - ☉
Shin	ש	XX Judgment - Fire
Tav	ת	XXI World - ♄

[Table 4: Attribution Key Derived from *The Book of Thoth* (Page 278)]

This concept of the "blind" is perhaps best summed up in the words of Maimonides, a medieval rabbi and philosopher living in Egypt during the early 2nd millennium CE:

> It is sometimes necessary to introduce such metaphysical matter as may partly be disclosed, but must partly be concealed: while, therefore, on one occasion the object which the author has in view may demand that the metaphysical problem be treated as solved in one way, it may be convenient on another

---

[20] See Case's *The Secret Doctrine of the Tarot*, and Waite's Introduction to Knut Stenring's *Sepher Yetzirah* translation.

occasion to treat it as solved in the opposite way. The author must endeavour, by concealing the fact as much as possible, to prevent the uneducated reader from perceiving the contradiction.[21]

According to Marvin Fox, author of *Interpreting Maimonides*, the rationale for using such occult blinds is derived from the fact that

> Rabbinic law... prohibits any direct, public teaching of the secrets of the Torah. One is permitted to teach these only in private to selected students of proven competence...[22]

Fox goes on to theorize that in spite of such rigid prohibition, it was in some cases considered expedient to circumvent the letter of the law in order to impart the truth to those who were able to receive it, albeit in such a way as to

> protect people without a sound scientific and philosophical education from doctrines that they cannot understand and that would only harm them, while making the truths available to students with the proper personal and intellectual preparation.[23]

One is left to wonder what kind of danger the "uninitiated" are being protected from. The possibilities are numerous, and we will not attempt to exhaust them here. The fact is the "blind" is a device that has become entrenched within occult literature. Unfortunately, this provides ample opportunity for self-styled "hierophants" to mislead the credulous: emboldened by the shield of the blind, they rest safe in the knowledge that in the event of objective scrutiny, the caveat of "intentional error" may be employed.

In contrast to the culture of secrecy growing up around the novel interpretations of ancient texts by charismatic fin de siècle occult leaders, one man stood apart – both in his depth of understanding regarding esotericism, and his desire to make such knowledge public. Poet, author, abolitionist, literary critic, lecturer, and amateur Egyptologist, Gerald Massey was known and respected by some of the most influential occultists of the 19th century.

In one of her letters to Massey,[24] Blavatsky praises him as a guru of Egyptology, and self-initiate of the "mysteries," bestowing "glory and honour" upon him, for what she considers to be his own hard-won corroboration of the esoteric doctrine she had purportedly been initiated into.

---

[21] Maimonides. *Guide for the Perplexed*. Page 10. Web <sacred-texts.com>
[22] Fox, Marvin. *Interpreting Maimonides*. Page 5. Chicago: University of Chicago Press, 1990. Ellipses added.
[23] Ibid.
[24] Blavatsky's 1887 letter to Massey; reprinted in the *Agnostic Journal* 10-3-1891.

CELESTIAL ARCANA

For his part, Massey vigorously rejected the secretive nature of esotericism, maintaining that all hidden knowledge should be made public. In his own words:

> I cannot join in the new masquerade and simulation of ancient mysteries manufactured in our time by Theosophists, Hermeneutists, pseudo-Esoterics, and Occultists of various orders, howsoever profound their pretensions. The very essence of all such mysteries as are got up from the refuse leavings of the past is pretense, imposition, and imposture. The only interest I take in the ancient mysteries is in ascertaining how they originated, in verifying their alleged phenomena, in knowing what they meant, on purpose to publish the knowledge as soon and as widely as possible.[25]

The fact that he was the Chosen Chief of the Most Ancient Order of Druids (from 1880-1906) seems – at least superficially – to be fairly contradictory with respect to his purported mission statement. One can only assume that in the event he was constrained by any oaths of secrecy,[26] Massey nonetheless publicly transmitted as much as possible within the given strictures. One of his most notable (yet rarely cited) contributions along these lines may have been his development of the celestial "heptanomis" concept.

According to Massey, the heptanomis was an ancient sevenfold division of the northern circumpolar region of the celestial sphere, with each division being the provenance of a discrete constellation. These seven stations of the pole were imagined as seven celestial mountains by the ancient Egyptians, and were each associated with a particular zootype. At any given moment within the vast 24-26,000 year precessional cycle, the northern celestial pole (NCP) of the earth can be found to point to one of these 7 constellations, or one of the interstitial "dark" regions separating each asterism. Massey theorized that these starless regions corresponded to intervals that were mythologically depicted as great floods or other catastrophic events by the ancients:

> Seven times over in the great year the typical catastrophe occurred. The station of the pole was changed. The island was submerged, the mountain was dislimned. ... In Revelation the heptanomis of seven astronomers is symbolized

---

[25] From Massey's remarks concerning Blavatsky's letter; reprinted in the *Agnostic Journal* 10-3-1891.

[26] It has been suggested that Massey's involvement with the Druidic Order did not involve a formal initiation. If this is the case, it may well be that no strict vows of secrecy were imposed upon him: "A misconception about Massey's religious beliefs stems from his connection with the Most Ancient Order of Druids to which he was elected Chosen Chief, an honorary position that he held from 1880 until 1906. The position might have involved some minor administrative duties, but it required no formal membership. To Massey, at least, it was not a religion and did not involve forms of initiation, ceremonial dress or attendance at active meetings at megalithic sites." Web. <gerald-massey.org.uk/massey/>

ORIGINS AND ATTRIBUTIONS

by the book of judgment sealed with seven seals. Seven seals are broken for the opening of the book. Seven angels sound upon seven trumpets. Seven thunders utter their voices. Seven plagues are loosed by the seven angels from the seven bowls of the wrath of God. Seven kings are overthrown, and seven mountains pass away, at this the final judgment of the great harlot [i.e., body of Draco] ... by those whose learning came to be unintelligibly interpreted and unintelligently abused by the ignorant fanatics of a later religious cult.

At the end of each three thousand seven hundred years in the cycle of precession the polestar changed, or, as represented, a star fell from heaven. Thus, when the second angel sounded, a mountain (one of the seven) sank down flaming to be quenched in the celestial sea. This was one of the seven mountains upon which the ancient harlot sat. At the same time a great star fell from heaven, which was one of the seven polestars. When the fifth angel sounded another polestar fell. The fall of the total seven has not been followed out one by one in stars. But the fall or wreck of the heptanomis piecemeal has been otherwise described; Enoch saw it as seven blazing mountains overthrown. Seven types of the over-toppling mount or station of the pole may be assigned approximately: (1) to the mount of the hippopotamus (or northern crown); (2) to the mount of the dragon; (3) the mount of the ape; (4) the mount of the jackal (or dog); (5) the mount of the bird (cygnus); (6) the mount of the tortoise (or lyra); and (7) the mountain of mankind.[27]

The significance of this fundamental thesis cannot be overemphasized, and as will be demonstrated throughout this book, Massey's theory of the heptanomis [Figure 2] contains a wealth of information that can be applied to the symbols of the Tarot.

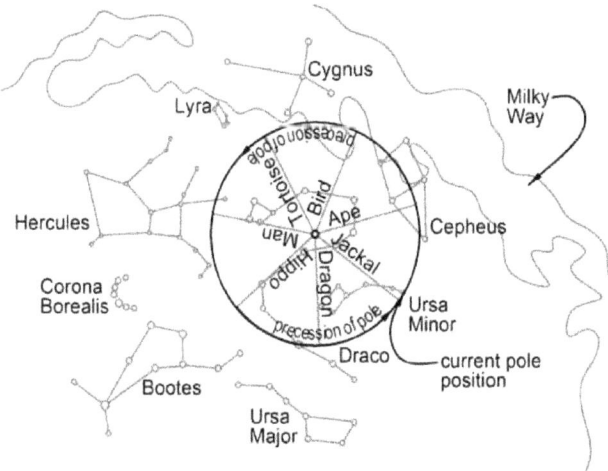

[Figure 2: Circumpolar heptanomis as seen from earth]

---

[27] Massey. *Ancient Egypt, the Light of the World.* Book 11. Page 701. <masseiana.org>. Bracketed comments and ellipses added.

CELESTIAL ARCANA

In the next several chapters, the celestial sphere and its structural components are demonstrated to be essential keys that can be used to unlock much of the hidden symbolism behind various esoteric constructs and pantacles, such as the Kabbalistic Tree of Life, the Nordic Yggdrasil, the Enneagram, the Great Seal, and the Porta Alchemica. The results of these investigations are then applied to the Major and Minor Arcana, and a new attribution system is proposed. This system is not intended to replace or "disprove" those preceding it, rather it is one that stands upon its own merits, and its implications may offer some insight into the rationale underlying previous systems, some of the original uses of the Tarot, as well as the so-called Secret Doctrine of Aeonic succession.

The Tree of Life

THE MAIN HEBRAIC mystical texts that will be investigated in an attempt to understand the method and motive behind the Yetzirac attributions conferred to the Major Arcana are the Sefer Yetzirah, Sefer Bahir, and Sefer Zohar.

Although the origins of these works are obscure, it has been suggested that their authorship can be traced to the 1st-3rd centuries CE, based on various considerations, such as writing style.[28,29,30] However, these texts were (and presumably still are) considered by some Jewish mystics and scholars to have been written much earlier than this, with the Sefer Yetzirah being attributed to Abraham himself.

As discussed previously, one of the inherencies in any study of early Hebraic mystical texts is that of the esoteric blind. Anyone who seriously reads the available texts with the aim of reconstructing the various cosmological forms – such as the Tree of Life, and its relationship to the cube of space – inevitably encounters some insurmountable barrier presented by one or several logical inconsistencies.

One is able to achieve a general understanding of the structures being described, but upon examination of the particulars, it becomes apparent that no single form can account for the various elemental contradictions. Exceptions to the rule must always be invoked in order for any particular structure to maintain logical coherence. Not even the best commentaries offer any real solution to this dilemma, nor should we expect them to, since they are all written under the constraints of rabbinic law. In fact, it is probably the case that there is no available printed version of any of these texts

---

[28] Benton, Christopher P. "An Introduction to the Sepher Yetzirah". Pages 1-2. Web. <maqom.com/journal/paper14.pdf>

[29] "The Bahir was first published around 1176 by the Provence school of Kabbalists, and was circulated to a limited audience in manuscript form. […] Most Kabbalists ascribe the Bahir to Rabbi Nehuniah ben HaKana, a Talmudic sage of the first century, and the leading mystic of his generation." Kaplan, Aryeh. *The Bahir: Illumination*. Weiser. 1979. Kindle Edition.

[30] Miller, Moshe. "The Zohar's Mysterious Origins". Web. <chabad.org/kabbalah/article_cdo/aid/380410/jewish/The-Zohars-Mysterious-Origins.htm>

(commentaries included) that has not been intentionally compromised. One is thus left to the impossible task of rectifying the contradictions as best one can.

The central cosmological structure of the Kabbalistic texts is the Tree of Life, or *Etz Chaim* [Figure 1]. The concept of the Tree of Life can be found in virtually every world culture, although the Hebraic version is perhaps the one most thoroughly developed in the West. While the details vary amongst the different cultures regarding this structure, there are several elements common to all versions that suggest an essential universality of cosmological function.

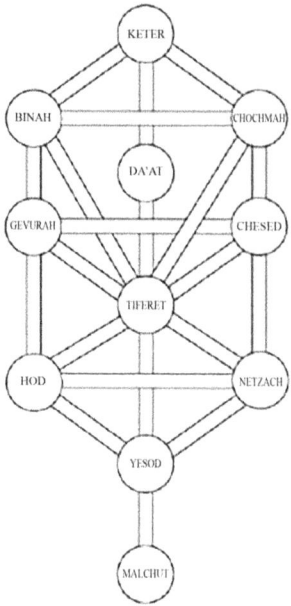

[Figure 1: One form of the Hebraic Tree of Life (*Etz Chaim*). Author's rendering (based on the "Kircher Tree" of 1652).]

Some of the most visually striking versions of the world tree can be found in the various 19th century depictions of the Nordic Yggdrasil, which have been derived from the poetic and prose Eddas. Examination of these illustrations [Figures 2 & 3] reveals numerous parallels with the celestial sphere, and its attendant functional components.

The trunk of the tree can be seen to correspond to the ecliptic axis, with the Midgard Serpent representing the ecliptic itself [Figure 4]. The rainbow bridge (Bifrost) is variously described as spanning Asgard and Midgard, or Asgard and the lower realms. In the Heine illustration, Bifrost is shown to terminate on either side of the Midgard ecliptic, but in the Bagge version, the bridge extends down into the realms below.

THE TREE OF LIFE

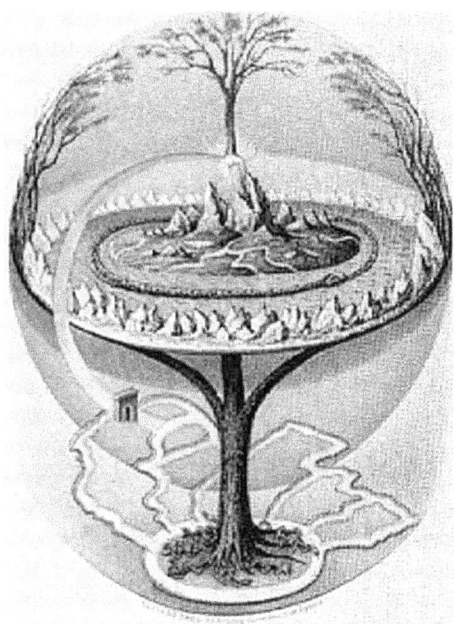

[Figure 2. Painting by Oluf Olufsen Bagge, 1847. From *Northern Antiquities*, by Paul Henri Mallet. Image source: Wikipedia.]

[Figure 3. Author's rendering of illustration by Friedrich Wilhelm Heine, 1886. From *Asgard and the Gods,* by Wilhelm Wägner. London: Swan Sonnenschein, Le Bas & Lowrey. Page 27.]

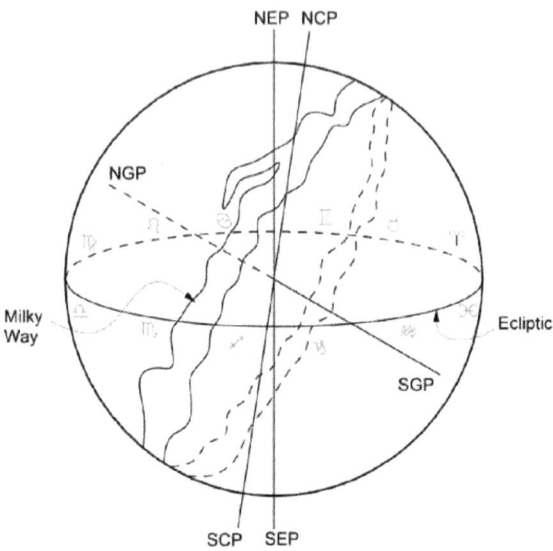

[Figure 4: The Celestial Sphere, oriented similarly to Bagge's illustration of Yggdrasil]

It is possible that Bagge is trying to indicate a connection between Bifrost and another bridge, Gjallarbru, which is said to span two rivers of the Nordic underworld. Regardless of the fine details of whether Bifrost was seen by the Norse to extend below Midgard or not, the resemblance of this bridge to the arch of the Milky Way, with its apex near the north ecliptic pole (NEP) is evident [Figure 5]. In addition, this nebulosity with respect to the "infernal regions" of the southern portion of the celestial sphere is to be expected from any culture originating so far north, as the majority of constellations south of the ecliptic can never be seen, or if so, only for short periods during various portions of the year.

By 1928 (and presumably much earlier), the concept of a parallelism between the sefirotic array on the Kabbalistic Tree of Life, and the nine Nordic worlds connected via Yggdrasil was being made. According to Manly P. Hall:

> The Scandinavian world-tree, Yggdrasil, supports on its branches nine spheres or worlds, which the Egyptians symbolized by the nine stamens of the persea or avocado. All of these are enclosed within the mysterious tenth sphere or cosmic egg – the definitionless Cipher of the Mysteries. The Kabbalistic tree of the Jews also consists of nine branches, or worlds, emanating from the First Cause or Crown, which surrounds its emanations as the shell surrounds the egg.[31]

---

[31] Hall, Manly P. *The Secret Teachings of All Ages.* Page 94. Web. <sacred-texts.com>

THE TREE OF LIFE

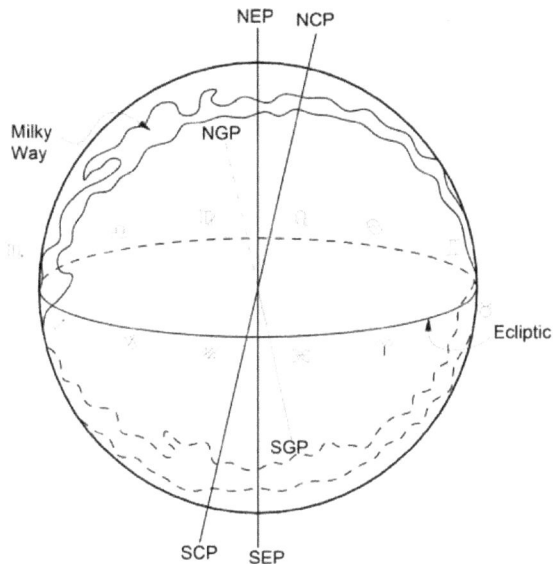

[Figure 5: The Celestial Sphere, oriented similarly to Heine's illustration of Yggdrasil]

This concept is illustrated in [Figure 6]. Here we essentially have a diagram of the celestial sphere, and if we apply the same rationale to any of the standard models of the Hebraic Tree of Life, the general congruencies become apparent.

The correlation between the central pillar of 4 sefirot and either the ecliptic or celestial axis appears to be borne out by various passages in the Hebraic source texts of Kabbalistic doctrine. It should be noted, however, that nowhere in any of the early Kabbalistic texts do we have a definitive statement as to whether the sefirot are even connected to the various branches, or "functionaries" of the Tree of Life, much less their specific configuration upon it.

In Section IV ("The Ten Sefirot") of the Bahir, we are told that the sefirot ("emanations") are so named, "because it is written (Psalm 19:2) 'the heavens declare (me-Saprim) the glory of God.'" This glory is subsequently broken down into a triad of "holies," the first of which is equated to the "highest crown," the second to the "root of the tree," and "the third holy is attached and unified in them all" (verse 128). The first and second "holy" clearly stand in a vertical relationship to each other, and it is therefore logical to place the third "holy" midway between the two, in order to portray its unifying nature in the most symmetrical fashion [Figure 7]. Whether each of these three "holies" should be ascribed to any particular sefirah is never specifically addressed, although the implication seems to have been codified within much of the derivative Kabbalistic literature stemming from the source texts.

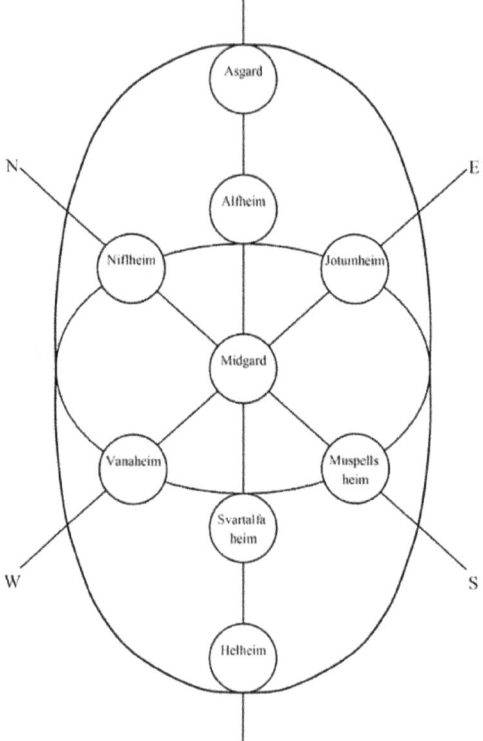

[Figure 6: "The Nine Worlds of the Odinic Mysteries." From Manly P. Hall's *The Secret Teachings of All Ages*.[32] Author's rendering.]

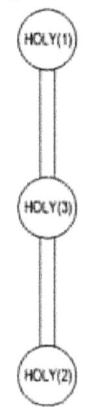

[Figure 7: Postulated relationship amongst the triplicity of "holies" that comprise the "glory of God," as presented in the Bahir.]

---

[32] Ibid. Page 28.

THE TREE OF LIFE

Thus, we see the third "holy" as corresponding to the central nexus of Tiferet, as depicted in the standard hermetic trees based upon the Kircher tree of 1652. Comparing these schematic Hebraic and Nordic trees, we see that Tiferet would correspond to Midgard, the geocentric core of the cosmological World Egg. That Crowley would identify this sefirah with the Sun is reasonable in an oblique sense, for further on in the Bahir, we find this sefirah referred to as the throne of the Blessed Holy One [v. 152]. This is the intersection of the ecliptic and celestial axes. As the ecliptic is predominantly solar in nature, and the celestial axis essentially terrestrial,[33] their intersection symbolizes the relationship between the solar ("divine") and terrestrial ("human"): it is the throne (i.e., physical seat) from which the celestial motions are observed as a testament to the will of the divine.

In verse 95, we are given a general schematic of the cube of space, within which is contained "the axis, the sphere, and the heart," as well as the Tree of Life. The logical inference is that the Tree of Life represents the various interconnections amongst these three unique components (axis, sphere, heart) placed within the cube of space [Figures 8 & 9]. Each of these components is said to possess 12 "functionaries." The resultant 36 functionaries are divided into 4 sets of 9, corresponding to the cardinal directions. This recalls the decan system of ancient Egypt, wherein the ecliptic was divided into 36 equal units of 10° each.

In his Astrotheology Lectures, Manly Palmer Hall speculates as to the significance and structure of this arrangement:

> From a standpoint of astrotheology, we have to assume that the ancients recognized the 12 signs of the zodiac, and the great northern and southern constellations, forming altogether 36 major patterns in the sky; that the ancients represented these or recognized them as the channels through which the great universal energies beyond man's conception flowed down into the mystery of the solar system, and into the mystery of his own corporeal life. Thus these constellations came to have certain meanings; certain distinct and special attributes, and it was believed that as the ancient Jewish people held that there were 12 tribes of Israel (Israel being an ancient generic term simply to signify the world, humanity, and kinds of life) that there were 12 orders of life that were nourished and fed by the energy which fell through the constellations through the zodiacal band and that these 12 were further differentiated by the northern and southern constellations until 36 forms ... could be recognized.[34]

---

[33] Further on, we will uncover another relationship between the ecliptic and celestial axes, where it is suggested that the former relates to Sirius, and the latter to the Sun.
[34] Hall, Manly P. *Astrotheology Lectures*. Web. <manlyphall.org/audio/astrotheology-2/astrotheology-part-2-of-5/>. 0:36:24-0:37:44. Ellipses added.

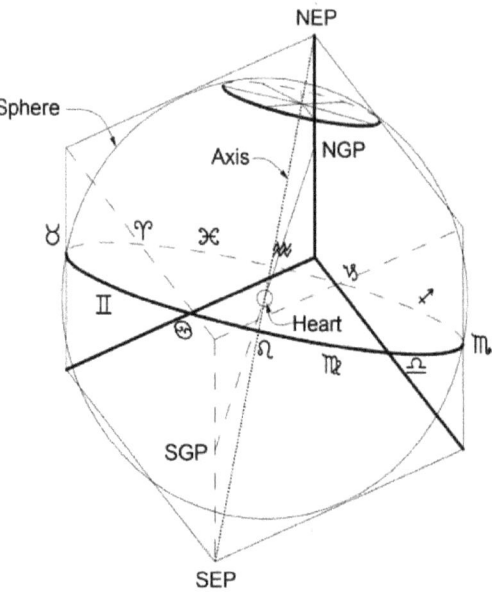

[Figure 8: Author's interpretation of the Cube of Space, Axis, Sphere, and Heart, as described in the Sefer Bahir.]

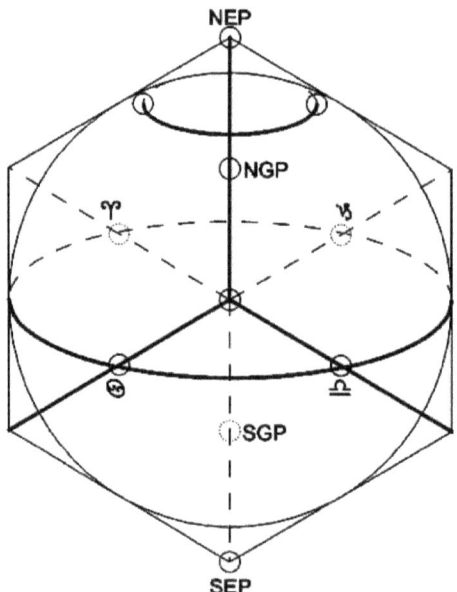

[Figure 9: Author's interpretation of one form of the Tree of Life, based on descriptions in the Sefer Bahir, Sefer Zohar, and Sefer Yetzirah.]

THE TREE OF LIFE

In addition, the Bahir states that there is another set of 36 functionaries placed above the first set, referencing a biblical verse in Ecclesiastes, which states that "one above another watches." Kaplan gives Ecc. 5:7 for this verse, but in the King James Version, it is actually 5:8: "If thou seest the oppression of the poor, and the violent perverting of judgment and justice in a province, marvel not at the matter: for he that is higher than the highest regardeth; and there be higher than they." The word used to describe these "watchers," or "officials" is derived from the Hebrew "gabah," which translates as "haughty," "high," "lofty," "taller," etc. If we postulate "gabah" in this context to reference the northern circumpolar region of the celestial sphere, the circle defined by the precessional motion of the NCP around the NEP presents itself as an obvious candidate for the location of this second "lofty" set of 36 functionaries [Figure 10].

Further on in verse 95 of the Bahir, however, it is suggested that there are actually 64 functionaries, with 8 "watching" above them. 7 of these "watchers" are ascribed to the classical planets, whereas the 8th appears to refer to a centrally located region described as the "throne of the Blessed Holy One," the "precious stone," and the "sea of wisdom." If we modify our original model accordingly, we arrive at the following structure, wherein the ecliptic axis is the central functionary, the circumpolar heptanomis completes the 8, who "above the others watch," and the remaining 64 are arranged equally along the ecliptic [Figure 11].

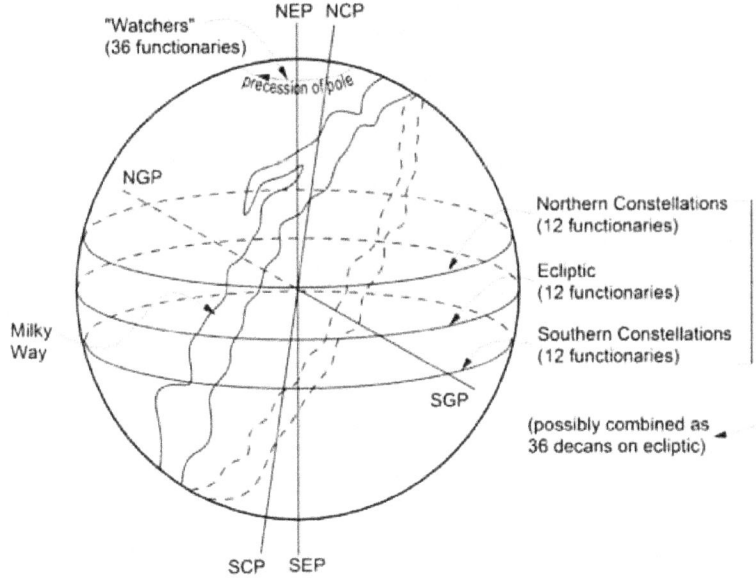

[Figure 10]

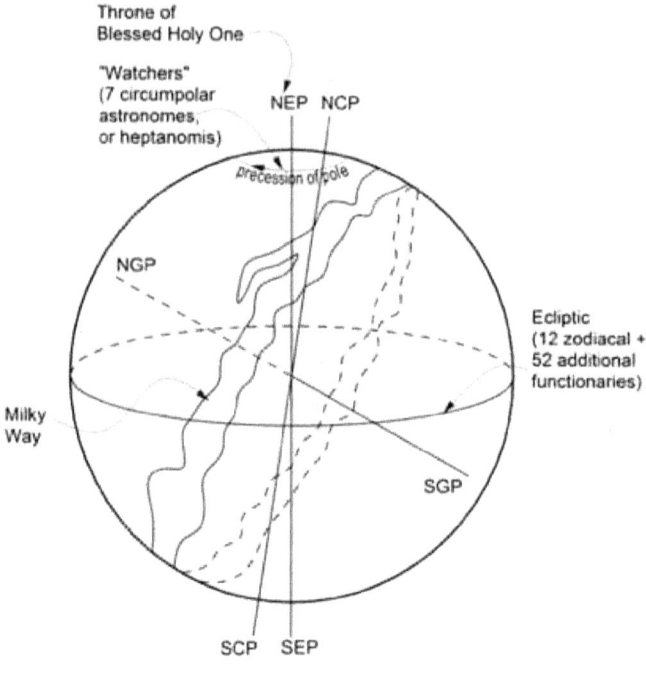

[Figure 11]

The locations of these heptanomic and zodiacal circles upon the celestial sphere is suggestive of the bronze vessel of Solomon, as illustrated in "The Lesser Key of Solomon," a grimoire compiled from British Museum source material by Crowley and S.L Macgregor Mathers [Figure 12]. According to Mathers and Crowley, this is the vessel Solomon was said to have constructed in order to contain the 72 spirits of the shemhamphorasch (or more properly Shem HaMephorash, "the full/explicit name of God"), which he subsequently invoked in order to construct his Temple.

[Figure 12: The Bronze Vessel of Solomon, as depicted in *The Lesser Key of Solomon*. (Page 51– Figure 158). Author's rendering.]

THE TREE OF LIFE

Another interesting corollary to this arrangement is that if we subtract the 12 zodiacal functionaries, we are left with 52, which happens to be the number of cards in the standard playing decks today. These 52 cards correspond to the same number of minor arcana in the Tarot, minus an additional 4 that appear there variously as the 4 suits of either a "knave," or "princess" court card.

As noted previously, the Bahir seems to indicate that the 7 planetary functionaries are to be associated with the northern circumpolar region corresponding with the Masseain heptanomis, with the eighth member of the "lofty" functionaries residing in the central region of the ecliptic pole. A very similar arrangement is made reference to in the Sefer Yetzirah, chapter IV verse 3:

> The three are one, and that One stands above. The seven are divided; three are over against three, and one stands between the triads. The twelve stand as in warfare; three are friends, three are enemies; three are lifegivers; three are destroyers...One above three, Three above Seven, and Seven above Twelve: and all are connected the one with the other.

If we organize our planetary functionaries around the central axis of the ecliptic according to the classical (Ptolemaic) domicile system,[35] the following arrangement presents itself, as shown in [Figure 13].

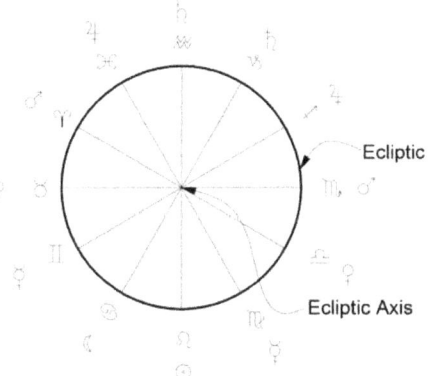

[Figure 13]

This configuration relates to the circumpolar heptanomis in the following way: As the vernal locus of the Sun moves in a retrograde fashion through the zodiac via the phenomenon of precession, there is a concomitant motion of the NCP through the circumpolar region of the heptanomis. Thus, for any

---

[35] The *Tetrabiblos* of Ptolemy is the earliest known reference to this system, although it is presumably much older, perhaps originating with the Babylonians, or Egyptians.

discrete precessional location along the ecliptic (i.e., the solar vernal point, or SVP), there is a corresponding point within the heptanomic circle that represents the celestial pole of the earth [Figure 14].

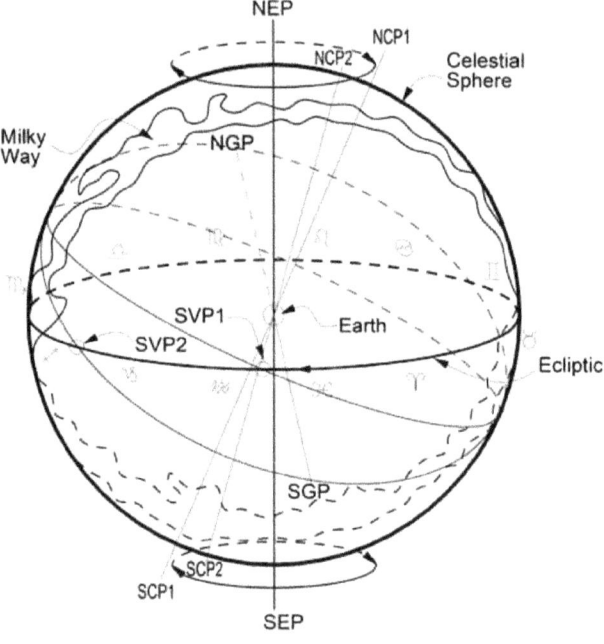

[Figure 14]

In book 9 of Massey's *Ancient Egypt, the Light of the World*, reference is made to a 6-fold division of this circumpolar region, presided over by the following asterisms: Hercules ("mankind"), Draco ("dragon"), Ursa Minor ("jackal"), Cepheus ("ape"), Cygnus ("bird"), and Lyra ("tortoise"). Massey subsequently posits a seventh circumpolar locus represented by Corona Borealis as "…the key-stone … to the conical mount of heaven [that] was first laid in the heptanomis as primary polestar of the seven which formed the circle of the crown…"[36]

Dividing the roughly 24,000-26,000 year precessional cycle by a factor of 7, we arrive at a figure of approximately 3,600 years, which corresponds to the time interval between each of these 7 circumpolar loci. [Table 1] shows the relationship between the circumpolar and ecliptic divisions of the precessional cycle. There is some degree of variance here, as the Hindu value for the duration of the Great Year differs from the Ptolemaic by a factor of 1,920 years (Hindu=24,000, Ptolemaic=25,920).

---

[36] Massey. *Ancient Egypt, the Light of the World.* Book 9. Pages 601-602. Brackets and ellipses added.

THE TREE OF LIFE

Aquarius	2100-4200 AD	0-3600 AD	Ursa Minor
Pisces	0-2100 AD		
Aries	2100-0 BC	3600-0 BC	Draco
Taurus	4200-2100 BC		
Gemini	6300-4200 BC	7200-3600 BC	Bootes
Cancer	8400-6300 BC	10800-7200 BC	Corona Borealis
Leo	10500-8400 BC		Hercules
Virgo	12600-10500 BC	14400-10800 BC	Lyra
Libra	14700-12600 BC		
Scorpio	16800-14700 BC	18000-14400 BC	Cygnus
Sagittarius	18900-16800 BC		
Capricorn	21000-18900 BC	21600-18000 BC	Cepheus
Aquarius	23100-21000 BC		

Relationship Between Circumpolar and Ecliptic Divisions of Precessional Cycle*

*Values approximate and based on a 25,200 year precessional cycle. This yields 2100 years/zodiacal sign (25,200/12), and 3600 years/pole station (25,200/7)

[Table 1: Relationship between circumpolar and ecliptic divisions of the precessional cycle]

Several possible models for dividing the circumpolar region of the precessional circle present themselves, but we confine ourselves to two for the present: one based on a six-fold division, the other on a sevenfold division [Figures 15 & 16]. In both cases, we see that the stipulation implied by the statement "three are over against three, and one stands between the triads" is met, in the sense that if we consider the solar element of the precessional circle to be central, three of the stations to one side are considered nocturnal rulers (Moon, Venus, Jupiter) of their respective signs, whereas the remaining triad (Mercury, Mars, Saturn) are considered diurnal rulers.

Such an arrangement calls to mind the system of Enneagram planetary or "essence" types introduced by Rodney Collin in *Theory of Celestial Influence*, wherein the Moon, Venus, and Jupiter are considered as feminine "types," while Mercury, Mars, and Saturn are masculine. The origins of this system have never been definitively verified, although as Collin was a student of Ouspensky, who was in turn one of Gurdjieff's foremost disciples, it seems reasonable to posit the latter as the ultimate source. Gurdjieff himself indicated that his own understanding of the Enneagram was developed subsequent to his initiation into the Sarmoung Brotherhood, an esoteric Sufi order purportedly based in central Asia. We shall see, however, that the words of Gurdjieff are by no means to be taken at their face value, and that even in those supposedly "objectively unbiased" instances of discourse on matters esoteric, it is an inevitable – and apparently even mandatory – stipulation that the grist for the acolyte must at all costs be "fortified" by some degree of obfuscation.

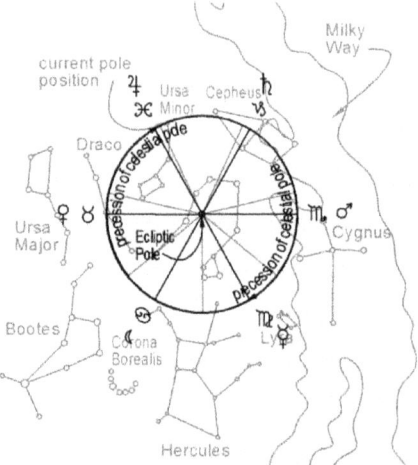

[Figure 15: 6-Fold Division of Circumpolar Precessional Circle]

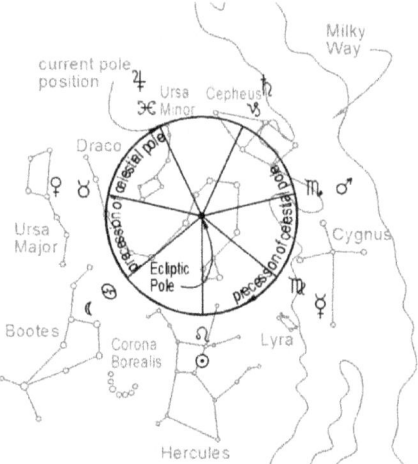

[Figure 16: 7-Fold Division of Circumpolar Precessional Circle]

A more prosaic explanation is that Gurdjieff developed his own system of astrological and planetary correspondences via perusal of the various sources of occult literature available to him during the late 19th century. First of all, he was by no means indisposed towards lying to those "asleep," in order to achieve his own aims, and furthermore, he unquestionably spent a good deal of time studying the occult literature of his time. He had certainly extensively perused Blavatsky's bulky tomes, and admitted to expending an enormous amount of effort in order to "track down the erroneous statements in…*The*

THE TREE OF LIFE

Secret Doctrine, and delighted to tell a completely mythical story of how its author had fallen in love with him."[37] Although he was far too cunning and evasive to ever reveal his true sources, it is noteworthy that the 17th century astrologer and herbalist Doctor Nicholas Culpeper (1616-1654 CE) presents an analogous scheme of complementarity between planetary types as what was subsequently elucidated by Rodney Collin in *Theory of Celestial Influence*.

It is unclear whether either Gurdjieff or Ouspensky personally sanctioned the specific planetary type correspondences that Collin presents in his book; nonetheless, he supposedly received much of the information contained therein through some means of channeling beyond the grave from Ouspensky,[38] who in turn had been psychically linked to *his* teacher, Gurdjieff.[39] The implication here of course is that there was some unbroken yet deliberately occluded chain of information transmission from teacher to student, and therefore Collin's pairing of Enneagram essence types [Figure 17] ultimately derives from a system espoused by Gurdjieff. This may or may not be the case; the least that can be said is that Collin's arrangement is almost identical to Culpeper's complementarity, as implied in the following passage from his *Complete Herbal*:

> You may oppose diseases by Herbs of the planet opposite to the planet that causes them: as diseases of Jupiter by Herbs of Mercury, and on the contrary; diseases of the luminaries [i.e. Sun and Moon] by the Herbs of Saturn, and the contrary; diseases of Mars by Herbs of Venus and the contrary.[40]

The Tree of Life can thus be seen as an occult representation of the celestial sphere, the trunk of which symbolizes the ecliptic axis. This interpretation is supported by a consideration of the Sefer Yetzirah, Bahir, and Zohar, as well as certain representations of the Nordic world tree, Yggdrasil. In this context, the various sefirot are considered to represent key functional components of the celestial sphere, relating to the heptanomis, the zodiac, and the ecliptic, celestial, and galactic axes [Figures 8 & 9].

The 7 stations of the heptanomis are related to the 7 classical planets according to the Aeonic rulership associated with each station of the celestial pole, with respect to the Ptolemaic domicile system. It therefore follows that these stations can also be connected to the 7 "blinded" planetary trumps; a concept that will be explored further on in this book. The complementarity of

---

[37] Webb, James. *The Harmonious Circle*. New York: G. P. Putnam's Sons, 1980. Print. Page 36.

[38] Ibid. Page 488.

[39] Ouspensky, P. D. *In Search of the Miraculous*. New York: Harcourt, Brace & World, Inc., 1949. Page 262.

[40] Culpeper, Nicholas. *Complete Herbal*. Chapter entitled "Original Epistle to the Reader". Bracketed comments added.

the heptanomic arrangement of these Aeonic planetary rulers is similar to that of the Enneagram "essence type" arrangement: in both cases, the solar element is central, with the remaining 6 planets being arranged in a bilaterally symmetrical fashion according to nocturnal/feminine and diurnal/masculine types. We will explore at greater length how the Enneagram is related to the celestial sphere, the zodiac – and by implication, the Tree of Life – in the chapter entitled "The Enneagram, Zodiac, and Celestial Sphere."

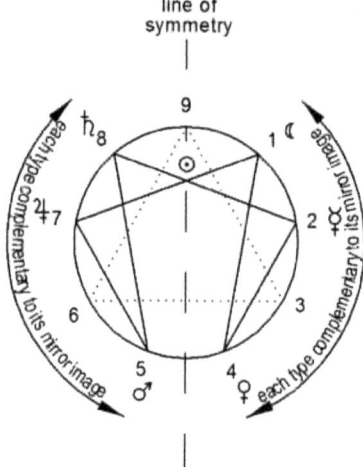

[Figure 17: Complementarity of Planetary Types per Rodney Collin.]

The Domicile System and World Ages

THE CLASSICAL PLANETARY domicile (or house) system ostensibly originated with Ptolemy via his *Tetrabiblos*, written circa 145-168 CE,[41] although the particular rulerships ascribed therein suggest that the source can be traced back to an Egyptian origin as early as sometime around 4000-2000 BCE, during the Ages of Taurus and Aries.

In Chapter 20 ("Houses of the Planets") of his astrological master work, Ptolemy begins by ascribing the Sun and Moon to Leo and Cancer, respectively, as these are the largest and brightest luminaries visible from Earth, and therefore according to a vague sort of logic of similitude, are to be most fittingly associated with the two constellations that "approach nearer than the other signs to the zenith"; i.e., the summer solstitial constellations. During Ptolemy's era (100-170 CE; the Age of Pisces), said constellations were in fact Gemini and Cancer, owing to the phenomenon of precession; a detail to be further expounded upon shortly.

Continuing his rationale of analogies, Ptolemy next ascribes Saturn to the winter signs of (tropical) Capricorn and Aquarius, as Saturn is the most distant classical planet from the Earth, associated with "coldness," and considered an astrological "malefic." He subsequently works inward from the outer circle of Saturn, applying similar logic to each successive planetary sphere from Jupiter to Mercury in a bilaterally symmetrical fashion, such that each planet (excluding the Sun and Moon) is ascribed rulership to two opposite signs – one considered a diurnal house; the other nocturnal.

The logic for such a house system is questionable. Additionally, it can hardly be expected that any accuracy can be attributed to an astrological system which has for the past 2,500 years (more or less) failed to account for precessional drift. In so doing, the whole of tropical astrology has been thrown into a confusing morass of semantics from which there is no logical way out other than what amounts to an arbitrary fixing of the zodiacal houses

---

[41] The actual date the *Tetrabiblos* was written is unclear, but it is known that it was written after the *Amalgest* (145 CE), and Ptolemy died in 168 CE.

with the vernal point at 0° Aries (even though it is now currently at approximately 5° Pisces), and arguing for some abstract notion of houses that is completely at odds with the current incontestable sidereal facts.[42]

The most that can be said for the tropical system is that it adheres to whatever unchanging seasonal effects of declination the various celestial bodies may have upon terrestrial phenomena. In other words, no matter what constellation the Sun may truly be in during the summer (which considered tropically is Cancer, but in reality – i.e. sidereally – is Gemini) it will always be a fact that this is the season in which the angle of incidence of that luminary's radiations are the most perpendicular to the Earth's atmosphere [Figure 1].

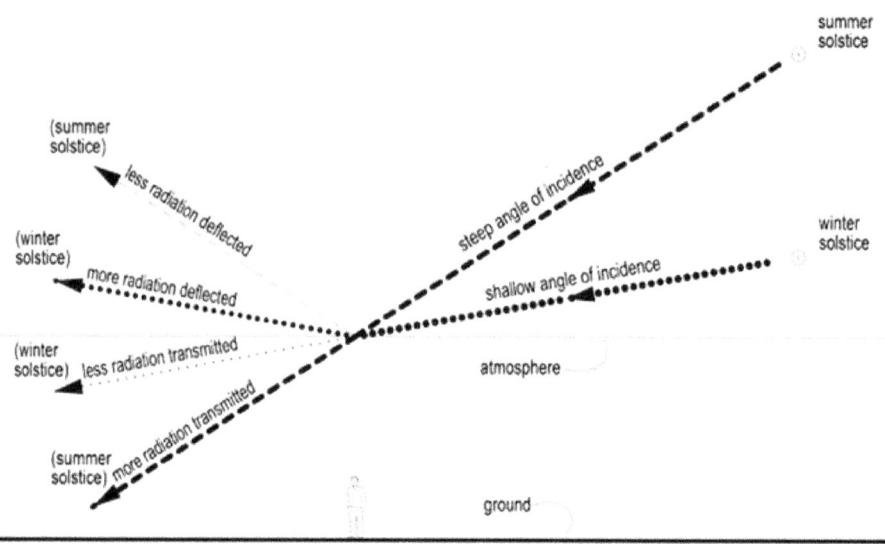

[Figure 1: Relationship between the angle of incidence of celestial radiations, and the degree to which they are absorbed into – or deflected by – Earth's atmosphere.]

Although there must undoubtedly be great significance to such declinational effects, it must be conceded that the tropical system can no longer lay legitimate claim to zodiacal characterizations of Sun signs that have any basis in a reality other than the arbitrarily frozen anachronisms of the Age of Aries, to which may have been added throughout the course of the past 2,500 years legitimate psychological attributes based upon more or less

---

[42] This does not exclude the *possibility* of a scientifically verifiable Astrology. Both the Sun and Moon exert incontrovertible effects on the behavior and biorhythms of all terrestrial life, and one does not have to move too far theoretically from these facts to postulate some veritable system of astrology that accounts for the other luminaries, as well.

THE DOMICILE SYSTEM AND WORLD AGES

objective observations. This approach can only lead to a hopeless admixture of old and new astrological traits, and the only way out is through a basic overhaul of such traits based on rigorous observation and appropriate application of the results, according to a sidereal system that never loses sight of the precessional variable.

In the blunt words of Irish siderealist Cyril Fagan:

> One wonders how many tropical astrologers fully realize that the tropical versions of the zodiac were all conceived and born in ERROR? No amount of sophistry, theo or otherwise, can metamorphose a blatant blunder into a truism ...
> If astrologers really understood the genesis of the tropical zodiacs and all their implication, they would drop them immediately as they would a viper.[43]

The question necessarily arises as to whether there is thus any need for a redefining of the houses of the planets to account for the effects of precession. The Ptolemaic house system appears to derive from the Age of Taurus, wherein the summer solstice (and hence the house of the Sun) truly corresponded with the constellation Leo, and the winter solstice (corresponding to the cold, distant Saturn) with Aquarius. The following passage from Massey's *The Natural Genesis* serves to illustrate in part the concept of the precessional shifting of the solstitial loci, and appears to suggest the concomitant necessity of a periodic redefining of the planetary houses:

> It is not improbable that some astronomer in the future who masters the mythological astronomy of the past, will discover that one form of the imaginary Meru, the inverted cone or sugarloaf [Figure 2], the "Lotus of Immensity" as it was called, is finally a figure of the circle of precession; the reversed cone or sugarloaf shape which is described in space by the axis or pole of the earth in the course of 25,868 years.
>
> Meru is also described as being intersected by six parallel ranges running east and west. In the Puranas, Meru or Jambu-Dvipa is encompassed about by six other Dvipas and seven oceans. In like manner the Chinese "posterior heaven" assigned to King Wan was represented by the hexagonal figure; the Hebrew קצוות שש or space in six directions. The Yi King (Book of Changes), consisting of sixty-four hexagrams, is related to this change. Such figures may have become mere arithmetical puzzles where their primary significance has been lost, but they did not originate as intentional enigmas. These six parallel ranges across Meru the present writer takes to represent the six divisions through which the planets were considered to file, seven as six (compare the seven-headed dragon that became six-headed). In the language of astrology, which was the ancient astronomy, each of the seven had one house on either

---

[43] Fagan, Cyril; Firebrace, Roy. *Primer of Sidereal Astrology*. Tempe, AZ: American Federation of Astrologers, 1971. Print. Page 136. Ellipses added.

side of the zodiac, excepting the Sun and Moon; these were the male and female of light, now reckoned as two aspects of the biune one. Thus –

Aquarius	Capricorn	Sagittarius	Scorpio	Libra	Virgo
[Saturn]	[Jupiter]	[Mars]	[Venus]	[Mercury]	[Sun]
WINTER SOLSTICE				SUMMER SOLSTICE	
[Saturn]	[Jupiter]	[Mars]	[Venus]	[Mercury]	[Moon]
Pisces	Aries	Taurus	Gemini	Cancer	Leo

Among the Egyptian coins of Antoninus Pius there is a series of twelve belonging to the eighth year of his reign (AD 146), which shows that the house of the Sun was then in the sign of the Lion and that of the Moon in the Crab; Mercury's double-house was in the Twins and Virgin; that of Venus in the Bull and Scales; Mars in the Ram and Scorpion; Jupiter in the Fishes and Archer; Saturn in the Sea-goat and Aquarius, thus –

Seagoat	Archer	Scorpion	Scales	Virgin	Lion
[Saturn]	[Jupiter]	[Mars]	[Venus]	[Mercury]	[Sun]
WINTER SOLSTICE				SUMMER SOLSTICE	
[Saturn]	[Jupiter]	[Mars]	[Venus]	[Mercury]	[Moon]
Waterman	Fishes	Ram	Bull	Twins	Crab

The sign of the solstice had changed. For these coins were struck at the end of a Sothiac cycle when the reckonings of the vague year were once more readjusted to the fixed year.[44]

The Sothiac (Sothic) cycle here refers to the ancient Egyptian system of timekeeping wherein the beginning of each year was marked by the heliacal rising of Sirius. During the Age of Taurus, this occurred during the summer months – corresponding to our own June, July, and August – and the summer Sun was to be found roughly in the constellations of Cancer, Leo, and Virgo, respectively.

As each civil year in the Egyptian calendar contained exactly 365 days (the "vague year" Massey mentions), as opposed to the 365.25 days in a Sothic year (Massey's "fixed year"), every 4 civil years, the heliacal rising of Sirius was observed to occur one day "later" than the previous year. It was thus observed that the civil and Sothic years diverged more and more, reaching their maximum difference of 182.5 days every 730 civil years, finally coming back into precise alignment after 1,460 years, or a so-called Sothic cycle [Figure 3].

If during the Age of Taurus the house of the Sun was considered to be in the constellation Leo by virtue of its solstitial locus, by the time of Antoninus Pius (146 CE), that same solar maximum elevation would have shifted some 2 signs to roughly the constellation Gemini, and in our present Age, it is to be found between Taurus and Gemini, precisely within the grasp of the upstretched arm of Orion.

---

[44] Massey, Gerald. *The Natural Genesis*. Book 9. Pages 53-54. Bracketed comment added.

THE DOMICILE SYSTEM AND WORLD AGES

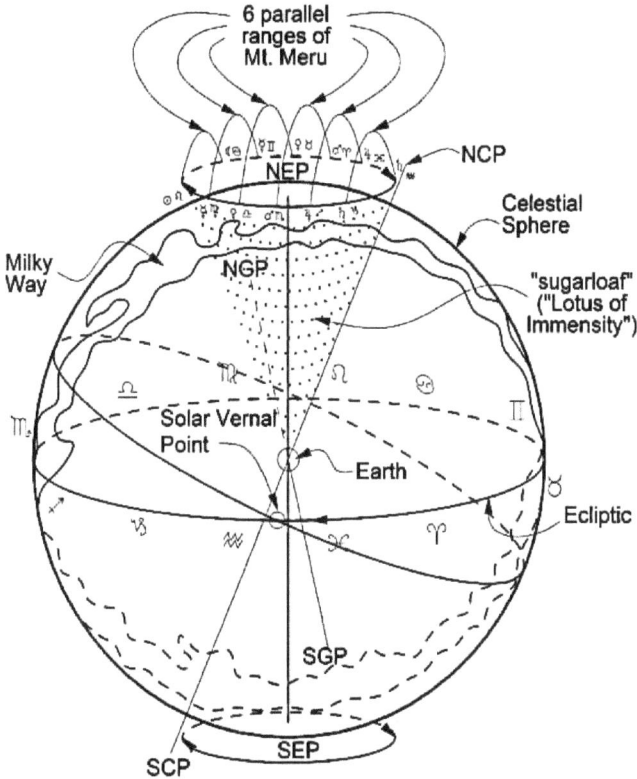

[Figure 2: Author's Interpretation of Mount Meru Based on Massey's description.]

[Figure 3: The 1,460 year Sothic Cycle.]

Framed in this way, Massey's planetary house and solstice diagram for 146 CE is a sign or so off, but this is of no real consequence to the discussion at hand. The overall suggestion appears to be that the planetary houses shift according to precession, which would give the following order of rulership for our current Age of Pisces/Aquarius:

Scorpio	Libra	Virgo	Leo	Cancer	Gemini
[Saturn]	[Jupiter]	[Mars]	[Venus]	[Mercury]	[Sun]
WINTER SOLSTICE				SUMMER SOLSTICE	
[Saturn]	[Jupiter]	[Mars]	[Venus]	[Mercury]	[Moon]
Sagittarius	Capricorn	Aquarius	Pisces	Aries	Taurus

Even though Ptolemy's own rationale for his domicile system should hardly be convincing to modern eyes, and analysis of his attributions of individual stars within each constellation to specific planets – based on a purported similitude of effects – will clearly show that such effects taken together can never equate to the house rulership system he proposes, if one were to accept Ptolemy's arrangement – in the interests of symmetry and antiquarian deference – there is in fact a way in which to frame such a system wherein it is not at all necessary to shift the rulerships according to the inexorable march of precession, and that is by expanding the seasonal connotations of the domiciles to embrace the cycle of the Great Year itself.

The Hindu model of Yugas as presented by Sri Yukteswar,[45] and the Greek concept of the 4 (or 5) Great Ages are both precessionally-based concepts or doctrines, which suggest that this 24-26,000 year cycle can be conceived of as analogous to the yearly cycle of the Earth's revolution about the Sun, with the seasonal changes at the smaller timescale corresponding to variations in – amongst other things – the level of consciousness or spiritual awareness present in humanity as a whole. The Great Year itself thus can be seen to have its own 4 seasons of Summer, Fall, Winter, and Spring.

In Yukteswar's system, the Summer of the Great Year – wherein the consciousness of humanity is at its highest naturally occurring level, people in general enjoy a state of relative spiritual enlightenment, conditions on Earth are more or less peaceful, etc. – occurs during the Age of Virgo, i.e. when the vernal equinoctial colure is to be found in that constellation.

The Golden Age of the Platonic Great Year appears to be centered in the Age of Cancer, Leo, or Virgo, depending on the sources one uses, and the inferences drawn from them. Massey suggests the Golden Age to have its primary locus in vernal point Leo, and this seems to be the Age many occult

---

[45] See Swami Sri Yukteswar's *The Holy Science*. First published in 1894, CE.

and New Age authors have gravitated towards for various reasons in order to frame their own doctrines of lost gnosis, past advanced civilizations, etc.

According to Yukteswar, the Hindu conception of the cyclical rise and fall of civilizations stems from the idea that our Sun orbits some other celestial body with a periodicity of 24,000 years, and that during the darkest Ages, this body is positioned between the Sun and the Galactic Center, effectively blocking the enlightening radiations emanating from it, whereas in the diametrically opposed Golden Age, no such obstruction exists, and said emanations are capable of being focused in some way by our Sun upon the Earth [Figure 4].[46]

If we extend this concept further to note that during the Ages of Virgo, Leo, and Cancer, the Galactic Center – localized roughly between Sagittarius and Scorpio – occupies the summer solstice region of the ecliptic, and thus (similar to the Sun in that season) presumably achieves a greater "radiational impingement" due to a higher angle of incidence more perpendicular to the atmosphere, it becomes clear how it may be considered fitting to conceive of those three Ages as the Summer Season of the Great Year.

Applying Ptolemy's planetary house system to this larger time cycle thus provides a framework within which it is possible to maintain a fixed rulership; not in spite of, but by virtue of the very mechanism that had originally called it into question.[47] Unless noted otherwise, any future references to planetary house rulerships in this book should be taken to imply just such a framework.

It may understandably be objected that the celestial mechanism thus proposed to explain the notion of "seasons" of the Great Year is wholly contingent upon latitude, and therefore entirely biased towards a Northern Hemisphere perspective. Notwithstanding the fact that we can certainly make no pretensions towards anything remotely approaching an empirical scientific approach towards astrology, we must concede that such an objection may not be without merit, while suggesting that in the end it may be a moot point.

In the first place – yes, such a mechanism does on the surface appear to have a "northern bias," but the very same must also be said concerning Ptolemy's own house system. In other words, if the Sun is the anachronistic ruler of Leo by virtue of it at one time having been in that constellation during the summer solstice, this could only be true for the Northern Hemisphere, as in the Southern Hemisphere, the very same conditions (Sun in Leo) would have applied to that region's winter solstice. Using Ptolemy's

---

[46] Swami Sri Yukteswar, *The Holy Science*. Introduction.

[47] It is purely speculative, but Ptolemy's fixing of the vernal point at 0° Aries can be interpreted as an occult reference to the conception of precessional World Ages – with a Golden Age (i.e. the "Summer Season" of the precessional cycle) occurring during the passage of the vernal point through the constellation Cancer.

logic, we should therefore have a situation in which the Sun ruled Leo in the Northern Hemisphere, and Saturn ruled it in the Southern Hemisphere.

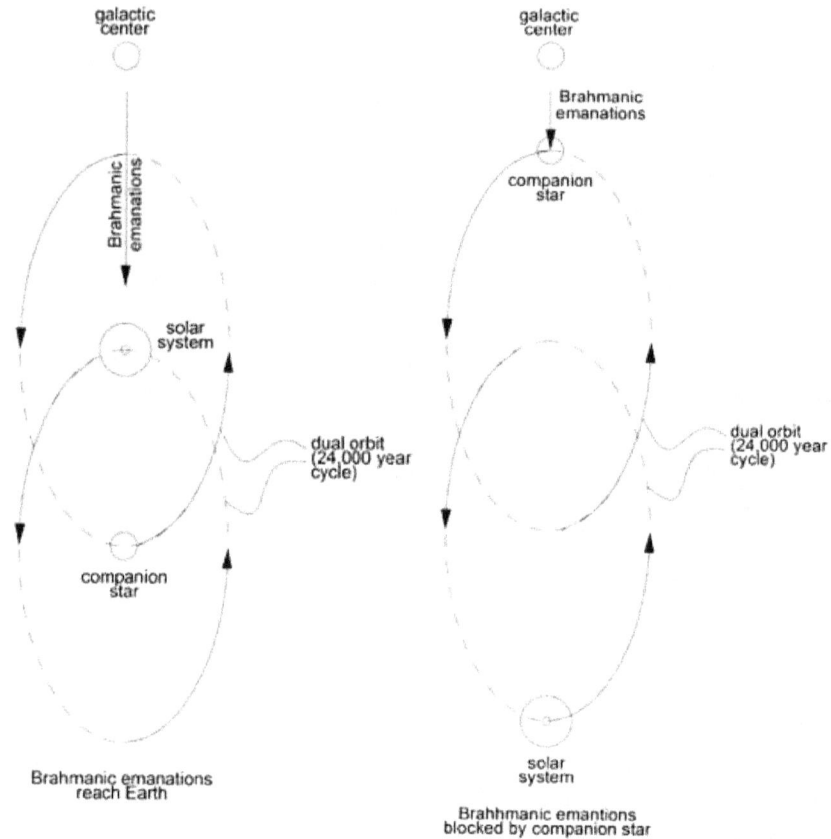

[Figure 4: Effect of the position of our Sun's "dual" on the transmission of enlightening radiations to Earth. Based on Yukteswar's model as described in *The Holy Science*.]

Similarly, the "seasonal" variations associated with the various Ages of the Great Year would also be constrained to a sort of "hemispheric bias" if the declination of the Galactic Center as focused onto Earth by some type of lensing effect from the Sun was the prime variable in determining the relative level of nascent spiritual awareness, in a global sense. This would suggest that we should expect those living in the Southern Hemisphere to be currently enjoying an Age of naturally occurring spiritual enlightenment, whereas those of us in the Northern Hemisphere are entrenched in the depths of a spiritual Dark Age.

Clearly the situation is more nuanced than this. In the first place, if there is any veracity to Yukteswar's Yuga theory, it must also be taken into account.

THE DOMICILE SYSTEM AND WORLD AGES

Additionally, there is a notion of a kind of "esoteric inversion" prevalent within occult literature wherein what is exoterically considered to be the current Age is in fact the very opposite of the "true" and "esoteric" Age. This concept will be explored further elsewhere, but in the present case it should be noted that the mechanism of the Great Year's "seasonal cycles," taken in combination with Yukteswar's Yuga model provides a richer framework within which to view this so-called "esoteric inversion" than either of the two concepts taken alone.

In the case of Yukteswar's model, the situation is relatively straightforward, and globally uniform. In other words, the Dark Ages are caused by the hypothetical that the Sun's "dual" blocks the Galactic emanations during the precessional segment of time where it stands between the Galactic Center and the Sun. 12,000 years later, at the diametrically opposed locus of the Sun's orbit, its dual is no longer positioned between it and the Galactic Center, thus allowing unobstructed free transmissions, resulting in a global "Golden Age."

If we consider this Golden Age to occur as the vernal equinoctial colure passes through Virgo, Leo, and Cancer – for the sake of simplicity considering it to be centered in Leo – it will become apparent that as we are now entering (or according to some sources, have already entered) the Aquarian Age, it must necessarily follow that we are, relatively speaking, in the midst of the darker Ages, if not the darkest.

According to the aforementioned "inversion formula," however, we are actually occultly entering the Age of Leo. Realizing the fact that during the Age of Aquarius, the Galactic Center achieves its ecliptic apex, or greatest angle of declination in the Northern Hemisphere during the summer solstice *at night*, this strange inversion where the exoteric diurnal sign is esoterically equated with its opposite and nocturnal sign begins to take on a more nuanced meaning. At least two layers of celestial influence can be seen at work: at one level, we have the global effects of the position of the Sun's "dual" in relation to it and the Galactic Center – is it inhibiting or permitting free transmission from the Seat of Brahma? Overlaying this global effect is the localized seasonal effect of the angle of incidence of the Brahmanic "logos" upon any particular terrestrial coordinate, and whether or not it is being focused by the lens of the Sun.

In our current Age, the "global" layer is characterized by the Galactic emanations being somewhat hindered by the Sun's orbital companion, whereas the "localized" layer connotes a solar lensing of the most potent angling of these radiations in the Southern Hemisphere. The Northern Hemisphere receives its solar focusing of the Brahmanic logos during the winter solstice, a time in which the angle of incidence of the radiations are at their lowest and thus least potent; however, nocturnally – during the summer solstice, said radiations are at their highest angle of incidence, and thus penetrate most effectively, albeit without the solar lensing effect. If the diurnal lensing effect is seen as influencing the outwardly visible, readily manifest (exoteric) side of life, the nocturnal condition may be considered as acting upon a more subconscious level that is comparatively invisible, yet just as potent.

As far as the effects of geographic location on the various qualities of celestial emanations reaching a particular person or culture within any given timeframe are concerned – whether one lives in the Northern or Southern Hemisphere – the situation is made infinitely more complex with our modern ease of travel and communication. Be that as it may, it is an inescapable fact that all of these things of which we speak – the Tarot; Eastern and Western astrology; the Hindu, Greek, and even Mesoamerican concepts of World Ages – are framed within the context of the Northern Hemisphere, and it is hardly any use to try and wrench ourselves free from this perspective when it informs the very thing we are attempting to investigate. Even if the questionable assertion that the Tarot originated in ancient Egypt is accepted, it must be realized that we are still dealing with a culture some 30° north of the equator.

There is yet another celestial influence in addition to that of the Galactic Center/Seat of Brahma that operates at the "localized," i.e. hemispherically biased layer, which perhaps should be mentioned at this point: that of Sirius.

The importance of this star in occult literature cannot be denied, although it is often the case that many of the allusions to it are for whatever reason deliberately occluded. We will not at this point address the various occult notions of the possible significance this luminary may have regarding the consciousness and evolution of humanity, but here confine ourselves to the observation that during our present Age, Sirius can be considered to exert its maximum influence in the Northern Hemisphere through a diurnal solar lensing effect during the summer months. Conversely, in the Southern Hemisphere, its maximum effects are realized nocturnally in the months of November, December, and January.

The Enneagram, Zodiac, and Celestial Sphere

IN *THE HARMONIOUS Circle*, James Webb presents a hypothesis that Gurdjieff's Enneagram is a composite affair, derived through various cullings from the esoteric literature that surfaced during the so-called Occult Revival of the Victorian Era.[48] He suggests it to be quite probable that at one level the Enneagram is intended to be a representation of the Kabbalistic Tree of Life, wherein the apex of the "informing triangle" corresponds to *Keter*, and the two loci forming the triangle's base correspond to *Chochmah* and *Binah*, respectively. *Malchut*, the tenth sefirot at the base of the Kabbalistic tree is posited to represent the bounding circle of the entire Enneagram, and the remaining 6 points are correlated as depicted in [Figure 1].[49]

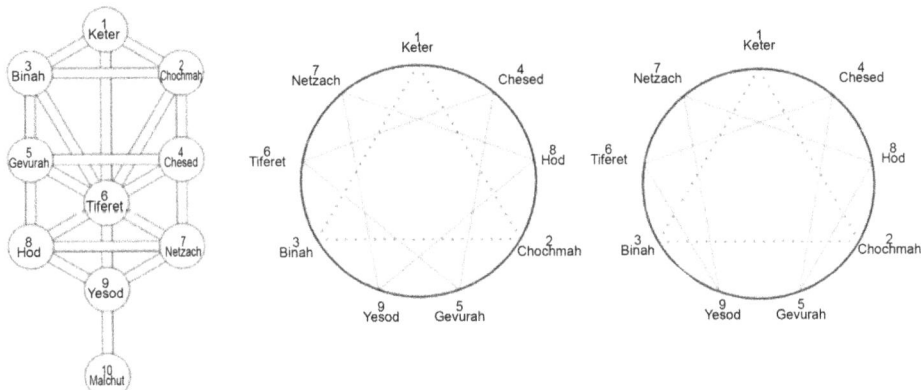

[Figure 1: Correlations between the Kabbalistic Tree of Life and the Enneagram, per James Webb. Author's rendering.]

The accuracy and significance of such a correspondence is a tentative and nebulous matter, but it can hardly be expected that one should be able to "nail down" a hard and fast rule set pertaining to the putative correlation in

---

[48] Webb, James. *The Harmonious Circle*. Page 531.
[49] Ibid. Page 516.

light of the rabbinic law mandating the use of the occult blind. Moreover, given his penchant for "burying the dog deeper,"[50] it is reasonable to assume that Gurdjieff also felt compelled to utilize such devices.

Nonetheless, if we stick to the general spirit of what has so far been revealed in our analysis of the Kabbalistic tree – namely that it is among other things an encoded diagram of the celestial sphere – and following Webb's lead, posit a similar case with respect to the Enneagram, we obtain some interesting results.

In his book *In Search of the Miraculous*, Ouspensky makes reference to a particular form of the Enneagram that Gurdjieff drew:

> There was yet another drawing of the Enneagram which was made under his direction in Constantinople in the year 1920. In this drawing inside the Enneagram were shown the four beasts of the apocalypse – the bull, the lion, the man, and the eagle – and with them a dove.[51]

Although Ouspensky provides no pictorial illustration for this version of the Enneagram, it seems likely that it must have looked similar to the version as depicted on one of the flyers for Gurdjieff's "Institute for the Harmonious Development of Man," as produced by one of his students, Alexander de Salzmann [Figure 2].

Here we have a tableau in which the "universal symbol"[52] is flanked by a demon and an angel with the 4 beasts of the apocalypse confined within the so-called "informing triangle," or "free trinity"[53] of the glyph. Astrologically speaking, these four beasts refer to the fixed zodiacal signs of the zodiac: Taurus (bull), Leo (lion), Aquarius (man), and Scorpio (eagle). The bounding circle of the Enneagram – sometimes referred to as the "circle of time" – is represented by the ouroboros, which is suggestive of the ecliptic.

All of these elements together are quite clearly indicative of an intended zodiacal reading of this particular version of the symbol, but how is one to explain the "illogical" composition of the four fixed signs, which one would expect to be arranged in an equidistant quadrangle along the perimeter of the ecliptic/ouroboros? What we see instead is a triangular arrangement, with the eagle's head (Scorpio) forming the apex, the bull (Taurus) and lion (Leo) forming the base vertices. The human or Aquarian element occupies the upper central crossing point, which Rodney Collin has ascribed to the Sun.[54]

---

[50] Bennet, J. G. *Gurdjieff: Making a New World*. Page 274.
[51] Ouspensky. *In Search of the Miraculous*. Page 295.
[52] See Ouspensky. *In Search of the Miraculous*. Page 294.
[53] Ibid. Page 289.
[54] Collin, Rodney. *Theory of Celestial Influence*. Page 90.

THE ENNEAGRAM, ZODIAC, AND CELESTIAL SPHERE

There are several ways to explain this arrangement. In the first case, if one refers to the celestial sphere as oriented in [Figure 3] – such that the region of the Galactic Center (Sagittarius/Scorpio) corresponds to the informing triangle's apex, and the point of crossing roughly to that of the NEP – it is apparent that the dove's location is essentially congruent with the NCP [Figure 4]. Through the transparency of the sphere, this polar locus is seen to be overlaid upon the constellation Columba ("the dove"), as well as Sirius – which according to Massey, is typified by the bennu-bird, phoenix, and dove.[55,56,57,58,59]

[Figure 2: Author's rendering of de Salzmann's flyer design for Gurdjieff's "Institute."]

---

[55] Massey. *A Book of the Beginnings*. Section 3. Page 133: "The dove was also a type of the genetrix, and bears her name of Tef."

[56] Ibid. Section 7. Page 308: "The Great Mother in her ancient type of the dove (Columbine [Columba]) and the Ancient of Days, the old father or pantaloon; the clown and harlequin are the two brothers Horus, the clown, kar-nu (Eg. inferior type) is the elder or child Horus, and harlequin is har, the younger, the spiritual type; har of the resurrection with the power of becoming invisible, or a spirit among mortals." Bracketed comment added.

[57] Massey. *Ancient Egypt, the Light of the World*. Book 9. Page 553: "It is also said to Osiris, 'Thy two sisters Isis and Nephthys come to thee, and they convey to thee the great extent (of the waters) in thy name of the great extender as lord of the flood.' These allusions show that there was an ark to which the two birds were attached as conductors. They are represented as hawks, but as the birds of east and west, or the earlier south and north, are equivalent to the dove of day and the raven of night in Semitic tradition. Isis was the lady or bird of dawn, and Nephthys the lady of darkness."

[58] Massey. *A Book of the Beginnings*. Section 18. Page 393: "The Egyptian bennu or phoenix was the constellation in which Sothis or Sirius (the Dog-star) was the chief star."

[59] Ibid. Section 1. Page 6: "About the time when the Nile began to rise in Lower Egypt, there alighted in the land a remarkable bird of the heron kind which had two long feathers at the back of its head. This, the Ardea purpurea, was named by the Egyptians the bennu, from 'nu,' a periodic type, and 'ben,' splendid, supreme—literally, as the crest indicated, tiptop. It was their phoenix of the waters, the harbinger of re-arising life, and was adopted as an eschatological type of the resurrection. The planet Venus is called on the monuments the 'Star of Bennu Osiris,' that is of Osiris redivivus."

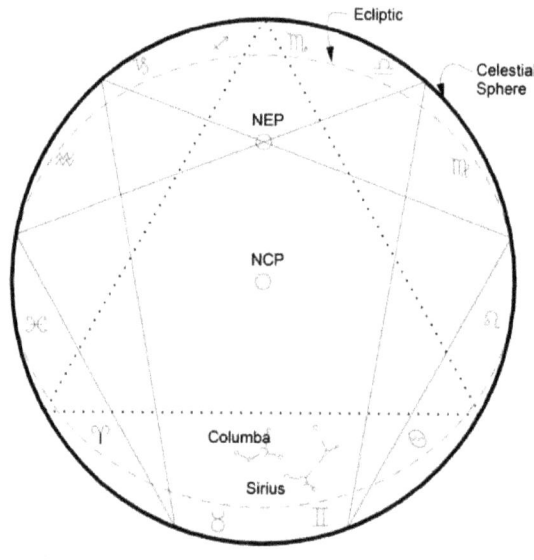

[Figure 3]

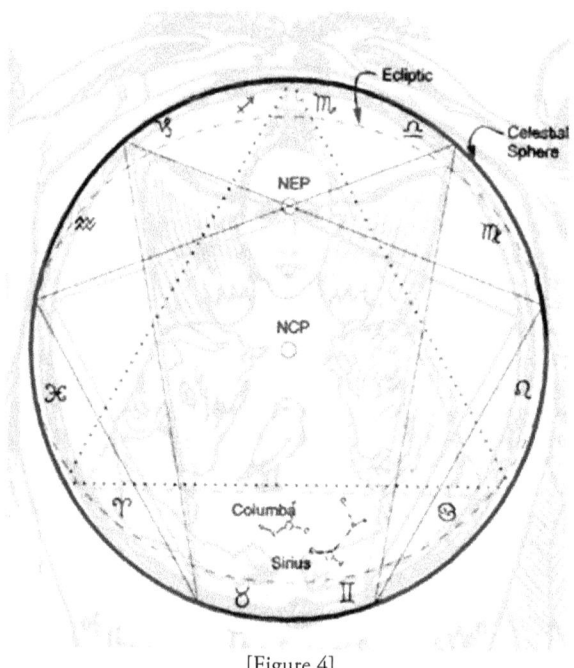

[Figure 4]

In this orientation, it is noteworthy that the solar point of crossing corresponds to the NEP, or axis of the ecliptic – elements of the celestial sphere functionally bound to the Sun. Also telling is the fact that this locus is illustrated so as to correspond with the center of the Aquarian figure's head [Figure 4], wherein the "seat of the soul," or the pineal gland, as well as the diencephalon region in general – both of which are known to be stimulated by solar radiation[60] – are located. Another implication involving the Sun and Aquarius is the fact that this constellation represents the vernal point of the impending Age – the exact inception of which is connected with the phoenix (i.e. dove-Sirius-Venus) cycle.

Taking the Galactic Center as apex of the "free trinity," and dividing the zodiac equally from either side of this central locus into 9 units, the resultant vertices of the equilateral triangle will be seen to correlate with Sagittarius (apex/Galactic Center), Aries (left lower vertex), and Leo (right lower vertex). Although not fitting precisely with the Scorpio/Taurus/Leo arrangement implied in de Salzmann's program illustration, we are nonetheless very close in this particular configuration, especially since Aries is known as the "ox goad," and the Galactic Center can be seen to comprise a portion of Scorpio, as well as Sagittarius.

Yet another way to orient the celestial sphere to correspond with this rendering of the Enneagram by de Salzmann is to slightly rotate the aforementioned configuration such that the NEP now corresponds with the apex of the informing triangle, and the NCP to that of the "point of crossing" [Figure 5]. In this orientation, the NCP overlaps the ecliptic, which passes underneath it, at its own point of crossing with the Galactic Equator; the base vertices remain associated with the signs of the ox goad and the lion, and the imputed association of the dove with Columba and Sirius still maintains.

In this case, the association of the solar point of crossing with the NCP is explained by the fact that this circumpolar locus is in a sense preeminently solar itself, as it serves among other things as the stellar demarcator of the passage of the Sun throughout the course of precession, and indeed both the polestar[61] and the Sun[62] are in the Egyptian Osirian mythos seen as representatives of Osiris.

---

[60] Walter E. Stumph & Thomas H. Privette. *The Steroid Hormone of Sunlight – Solitrol (Vitamin D) As a Seasonal Regulator of Biological Activities and Photoperiodic Rhythms.* Journal of Steroid Biochemistry. Molecular Biology Volume 39, No. 2. Pages 283-289, 1991.

[61] Massey. *Ancient Egypt, the Light of the World.* Book 9. Pages 574, 581: "The 'mooring-post,' which represents the pole, is designated 'the lord of the double earth in the shrine,' that is, Osiris as the power of the pole.... The pole and equinox are travelling pari passu, one in the upper circle of the heavens, the other in the larger lower circle of the ecliptic, and the shifting of the equinox was correlated more or less exactly to the changing of the pole-star. The power that presided over the pole as Osiris was given rebirth as Horus in the vernal equinox. The pole-star symbolized the lord of eternity."

[62] Ibid. Book 10. Page 648: "The Sun-god, whether as Atum-Iu (Aiu or Aai) or Osiris-Ra, is a mummy in Amenta and a soul in heaven. The imagery is quite natural: the nocturnal Sun became a mummy as a figure of the dead, and a soul or spirit in its resurrection as a figure of

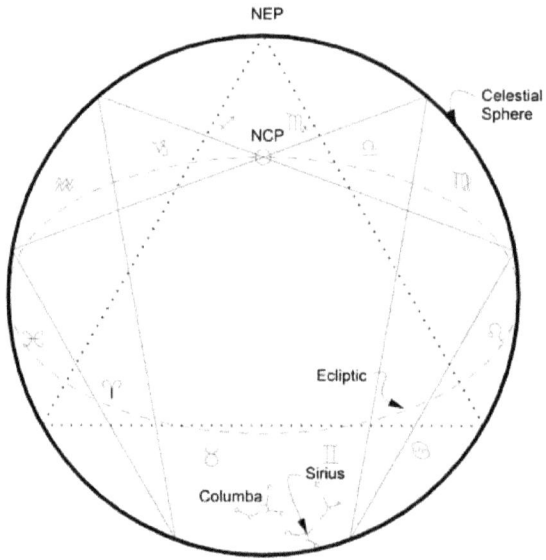

[Figure 5]

Although these two orientations of the celestial sphere are suggestive with respect to the depiction of the Enneagram under discussion, there is yet a third way in which this so-called "universal symbol" can be related to the zodiac and celestial sphere, and by extension, the Tree of Life. To do so, we refer to the following passages from Massey's *Natural Genesis*:

> Various figures of the put-circle of the nine gods, or the heaven of nine divisions over-arching the abyss that was hollowed out below by Ptah can be distinguished. The put are the nine that sat upon the waters of the quarter at present occupied by the three water-signs. The image of the put circle ... is a circle three-fourths filled in and one fourth hollow. Now, Diodorus Siculus tells us that the Chaldees figured the earth as a round boat turned upside down, with the hollow underneath. The boat or kufa still in use on the rivers Euphrates and Tigris is somewhat like a beehive with a considerable bulge in the middle. This figure of the earth corresponds to the Egyptian put-sign with its hollow underneath. The hollow, however, was the abyss that was founded or opened by Ptah for the sun to pass through the celestial waters and circumnavigate the globe. Various types of this formation of "the world" are extant, in addition to the put-circle and the inverted boat or beehive. The horseshoe figure is one. Hence its mystical value as a symbol of superstition. The headdress of Hathor has the shape of the horseshoe. The letter omega Ω is

---

the living. Atum, or Osiris, as the Sun in Amenta, is the mummy buried down in Khebt or Lower Egypt, and Iu in the one rendering, or Horus in the other, raises the mummy-god."

THE ENNEAGRAM, ZODIAC, AND CELESTIAL SPHERE

> another form of the same sign. Nine stones arranged horseshoe shape on the edge of the water, as at the "Nine Stone Rig," where stood the "headless cross," formed another. The "Headless Cross," or tau, is also an image of the three quarters, the fourth being the crossing, the abyss.[63]

And further on:

> Buddha was also portrayed standing within the horseshoe figure that is thus shown to be a type equivalent to the tree of nine branches under which he sits. The horseshoe symbol of good luck is thus connected with the nine months' period of gestation.
>
> The Hindu Golden City of the Gods, also called the 8-leaved Lotus, has eight circles and nine gates, in agreement with the eightfold Am-smen and Sesennu (Eg.) which passed into, or were followed by the put-circle of the nine. It is said the initiates know that living being which resides in the lotus with nine gates with three spokes and triple supports. Thus the lotus of nine gates rests on a threefold rootage in the waters, which is equivalent to the three water-signs. Also in a Hindu representation of paradise previously described there is a silver bell with nine precious stones surrounding the square of the four quarters. It comes to this at last. The four quarters represented by the put-circle, the tree and well or tree and bell are identical with the ankh-cross ♀ in a reversed position ☿ , with the feminine ru below and masculine tau above; and from this form of the figure was derived the well-known cross and circle or Imperial Globe as a symbol of worldwide supremacy. The legends relate that Gautama Buddha was reborn under the tree in the ninth incarnation of Vishnu, and that it was by means of the tree that he attained Nirvana, or passed into the divine circle of the gods, called the put pleroma in the Kamite mythos. Here also the number identifies the name of Buddha with the Egyptian put, for number 9, the circle of the nine, and with Putah, the founder of this circle of the nine gods. Buddha in China is Yu, Fo, Fot, or Boud, whose great work was the dividing of the land into nine parts after the deluge, which is identical with the work of Ptah, who founded the put-circle of nine gods upon the waters that were thus limited to one quarter of the four. The put pleroma of nine gods was likewise extant in China. In the third of the divine dynasties there was a company of nine brothers, who were the ruling powers, and during their reign, as in the time of Yu, the earth, the mountains, and the waters were separated into 9 divisions. Pure customs then prevailed, good government was established, human beings occupied one territory, and males and females originated food and drink. The Chinese have a sacred cap, exclusively consecrated to the emperor, styled "the orbicular Cloudy Court Cap of Nine Seams"; whereas the empress wears one called Seven Gems, or the White Water Lily … *The put-circle, then, established by Ptah, we have to look upon as a sort of zodiac of nine signs, imaged by the nine stones, nine branches, nine bridges, or other forms of the nine, and representing the nine*

---

[63] Massey. *The Natural Genesis*. Book 9. Pages 63-64. Ellipses added.

> *months which, together with an inundation, made up the earliest solar year, the fourth quarter being typified by the abyss (our three water signs) that the sun navigated in the passage fabled to have been created by Ptah.*[64]

In general, much of the Egyptian symbolism that Massey leads us through with an overwhelming series of details is to be seen as deriving from the Age of Taurus, when the polestar was localized in the region of the Great Mother and Genetrix, Ursa Major. In this Age, the summer solstice occurred when the Sun was in Leo, and the three diurnal "water signs," or signs of inundation corresponding to the 3 months of flooding of the Nile were therefore Cancer, Leo, and Virgo. The corresponding nocturnal "water signs," during which the Osiris as the Full Moon could be seen making safe passage through the underworld are thus the diametrically opposed signs of Capricorn, Aquarius, and Pisces. During this Age, the zenith, or point at which the Sun reached its ultimate height during the course of the year was in Leo, which was equated with the south. The nadir corresponded to the lowest point of the Sun during the winter in the sign of Aquarius. The three lower signs of Pisces, Aquarius, and Capricorn thus constituted the region of "the abyss" during the Age of Taurus.

According to Massey:

> In the "Chapter of vivifying the soul forever" the boat of the sun goes along "sounding the heaven at the great place," i.e., in the northern quarter, the abyss, and we read "The Heaven is open, the Earth opens, the South opens, the North opens, the West opens, the East opens, the Southern Zenith opens, the Northern Nadir opens." This describes the sixfold heaven which followed that of the four quarters, and the new figure was registered and represented in manifold ways.[65]

If we take only "the heaven of nine divisions over-arching the abyss that was hollowed out by Ptah," and connect the two ends of the "horseshoe" or "omega" such that it now forms a closed circle, we arrive at the put-circle proper, as illustrated in [Figure 6], in which the three signs of the abyss are notably absent. In this configuration, if we now connect the signs of Leo, Scorpio, and Taurus, an equilateral triangle corresponding to the "free trinity" of the Enneagram results [Figure 7].

It seems likely that this form of the zodiac was meant to be eschatological, as opposed to uranographic, since the ancient Egyptians were known to have excised the undesirable months of the year for magical purposes, thereby enabling their dead to live forever in the most favorable conditions, which

---

[64] Massey. *The Natural Genesis.* Book 9. Pages 67-68. Italics and ellipses added.
[65] Ibid. Pages 197-198.

constituted their "happy hunting grounds."[66] This removal of the three "feminine" months of the year may have come about in part as a result of the transition into a more patriarchal world view, with its predominantly solar chronometry.

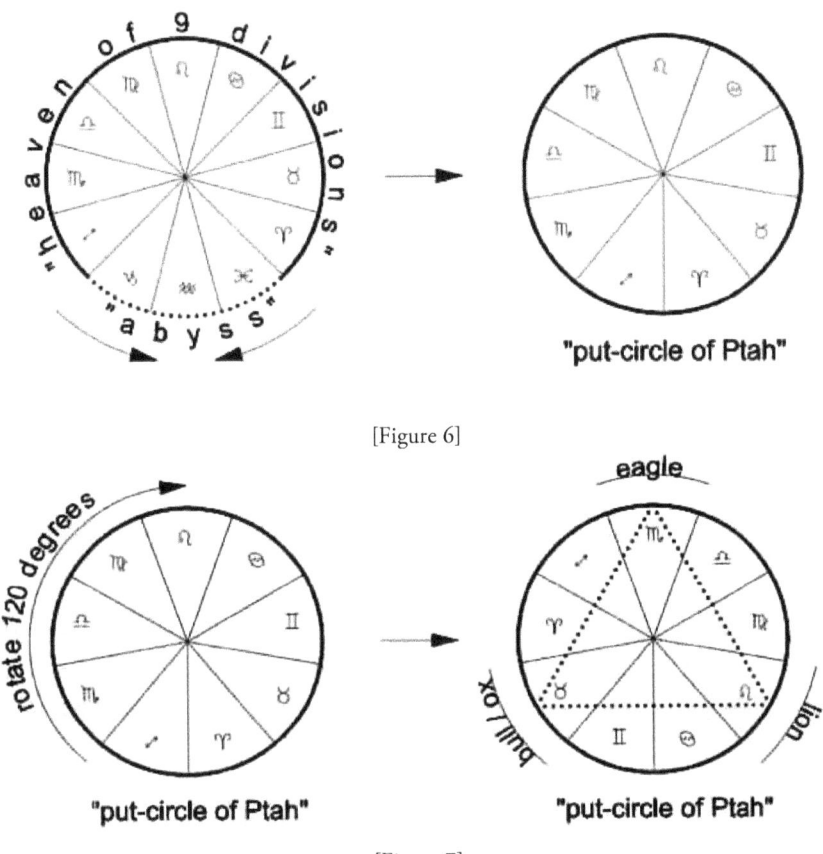

[Figure 6]

[Figure 7]

These three months of the Sun's hidden passage through Amenta have become equated with the proportion of days in a year associated with the "curse." As there are a total number of 13 lunar months in a year, if one takes seven days as a round figure to represent the complete duration of each menstruation, this yields a total of 91 days out of the year associated with the ancient mysteries of suffering and blood sacrifice. These 90 or so "taboo" days correspond roughly to one quarter of the days in a year; the same number of days that were grouped together as the invisible three signs of the abyss in the put-circle of Ptah, and it is notable that the Enneagram is devoid of the

---

[66] Fagan. *Astrological Origins*. Pages 50-51.

inverted yonic triangle, which one might typically expect to be present, in order to balance out the upright phallic one.

In regard to the path of "inner circulation" (in the order of 142857142..., etc.) much has been said about the fact that somehow, "mysteriously," any integer (that is not itself a multiple of seven) divided by seven results in a number whose decimal fraction consists of a remainder, the numbers of which inevitably occur in some sequence arranged according to this order [Figure 8].

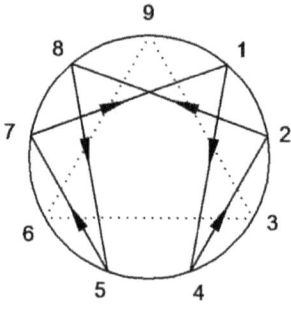

[Figure 8]

In a general sense, this pattern illustrates that every discrete phenomenon occurs through the interaction of a triad of unique forces, which together characterize the essential nature of that phenomenon. Moreover, in order for that single phenomenon to manifest completely, an additional complementary phenomenon – existing on the same level as the first, and also itself comprised of an analogous triplicity of forces themselves complementary to the first set of three – is required to "balance out" the first phenomenon [Figure 9].

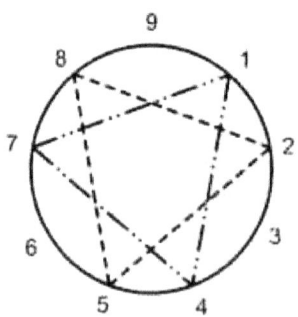

—···— first triplicity
- - - - second triplicity
(complementary to first)

[Figure 9]

THE ENNEAGRAM, ZODIAC, AND CELESTIAL SPHERE

Thus, in order for the first triad (1-4-7) to complete itself, assistance from the second triad (2-5-8) is required. Cycling through in the order of inner circulation, we start the first triad: 1-4... But before this triplicity can be completed, the complementary triad must feed on the energy supplied by the beginning of the first, yielding 2-5-8, thereby generating the necessary momentum for the completion of the first, moving from 4 to 7. The overall numerical sequence representing the interaction and mutual facilitation of these two triplicities is therefore 1-4-2-8-5-7.

This is not to imply a mutually annihilative relationship wherein the two complementary phenomena existing on the same plane essentially cancel each other out for a "net-zero sum," for yet a third phenomenon existing at a higher lever (i.e. more "spiritual"; comprised of a higher order of vibrations; with more "degrees of freedom," etc.) is involved in the mediation of the two lower complementary phenomenon; this is the so-called "informing triangle" of the Enneagram.

The net result is a kind of "nothingness," which may be indicated at one level by the circular portion of the "universal symbol" – the circle of time; the ouroboros; or quite simply, the number zero. But this is that ineffable nothingness, about which little can be said without distorting the truth of the matter. In order for this "nothingness" to be defined properly, entire universes are required.

Admittedly, this analysis of the inner circulation is somewhat novel, but according to Gurdjieff, there are as many different interpretations of the Enneagram as there are levels to an individual's understanding. And yet this symbol is seen as being an objective one; the fundamental hieroglyph, no less, of a universal language.[67]

In my opinion, there really never has been an adequate explanation of the pattern of inner circulation. In many ways, its significance seems to amount to little more than a fascination with the bilaterally symmetrical shape created by the mapping of the sequence 1-4-2-8-5-7 onto a circle divided into 9 sections. Perhaps one of its functions is to hold the fascination of the esoterically inclined – an intellectual fetish item, so to speak, designed to ensnare and befuddle the "formatory," or "mechanical" mind, as a means to exhaust it such that one is either spent in the impossible task of reaching a logical solution to its conundrum, or one rises above and escapes from the tenacious grasp of puny logic, entering into the new and mysterious realm of the intuitive and subconscious mind: that aspect of the higher mental function which Gurdjieff always maintained should be the proper locus of "waking consciousness."

---

[67] Ouspensky. *In Search of the Miraculous.* Pages 283-4; 294.

But this progression, generated in the decimal fraction of any integer indivisible by 7 when divided by 7, is not the only one that results in such a "mystical" pattern. If we take what I refer to as the "collapsed" Fibonacci sequence, it can be seen that there are at least 3 unique bilaterally symmetrical patterns centered upon the number 9, similar to the Enneagram, although surpassing it in complexity.

The Fibonacci sequence proper is obtained by starting with the number 1, and adding it to itself. The result, 2, is then added to the previous number, 1, yielding 3. This process is repeated ad infinitum, yielding the following sequence: 1, 1, 2, 3, 5, 8, 13, 21... To achieve the collapsed Fibonacci sequence, we convert all numbers to their "mystical" equivalents through the use of so-called theosophical addition, [68] which yields the following comparable sequence: 1, 1, 2, 3, 5, 8, 4, 3... Continuation of this process shows that this progression of numbers eventually (after 23 iterations) begins to repeat: 1, 1, 2, 3, 5, 8, 4, 3, 7, 1, 8, 9, 8, 8, 7, 6, 4, 1, 5, 6, 2, 8, 1, 9, [1, 1, 2, 3, 5...]. Taking this result, and mapping it onto the circle of 9 divisions generates the following pattern [Figure 10]:

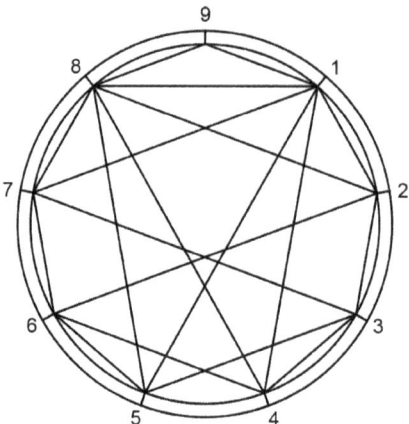

[Figure 10: Collapsed Fibonacci sequence: 112358437189887641562819...]

If we extend this concept beyond the strict Fibonacci sequence (which always starts with 1) and start with 2, adding 1 to that, following the previously mentioned formula, the following 24-digit repeating sequence is generated: 2, 1, 3, 4, 7, 2, 9, 2, 2, 4, 6, 1, 7, 8, 6, 5, 2, 7, 9, 7, 7, 5, 3, 8... Mapping this onto the circle of 9 divisions, we get the following pattern [Figure 11]:

---

[68] Theosophical addition is nothing more than converting numbers that go beyond single digits (i.e., any number after 9) back into a single digits by adding the several digits repeatedly, until a single digit number is obtained. Thus, 149 yields 1+4+9 = 14 = 5.

THE ENNEAGRAM, ZODIAC, AND CELESTIAL SPHERE

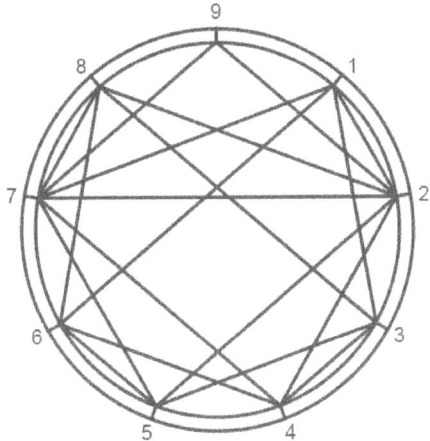

[Figure 11: Collapsed Fibonacci sequence: 213472922461786527977538...]

Similarly, the sequence 3, 1, 4, 5, 9, 5, 5, 1, 6, 7, 4, 2, 6, 8, 5, 4, 9, 4, 4, 8, 3, 2, 5, 7... yields [Figure 12]:

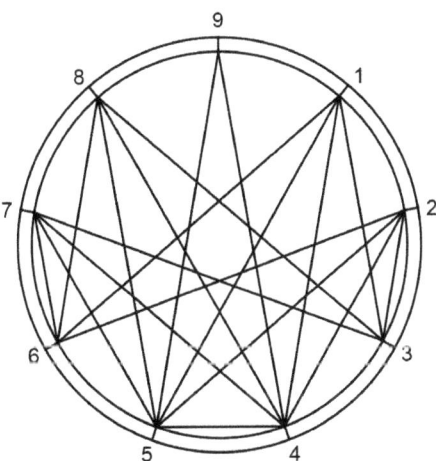

[Figure 12: Collapsed Fibonacci sequence: 314595516742685494483257...]

The subsequent iterations of this process are each duplicates of one of the previous three patterns. Thus, 4, 1, 5 6... yields [Figure 10]; 5, 1, 6, 7... is equivalent to [Figure 12], and 6, 1, 7, 8... generates [Figure 11]. Additionally, 7, 1, 8, 9..., 8, 1, 9, 1..., and 9, 1, 1, 2... each yield the same pattern as [Figure 10].

What are we to make of these "magical circles?" Are these also examples of universal hieroglyphs that can be used to unlock the mysteries of the universe as is purported of the Enneagram, or are they simply another example of the bilateral symmetry found in the infinite forms of natural phenomena we see all around us? One viewpoint does not exclude the other. This being said, I personally am quite skeptical of most of the explanations of the Enneagram's inner circulation that have been proposed. Rodney Collin attempts to link it to the unseen and mysterious processes involved in agriculture throughout the course of a year,[69] but this explanation seems to lack substance and credibility. The same could be said for Ouspensky's treatment of the inner circulation as an example of the ebb and flow of the arterial and venous blood in the human organism. Moreover, Ouspensky's distribution of planetary attributions along the Enneagram does not agree with Collin's, as the former arranges them according to the order of days in the week [Figure 13], whereas the latter organizes them in the classical geocentric pattern [Figure 14].

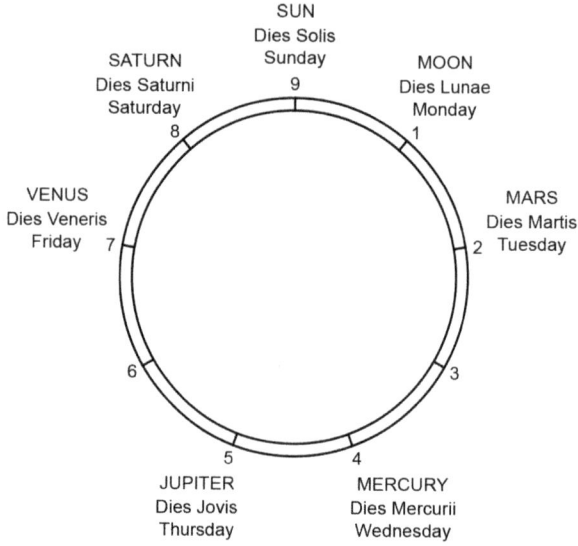

[Figure 13: Ouspensky's arrangement of planets along the Enneagram, from *In Search of the Miraculous*]

Perhaps one of the most reasonable explanations of the significance of the Enneagram does not even take into account the pattern of inner circulation. According to Gurdjieff, the Enneagram is above all a representation of the interaction between the two most fundamental and universal laws: the Law of Three, and the Law of Seven.[70] The Law of Three has been mentioned previously: every phenomenon consists of a triad of forces, each of which must be present for the phenomenon to manifest, or come to completion.

---

[69] Collin. *The Theory of Celestial Influence*. Page 168.
[70] Ouspensky. *In Search of the Miraculous*. Page 287.

THE ENNEAGRAM, ZODIAC, AND CELESTIAL SPHERE

One force is considered "active", one "passive", and one "neutralizing." The order of interaction of these forces determines the nature of the phenomenon. For example, any phenomenon manifesting from these forces in the order neutralizing-passive-active is said to represent the process of "regeneration", whereas the order active-neutralizing-passive corresponds to "decay."[71]

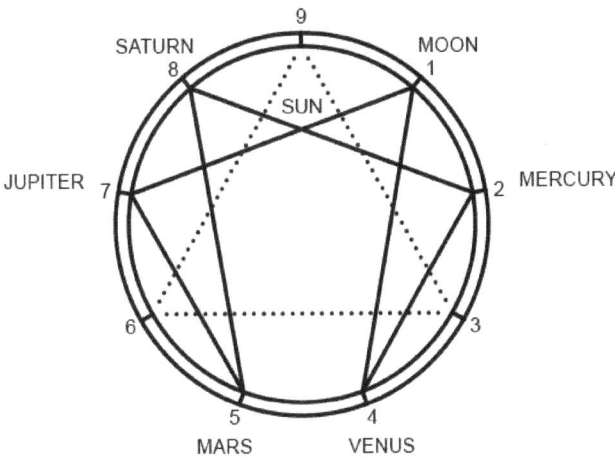

[Figure 14: Collin's arrangement of planets along the Enneagram, from *The Theory of Celestial Influence*]

The Law of Seven (also known as the Law of Octaves) refers to the idea that each phenomenon that comes to completion does so in seven discrete steps, with the eighth step being a repetition of the first. Moreover, this eighth step is either twice or half the frequency of vibrations of the first step, according to whether the phenomenon represents an ascending or descending octave.[72] Each "note" in the octave is separated from the previous one by a certain number of vibrations, constituting a "full step." This is the case for all of the notes of the octave, with the exception of the intervals between notes 3 and 4, and 7 and 8, which are each separated by only a "half step" [Figure 15]:

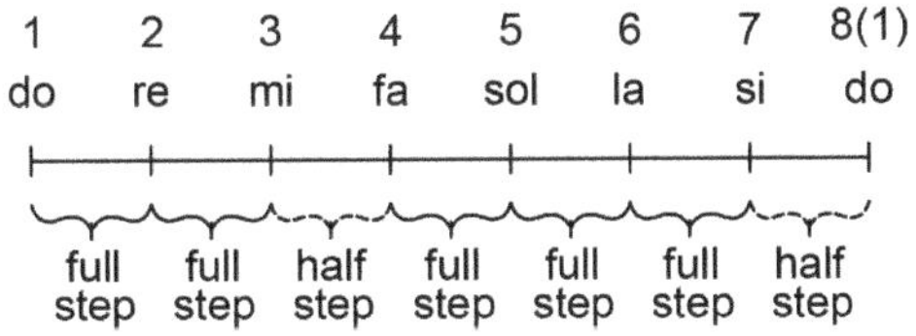

1	2	3	4	5	6	7	8(1)
do	re	mi	fa	sol	la	si	do
full step	full step	half step	full step	full step	full step	half step	

[Figure 15: An octave, comprised of 5 full steps and 2 half steps]

---

[71] Collin. *The Theory of Celestial Influence*. Pages 53-57.
[72] See Ouspensky's *In Search of the Miraculous*. Chapter 7.

In order for any octave to come to completion, a certain catalyst, or "shock" is needed to maintain continuity at the points where the half steps occur. If we consider the Enneagram as a representation of a single octave, we can see how the Law of Three and Law of Seven are related [Figure 16].

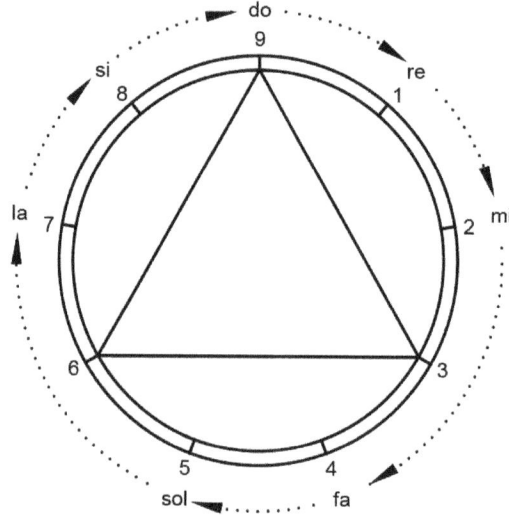

[Figure 16: Relationship between the Law of Three and Law of Seven as depicted on the Enneagram]

In this arrangement, point 9 of the informing triangle represents the initial and concluding "do," with points 3 and 6 marking where the shocks are necessary in order to pass through the two half step intervals. It should be noted that although the first shock at point 3 is in the expected location between "mi" and "fa" of our octave, the second one appears to be in the wrong place: between "sol" and "la", instead of "si" and "do." This apparent inconsistency was noted by Gurdjieff, but never specifically addressed. The following explanation for it is given, based on some additional clues left by Ouspensky:[73]

Beginning the octave, we pass from "do" to "re" to "mi," with point 3 of the informing triangle supplying the necessary shock to maintain the continuity of the octave through the "mi-fa" interval. This point 3 actually represents the beginning note of another octave, existing at a higher or lower level as the octave we began with at point 9. If the original octave is ascending, point 3 represents a shock from a higher octave, and if descending, vice versa [Figure 17].

---

[73] See *In Search of the Miraculous*. Pages 376-8.

THE ENNEAGRAM, ZODIAC, AND CELESTIAL SPHERE

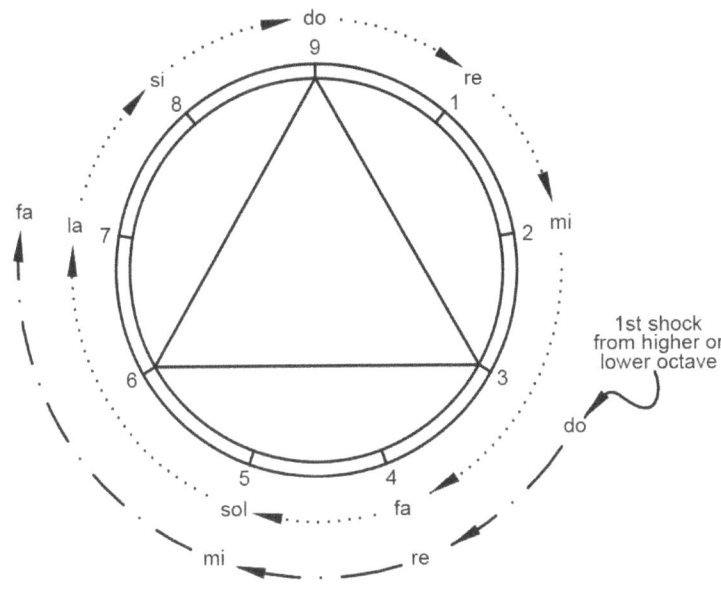

[Figure 17: The first shock of the original octave is supplied by the "do" of an additional octave]

At this point, we now have two different octaves interacting with each other, which also represent two of the three forces necessary to constitute the triad required for the completion of each discrete phenomenon. Examination of [Figure 17] shows that in order for the second octave to maintain continuity through its own "mi-fa" interval, an additional shock is now necessary at point 6. This is supplied by yet another "do" of a third octave (higher, or lower, according to the aforementioned formula) [Figure 18]. Finally, the "mi-fa" interval of the third octave beginning at point 6 is filled by "do" corresponding to the analogous "do" of either half or double the frequency of the original note of the beginning octave. The process is now complete, and we have arrived at exactly one octave higher or lower to our original starting point. Thus, the culmination of each discrete phenomenon can be represented by the sequence of notes in an octave (Law of Seven), and it requires the interaction of "dos" from a triplicity of octaves, represented by points 9, 3, and 6 of the informing triangle (Law of Three).

If we apply these concepts to the cycle of a year, as represented by the put-circle of Ptah (from the perspective of the Age of Taurus; see [Figure 19]),[74] we see that our octave begins in Scorpio, corresponding to the autumnal

---
[74] Note that this view is a mirror image of those portrayed in Figures 2, 3, 4, 5, and 7. This is necessary in order to take the point of view of the human anthropomorph in Figures 2 and 4, thereby illustrating the course of the year as clockwise as opposed to anti-clockwise.

equinox. Notably, the Jewish New Year (Rosh Hashanah) is celebrated around the first new moon of the autumnal equinox,[75] so it is possible that this tradition could have originated from an earlier Egyptian one, as this time of year represented the end of the inundation, and the beginning of the planting season. In the Egyptian mythology, the Sun at this point was anthropomorphized as the dying Osiris, entering the underworld of Amenta; the seed of a future life and sustenance, buried in the earth.

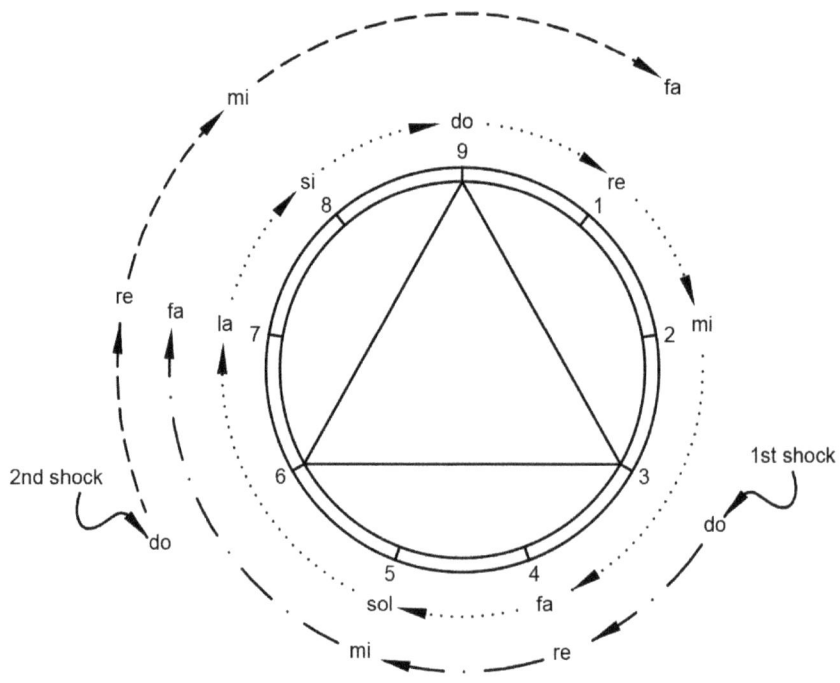

[Figure 18: The first shock of the second octave is supplied by the "do" of a third octave]

Next, we pass through points 1 and 2, corresponding to late fall/early spring, and arrive at point 3, representing the spring equinox, with the Sun in Taurus. Osiris is here reborn, having "magically" escaped the suffering of Amenta represented by the three excised taboo months that were notably anthropomorphized as Harpocrates, the "silent god."[76] The second octave thus begins, catalyzing the continuation of the year at point 3, through the increased solar radiations of the spring.

The third octave enters at point 6, corresponding to the summer solstice, with the Sun in Leo. Here Osiris is at his height of power, and is represented

---

[75] Massey. *Ancient Egypt, the Light of the World.* Book 10. Page 661.
[76] Massey. *A Book of the Beginnings.* Section 7. Page 270; Section 8. Page 353.

as the conqueror who bruises the head of the Apap dragon of drought (constellation Hydra) under his foot, via the inception of the summer inundation.[77] Here the cycle of the year receives the second shock that catalyzes its full completion, passing through points 7 and 8, and finally back to point 9, which represents the place of judgment, or *maat*, of the year. This is a time of reflection upon the previous year, leading to an assessment of where progress has been made and where further work needs to be done in the year to come.

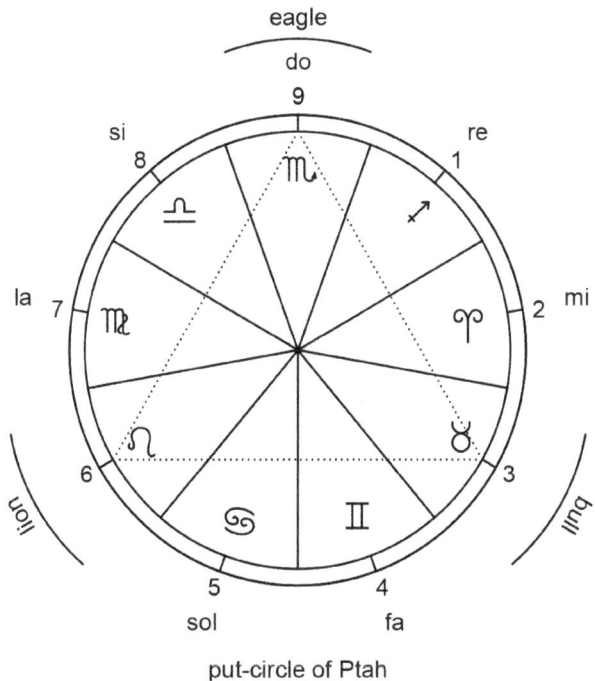

[Figure 19: Yearly cycle represented as an octave on the Enneagram/put-circle of Ptah]

---

[77] Massey. *Ancient Egypt, the Light of the World.* Book 5. Page 270.

The Great Seal

IN ADDITION TO the previously discussed connections amongst the Enneagram, zodiac, and celestial sphere; where particular emphasis is given to the 4 fixed signs (especially Aquarius) as well as the Sirius/Columba region of the heavens, there is yet another suggestive example of how this "universal symbol" can be seen to bear a striking geometrical resemblance and hint at an as of yet undisclosed significance to a particular symbol – the Great Seal of the United States.

One of the original design proposals for the reverse of the Great Seal depicted Moses parting the waters of the Red Sea, in order to turn back the Pharaoh in pursuit of the Israelites.[78] This symbolism derives from an earlier Egyptian source, and is fundamentally connected to Sirius. According to Massey, Moses is a figure of the Egyptian Shu-Anhur, constellated as Cepheus.[79] During the Age of Taurus, this circumpolar asterism could be seen rising achronycally as the setting Sun entered Leo/Hydra. It was at this point that the "dragon of drought" (i.e. Hydra, anthropomorphized as the pursuing Pharaoh) was "slain" as the floods commenced.

The Great Seal in its present form was adopted in 1782, but July 4, 1776 was the date that the first committee for its design was established.[80] The obverse of the seal depicts the familiar eagle behind the shield, clutching the olive branch and arrows; its head facing the direction of the peace offering, possibly as an indication of the United States' preference towards a non-violent approach to foreign relations.[81] On the reverse, we see the unfinished pyramid with the so-called "Eye of Providence" observing from on high.

There is a certain intuitive correspondence suggested by the eagle with its clutches and the triangular motif of the pyramid, for on the obverse we can

---

[78] Case. *The Great Seal of the United States: It's History, Symbolism and Message for the New Age.*

[79] Massey. *Ancient Egypt, the Light of the World.* Book 10. Page 660.

[80] *The Great Seal of the United States.* U.S. Department of State. Bureau of Public Affairs. Web. <state.gov/documents/organization/27808.pdf>

[81] Case. *The Great Seal of the United States: It's History, Symbolism and Message for the New Age.*

quite easily make out the familiar and unique 6-pointed figure of the Enneagram's "inner circulation," whereas in the case of the pyramid on the reverse, the "informing triangle" portion of the Enneagram is the obvious correlation [Figure 1].

It is not being suggested that the Great Seal is a derivative of the Enneagram, although if one is to take Gurdjieff's indication that the latter is an ancient (i.e. thousands of years old) symbol which only became accessible publicly due to his own dispensation, this would certainly be within the realm of possibility. The earliest known published version of an Enneagram-like diagram is perhaps Raymond Lull's "Figures 'A' and 'T'" from *Ars Generalis Ultima*, circa 1305 [Figures 2 & 3], and much later in 1665, Athanasius Kircher published a very striking version (albeit depicted without the path of "inner circulation") as the frontispiece to his *Arithmologia*, wherein something very closely resembling the Eye of Providence of the Great Seal occupies the same relative position as the central "informing triangle" of the Enneagram [Figure 4].

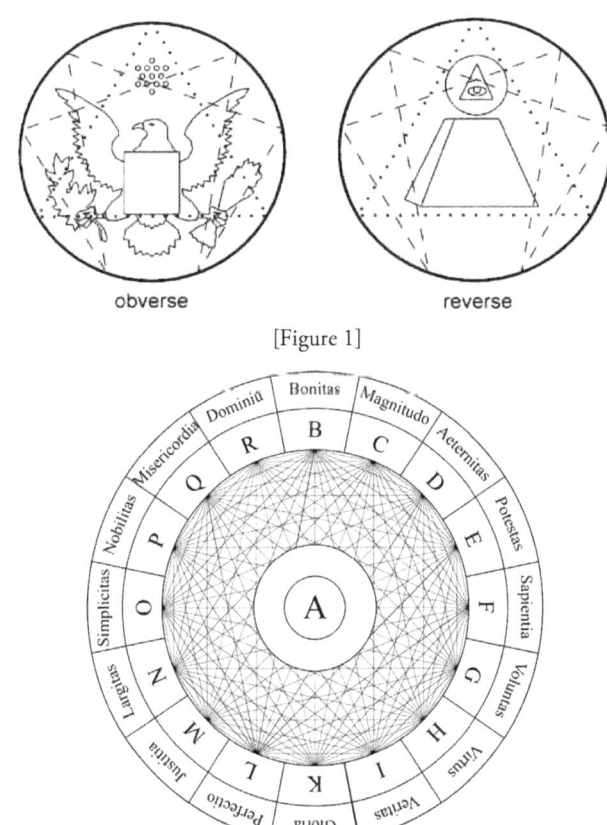

[Figure 1]

[Figure 2: "Figure 'A'" from Raymond Lull's *Ars Generalis Ultima*. Author's rendering.]

THE GREAT SEAL

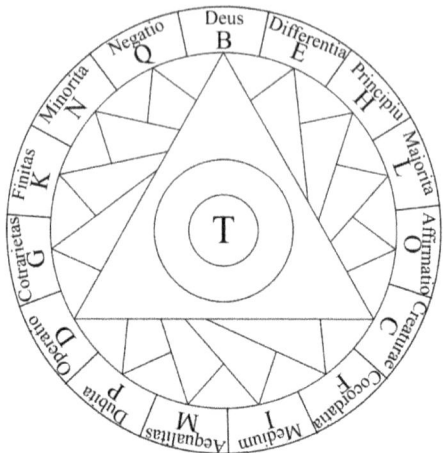

[Figure 3: Partial reconstruction of "Figure 'T'" from Raymond Lull's *Ars Generalis Ultima*. Author's rendering.]

[Figure 4: Frontispiece to Kircher's *Arithmologia*. Author's rendering.]

The Enneagram can be seen as a kind of "pantacle," in the sense that Eliphas Levi used the term:

> The Pantacle, being a complete and perfect synthesis expressed by a single sign, serves to focus all intellectual force into a glance, a recollection, a touch. It is, so to speak, a starting-point for the efficient projection of will.[82]

This refers to the constructive use of distilled symbolism as a means of organizing Thought so as to direct Will towards some useful purpose – in which case, all of the various elements of the talisman, pantacle, hieroglyph, etc., are fully understood, with the condensed symbolic shorthand serving as an efficient template and organizing influence for the intellectual and emotional energies which are constantly volatizing outwards due to the infinitude of stimuli which are ever deflecting the realization of one's original aims. This constructive use of course presupposes that one fully understands the symbolism employed, and that one's intentions are benign. The talisman loses its constructive potency when true knowledge of its symbolism is lost or perverted, and the aims of those individuals employing them become essentially selfish. In this case, the pantacle becomes in many ways its own opposite, and has correspondingly negative effects both individually and collectively. Regarding this, Levi says the following:

> Superstition is derived from a Latin word which signifies survival. It is the sign surviving the thought; it is the dead body of a Religious Rite. Superstition is to initiation what the notion of the devil is to that of God. This is the sense in which the worship of images is forbidden, and in this sense also a doctrine most holy in its original conception may become superstitious and impious when it has lost its spirit and its inspiration. Then does religion, ever one, like the Supreme Reason, exchange its vestures and abandon old Rites to the cupidity and roguery of fallen priests, transformed by their wickedness and ignorance into jugglers and charlatans. We may include among superstitions those magical emblems and characters, of which the meaning is understood no longer, which are engraved by chance on amulets and talismans. The magical images of the ancients were pantacles, i.e. kabalistic syntheses. Thus the wheel of Pythagoras is a pantacle analogous to the wheels of Ezekiel; the two emblems contain the same secrets and belong to the same philosophy; they constitute the key of all pantacles, and we have made mention previously of both.[83]

Here Levi provides us with a diagram of the Star of Zion with the four apocalyptic beasts that very closely resembles the deHartman flyer for

---

[82] Levi, Eliphas. *Transcendental Magic, Its Doctrine and Ritual*. Chapter 10. Translated by A.E. Waite, 1896.
[83] Ibid. Chapter 18 – Charms and Philtres.

THE GREAT SEAL

Gurdjieff's "Institute," and it is quite likely that the latter is derived directly from the former [Figure 5].

[Figure5: An Example of a Pantacle, from Levi's *Doctrine of Transcendental Magic.*]

The similarities noted between Gurdjieff's Enneagram and the Great Seal may be dismissed as coincidence, but if there is any doubt concerning the implicit occult references in what amounts to the use of pantacles on the fundamental unit of America's paper currency, the following remarks by one of the foremost occult scholars of our era and 33rd degree Freemason, Manly P. Hall, should provide ample reason to reconsider:

> Not only were many of the founders of the United States Government Masons, but they received aid from a secret and august body existing in Europe, which helped them to establish this country for a peculiar and particular purpose known only to the initiated few. The Great Seal is the signature of this exalted body--unseen and for the most part unknown--and the unfinished pyramid upon its reverse side is a trestleboard setting forth symbolically the task to the accomplishment of which the United States Government was dedicated from the day of its inception.[84]

Elsewhere, in the same text, he states:

> As an imperishable reminder of their *sub rosa* activities, the Rosicrucians left the Great Seal of the United States. The Rosicrucians were also the instigators

---

[84] Hall. *Secret Teachings of All Ages*. Ch. 19 – Fishes, Insects, Animals, Reptiles, and Birds; Part II.

of the French Revolution, but in this instance were not wholly successful, owing to the fact that the fanaticism of the revolutionists could not be controlled and the Reign of Terror ensued.[85]

One is left to wonder what this secret purpose is, and whether the presumably Initiated Founding Fathers took into account the laws of equilibrium and Justice in their tacit acceptance of the mass genocide perpetuated upon the Native Tribes of this continent under the guise of Manifest Destiny.

Charles Adiel Lewis Totten, 19th Century American Military tactician and proponent of Anglo-Zionism states the following in reference to the Great Seal:

> Such is the oneness or harmony of the Great Seal of the United States of America, that there is not an emblem or motto in its whole concert that is not directly related to the instrument as a whole, and at the same time severally to all of its parts, and that the governing arithmography of the design from general to particular is couched in the terms of the very same physical factors of modern science that have already been pointed out as veiled in the numerical language of inspiration itself. In each case men have wrought wiser than they knew. There now can be no doubt of this, and consequently Providence alone can have overruled the results, and lo, at the very end of the Age permits their interpretation, or revelation, i.e., their discovery and unveiling, in order to beget belief from any who have preserved the seeds of faith in such an Age of waning belief as this in which we live.
>
> And it will therefore be perceived how necessary it was to separate our own discussion of the Great Seal into two parts or volumes. By so doing, we have divided the letter as it were, from the spirit, as much as possible, and recognized the fact that Manasseh himself is a dual tribe, and that his elements as yet are not all "wise" (Matt. XXV. 2).[86]

In this case, the Tribe of Manasseh being referred to is the so-called "13th Tribe" of Israel, which Totten equates to the United States of America. One must assume that the implication here is that this "tribe" consists of the descendants of the Hebraic King David through some Providential lineage. Further on in the same book, Totten engages in some rather questionable and in some ways simply preposterous Gematrial gymnastics, eventually arriving at what he refers to as "the M-ography of our topic":

> The Hebrew letter *mem* (מ = m) which is the initial letter of Manasseh's name is, like himself, and his tribe in the sequence of adoption, the 13th letter in the

---

[85] Ibid. Chapter 32 – Rosicrucian Doctrines and Tenets.
[86] Totten. *The Seal of History: Our Inheritance In the Great Seal of "Manasseh", The United States of America: Its History and Heraldry; and Its Signification Unto "The Great People" Thus Sealed.* 1897. Pages 130-131. Nabu Public Domain Reprint.

alphabet. Its significance is that of *eldership, number,* and *many* as shown by Dr. John Lamb, in his Hebrew characters derived from Hieroglyphics (London, 1835). We have already shown that it was the root of the word Tom or *Twin* (p. 291, Study Number Eighteen). But Manasseh, while an elder in Joseph was a twin, or double in his own tribal organization, which was unique in Israel and consisted in two "half-tribes" that were undoubtedly somewhat independent, or divided in partisanship which is the root and safeguard of a Republic.

His initial letter (M) is a letter that is peculiarly sacred to all languages and religions: it was the symbol of a stream, or of stripes as on our flag (MMM) significant in hieroglyphics of waves of water, which is the familiar emblem of Aquarius (M) the "Water Bearer," whom some regard as Manasseh's Sign upon the zodiac. It was primarily an emblem for peoples, nations and tongues, to wit, as before noted of multitudes, a Josephetic promise of increase.[87]

If, as Hall suggests, the Founding Fathers did indeed have some connection to a hidden and presumably enlightened governing body, and they left their "signature" encoded within the Great Seal, what implications may we derive from it regarding the future of the nation they have so marked? Hall indicates that the eagle on the obverse of the Great Seal is in reality a phoenix:

> Only the student of symbolism can see through the subterfuge and realize that the American eagle upon the Great Seal is but a conventionalized phoenix, a fact plainly discernable from an examination of the original seal… Masonry will be in a position to solve many of the secrets of its esoteric doctrine when it realizes that both its single- and double-headed eagles are phoenixes, and that to all initiates and philosophers the phoenix is the symbol of the transmutation and regeneration of the creative energy – commonly called the accomplishment of the Great Work. The double-headed phoenix is the prototype of an androgynous man, for according to the secret teachings there will come a time when the human body will have two spinal cords, by means of which vibratory equilibrium will be maintained in the body.[88]

This double-headed stage of evolution is clearly not meant to be taken literally, and quite likely refers to the activation of the diencephalon (literally "two-headed") region of the brain wherein resides the pineal gland, and the notion that the future "enlightened" humanity will have through some means permanently awakened this "third eye" or "ajna chakra," thereby achieving the state of consciousness which has been known by mystics throughout the Ages as humanity's "birthright."

---

[87] Ibid. Pages 135-136.
[88] Hall. *Secret Teachings of All Ages.* Ch. 19 – Fishes, Insects, Animals, Reptiles, and Birds, part 2. Web. <sacred-texts.com>

CELESTIAL ARCANA

The symbolism of the phoenix on the Great Seal, Levi's pantacle of Ezekiel, Gurdjieff's modified Enneagram, and the zodiacal and precessional connotations inherent within each of them suggests a possible reference to the so-called phoenix cycle, a concept deriving from ancient Egyptian observations of the cyclical heliacal risings of Sirius as harbinger of the yearly inundation.

Canis Major and Orion were collectively referred to as the Ship of the Bennu-Asar – "Bennu" referring to the bennu bird, phoenix, or Sirius; "Asar" being an appellation of Osiris, or Orion. When Venus rose along with Sirius (Isis) and Orion (Osiris), the planet was referred to as "the star of the Ship of the Bennu-Asar."[89]

Considering the preeminence ascribed to Venus when proximal to Sirius, and the fact that both luminaries had a dove as one of their types (the dove itself being another form of the phoenix), it seems reasonable to posit that the condition of their simultaneous heliacal rising may have in some way been related to the ancient Egyptian phoenix cycle, perhaps serving as one of its units of division, if not a type of phoenix cycle in its own right. There are various lengths of time (i.e. 250, 500, 1000 years) that have been ascribed to this cycle, which we shall explore in more detail in the chapter on the Devil trump. In the present case, 243 years – the amount of time it takes Venus to return to the same position relative to the Earth and Sun – will serve as our unit of measurement.

If these cycles were calculated by the ancient Egyptians with predominance ascribed especially to the simultaneous heliacal rising of Sirius and Venus as the fixed point of reference, it is interesting to note that during our present Age, if one is to consider the conjunction proper of the Sun and Sirius – along with the heliacal proximity of Venus – to comprise the alpha and omega points of any given cycle, the date which presents itself as the ideal candidate for such consideration is none other than July 4.

Currently, July 6 marks the exact conjunction (longitudinally) of the Sun with Sirius. Given that each day of a 365-day year comprises .986° of a circle, and using the formula of 1° of precessional drift per 72 years, it follows that the precessional motion of the Sun along the ecliptic in terms of "daily quanta" occurs in units of 70.992 years (72x.986). Thus, if currently the Sun is longitudinally conjunct Sirius roughly around July 6, it follows that 243 years (one putative phoenix cycle) prior, such a condition occurred 3.42 days earlier, or around July 2-3.

Continuing with this logic, it also follows that the period during which the annual conjunction of the Sun with Sirius occurred on the summer solstice

---

[89] E.A. Wallis Budge. *The Gods of the Egyptians*, Volume 2. Ch. XIX – Miscellaneous Gods.

THE GREAT SEAL

(June 21) – thereby engendering the most potent "lensing" effects of whatever consciousness affecting radiations purportedly emanate from this star by virtue of its high solstitial angle of incidence – falls somewhere around 949 CE. This period marks the height of the maturity of the Monastic Christian civilization, according to Rodney Collin,[90] the year 948 CE being the beginning of Abbot Maieal's tenure at the Abbey of Cluny.

In a sense, this period can be seen to represent the "height" of the Piscean Age, and it is interesting to note that the antipodal era, some 12,956 years prior (circa 12,000 BCE) corresponds to the Age of Virgo, during which period the polestar was Vega, and presumably an "exoteric" Golden Age of sorts was being enjoyed, as the radiations from the Galactic Center were during this period pouring forth into the Earth from Sagittarius' high angle of incidence during the summer solstice.

Analyses of the horoscopes for both the American Declarations of War and Independence ratified on July 6, 1775 and July 4, 1776, respectively, reveal some interesting correspondences when considered from a sidereal astrological perspective [Figures 6 & 7]. Examination of the first horoscope shows the Moon to be simultaneously square the Sun and Pluto. According to Fagan, "When the Moon is configured with Pluto – the Ejector – the native tends to become antisocial and a rebel against the existing order of things."[91] Additionally, "Should the Moon be configured with the Sun, the native will seek to be esteemed, honored, feted and flattered, feeling very conscious of his dignity and importance."[92]

With regard to the second horoscope – that of July 4, 1776 – we see a number of very significant planetary configurations. The opposition of mercury to Pluto provides an influence that engenders actions of rebellion and separation (Pluto) as expressed through the use of words (Mercury). The squared configuration of Saturn with Mercury, the Sun, Jupiter, Venus, and Mars makes this "malefic" planet potentially a key influence in this particular horoscope, although as there is no official time recorded for the signing of the Declaration, determination of the angularity of the planets for the horoscope under consideration is difficult, if not impossible. Nonetheless, as to the effects of Saturn, Fagan says the following:

> It has been repeatedly pointed out in these pages that millionaires and successful businessmen usually have Saturn angular in their horoscopes, and frequently configured by conjunction, opposition or square aspect with the luminaries and otherwise well buttressed by the testimony of the benefics: for

---

[90] Collin, Rodney. *Theory of Celestial* Influence. Appendix 8: The Cycle of Civilizations. Page 363. New York: Penguin Arkana, 1993.
[91] Fagan. *Solunars Handbook*. Page 55.
[92] Ibid. Page 54.

Saturn is the significator of want, acquisitiveness, covetousness, greed, jealousy and malice.

The author of the Tetrabiblos [Ptolemy] informs us, "If Saturn alone is ruler of the soul and dominates Mercury and the Moon, if he has a dignified position with reference to the universe (i.e. zodiac) and the angles, he makes his subjects lovers of the body, strong-minded, deep thinkers, austere, of a single purpose, laborious, dictatorial, ready to punish, *lovers of property, avaricious*, violent, *amassing treasure*, and jealous..."[93]

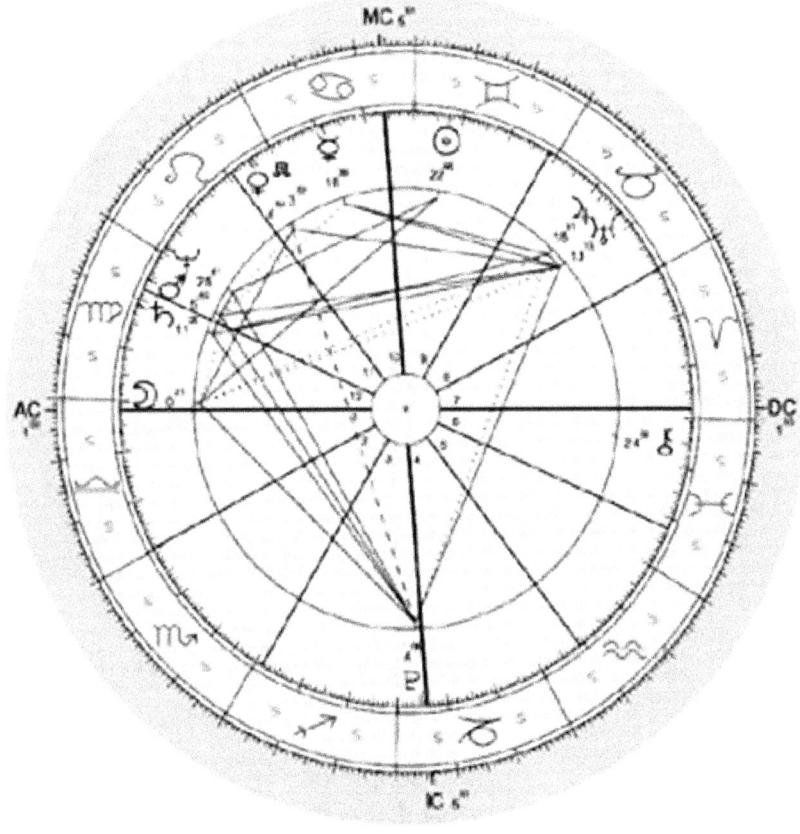

[Figure 6: Horoscope for July 6, 1775]

It is not being suggested that the planetary configurations of either of these horoscopes and their angularity at any particular moment in time are to be

---

[93] Ibid. Page 65. Italics Fagan's.

read in the same way as the horoscope for an individual person, for according to Fagan:

> It is probable that the fixed stars and the degrees of the zodiac, being relatively static, hold dominance over empires, countries, cities and other things – but never the planetary bodies which are in constant motion.[94]

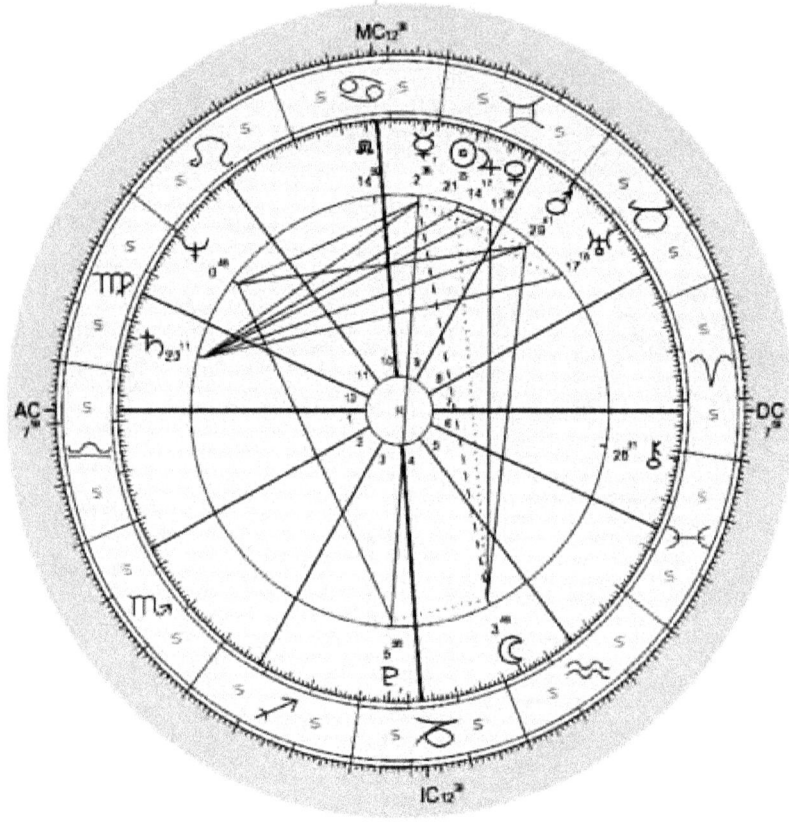

[Figure 7: Horoscope for July 4, 1776]

Of course, the fixed star of primary importance in each of these horoscopes is Sirius, which is essentially conjunct the Sun in both cases, and it may be that the coordination of these momentous events with the heliacal rising of this star was a deliberate and calculated gesture on the part of the

---

[94] Ibid. Page 18.

Founding Fathers. Moreover, the various planetary configurations involved in each chart are most certainly to be seen as affecting individuals, and if there is any objective reality to be ascribed to astrological influences, the groups of individuals involved in the ratification of both Declarations must have been affected by – and possibly attempted to exploit – the various planetary and stellar influences unique to the specific dates under consideration.

The Porta Alchemica

ACCORDING TO LEGEND, the Porta Alchemica in Rome, Italy [Figure 1] was conceived as a memorial to a successful alchemical transformation achieved in Queen Christina of Sweden's alchemical laboratory.[95] The pediment of this "door to nowhere" bears a circular glyph, which is similar in certain key respects to Massey's description of the put-circle of Ptah.

[Figure 1: The Porta Alchemica. Author's rendering]

---

[95] Susanna Åkerman. *The Porta Magica and the Italian Poets of the Golden and Rosy Cross.* Web. <levity.com/alchemy/queen-christina.html>

In contrast to the ninefold divisioning of the put-circle, here we have a sixfold scheme, arranged according to the familiar seal of Solomon, a similar motif being found in the 13 stars above the eagle's head in the Great Seal [Figure 2]. Prominent within this glyph is a representation of the Imperial Globe, which according to Massey refers to the two solstitial periods of the year, with the three upper arms of the cross representing the summer season, and the circle below signifying the wintery "abyss" [Figure 3]. Here at one level then, we simply have a depiction of a zodiac of sorts. The upward-pointing triangle can be read as a masculine lingaic element, corresponding to the Sun, summer, and day – with the inverted triangle naturally indicative of a yonic element recalling the Moon, night, and winter.

[Figure 2: stars of the Great Seal Author's rendering]

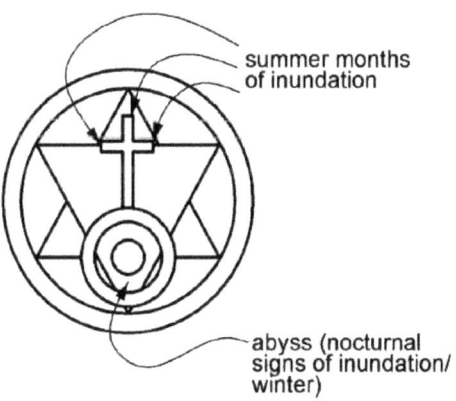

[Figure 3: Pediment of Porta Alchemica. Author's rendering]

Various planetary glyphs and inscriptions are found to either side of the portal, which is itself flanked by statues of the pygmy god Bes, who according to Massey, "…is a figure of Child-Horus … He comes capering into Egypt along with the Great Mother, Apt, from Puanta in the far-off south. In reality, Bes-Horus is the earliest form of the Pygmy Ptah."[96] If we consider Ptah as the "excavator of Amenta" and "opener of the way," an association with the eastern and western horizons becomes evident, as he was seen to have hollowed out the underworld for the safe passage of the Sun, or Osiris, through it at night [Figure 4].

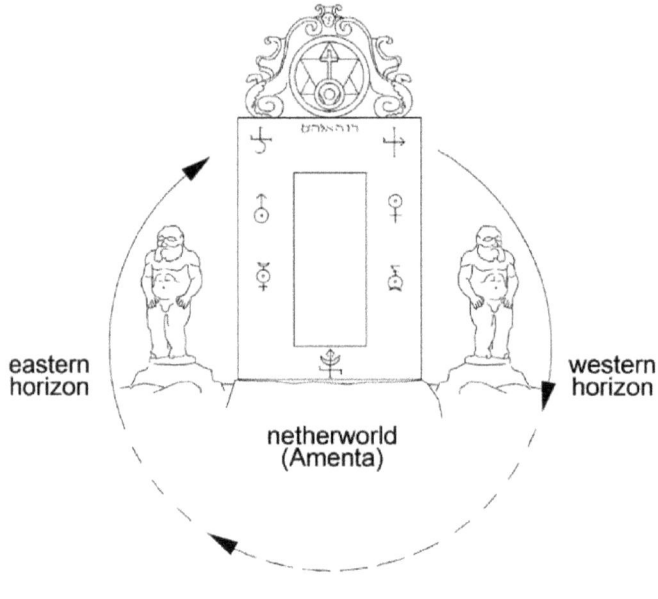

[Figure 4]

Regarding the symbols arranged on either side of the portal, and the 7th symbol at the base, it is clear that at least 5 of them are planetary, depicting – starting from the lower left and working clockwise – Mercury, Mars, Saturn, Jupiter, and Venus. The other two symbols have variously been labeled as either the Sun and Moon, or else the alchemical substances antimony and vitriol, but these designations have no authoritative basis, and in many ways seem inadequate.

With respect to the proposed ascription of the Sun and Moon to these two "mystery glyphs," one is hard-pressed to imagine that these complex symbols would be used instead of the much simpler ones which are familiar to all (☉, ☽), in light of the fact that the other 5 planetary glyphs are essentially congruent with the standardized versions. If anything, the lower right so-called "solar" glyph, and the bottom "lunar" glyph of the portal's base more closely resemble the planetary sigils of Uranus and Neptune, respectively (♅,

---

[96] Massey. *Ancient Egypt, the Light of the World.* Book 5. Page 250. Ellipses added.

♆). Similarly, the standard symbol for antinomy (♁) is simply an inverted sign of Venus, which calls into some doubt the attribution of that alchemical compound to the lower right sigil, which exhibits an additional pointed crook-shaped element at the top, and an inverted lunar crescent at the bottom.

The so-called vitriol symbol at the base more closely resembles the Calx Vive symbol as given in Valentine's *The Last Will and Testament*, as seen in this reproduction of the table of alchemical characters from Manly P. Hall's *Secret Teachings of All Ages*: [Figure 5]. This Calx Vive, or "Living Stone" is none other than the Philosopher's Stone. The metal antimony, and its associated crystalline species known as the Star Regulus, represent an intermediate stage within the overall alchemical process. If there is to be any antimony symbol proper ascertained within the various signs engraved upon the Porta Alchemica, it is clearly to be found gracing the topmost lintel portion spanning the lingaic and yonic triangles of the Star of Zion [Figure 6].

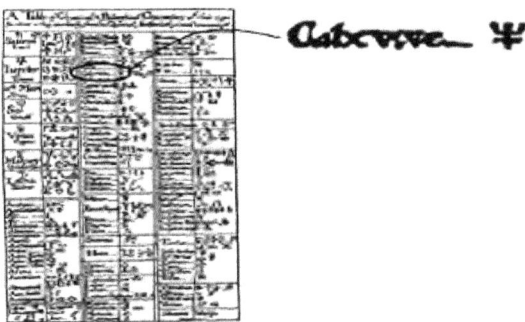

[Figure 5: Alchemical Table from Valentine's *Last Will and Testament*. Author's rendering.]

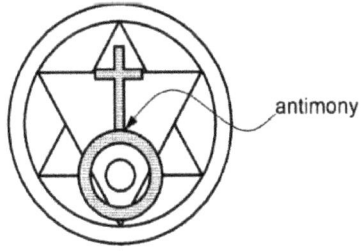

[Figure 6: The Antimony Symbol of the Porta Alchemica. Author's rendering.]

If we consider the portal as a whole to represent a type of zodiac with the topmost portion representative of the 3 summer months of inundation as codified in the ancient Egyptian Sothic mythos – namely Virgo, Leo, and Cancer, with Leo (and thus, Regulus – doubling as the "Star Regulus," or antinomy, of alchemy) occupying the solstitial apex – the lower portion of the abyss, designated by the symbol at the base resembling that of Neptune would correspond to the nocturnal "water signs" of inundation: Aquarius, Pisces, and Capricorn.

Of course the existence of Uranus and Neptune was supposedly not known of during the era under consideration, and it is not being suggested that the signification of the actual planets was intended for the two glyphs at the lower right and base of the portal. Let us take the lower right symbol to be indicative of Uranus, or the ancient Egyptian Urnas, who according to Massey was representative of "the celestial water out of (nas) which all came at first...a river that runs through the fields of the Aahlu (Elysium) cultivated by the osirified deceased."[97] This is the nocturnal firmament of the underworld, typified by the Milky Way. Viewing the Porta Alchemica as a representation of the celestial sphere oriented such that the topmost portion corresponds to the ecliptic pole region centered in Draco, and looking from within the sphere such that this pole is centrally justified, passing down through the constellation Leo which occupies the apex of the ecliptic arch as the summer solstitial keystone and "regulus" of the year, the following arrangement presents itself [Figure 7]:

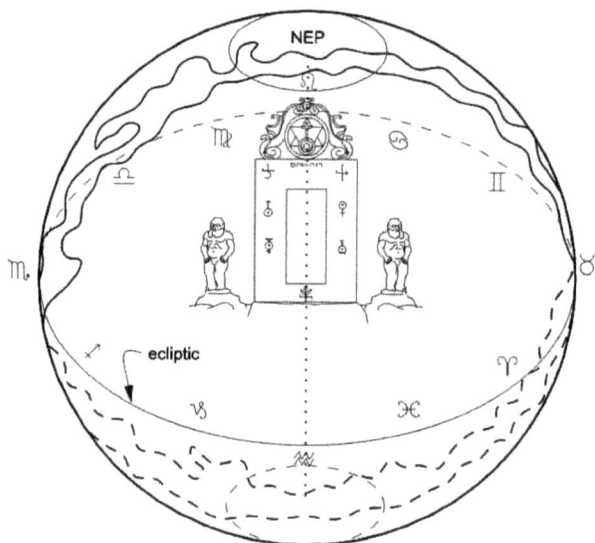

[Figure 7]

As previously noted, the figure of Bes which flanks each side of the portal is symbolical of either horizon, and calls to mind the functionality ascribed to this god by the Ancient Egyptians as the hollower out of Amenta. That such a reading was intended to be applied at some level to the portal is corroborated

---

[97] Massey. *The Natural Genesis*. Book 9. Page 26.

by the Latin inscription at its base, which translates roughly as "It is an occult work of wisdom to open the earth, so that it may generate salvation for the people."

The symbol at the base – if read as indicative of Neptune – refers not to the planetary god, but to the Roman Neptune as god of the nocturnal waters of the abyss or underworld, "…the god who completed the circuit round, the protector by night, the seer unseen."[98] In our current orientation of the celestial sphere, this refers to the region of the Duat containing the extra-zodiacal constellations of Columba, Phoenix, Carina, and Canis Major. This is roughly the region of the South Celestial and Galactic Poles, and here we must call attention to the associations of Isis with the dove (Columba), phoenix, Canis Major, and the star Sirius. In fact, this entire celestial region of Carina and Canis Major is known collectively as the barque of Isis, i.e. the stellar matrix within which this resplendent blue Goddess traverses the night skies.

The Hebraic inscription above the portal's lintel translates to *Ruach Elohim*, or "Spirit of the God(s)." According to E.J. Langford Garstin:

> The Throne of God, under which imagery our Fire is presented to us, is the Briatic World. And this is composed of the Sefiroth Chokmah and Binah, the Father and Mother, Wisdom and Understanding, Chiah and Neshamah, the positive and negative aspects of the Divine Mind of the fully initiated and regenerated man, which form, as it were, the Mercava or Chariot of Yechidah, their synthesis and the Divine Self.
>
> Of these, Chiah as we have pointed out, is, as it were, the living vital principle, so that they correspond almost word for word with our quotation "life and understanding." They form the soul of man in his higher aspect, as Sendivogius tells us, whereby man is distinguished from other creatures and resembles his Creator; and its fiery nature is indicated in that *Ruach Elohim*, whose symbol is the letter Shin, the Mother Letter of the element of Fire.[99]

In this connection, it is interesting to refer to the following sigil [Figure 8], which Eliphas Levi presents without any overt explanation in his chapter on Transmutations in *Dogma et Ritual de la Haute Magie*:

[Figure 8: A sigil from Levi's *Dogma et Ritual de la Haute Magie*.]

---

[98] Ibid. Book 8. Page 526.
[99] Garstin. *The Secret Fire*. Chapter IX. Italics added.

THE PORTA ALCHEMICA

Here again we have at one level a zodiacal diagram, or encoded representation of the celestial sphere. The triple tau motif in the top half of the circle has been shown by Massey to signify the 3 summer months of inundation, and if we reference the original Taurean epoch of ancient Egypt from whence such symbolism apparently derives, we thus may ascribe the zodiacal signs of Virgo, Leo, and Cancer to each of the taus, corresponding to the O, T, and A.[100] The chalice below corresponds to the watery Abyss; the Underworld; the land of Amenta. If we consider the top portion of the circle to correlate with the circumpolar region of the celestial sphere, it becomes evident that the central axis represented by the solar ecliptic pole is associated with the fiery element, represented by the Hebrew letter *Shin* [Figure 9].

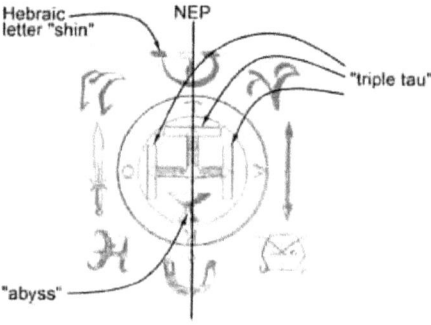

[Figure 9: Levi's sigil interpreted with respect to the celestial sphere.]

In this context, *Shin* relates to the sefirot *Chochmah* and *Binah*, but it should be noted that both of these "briatic" sefirah are united at a yet higher "atziluthic" level via the sefirah *Keter*, and in a sense, it is *Keter* which can be seen to most fittingly correspond to the infinite upward extent of the North Ecliptic Pole; centered in the Cat's Eye Nebula, in the constellation Draco. Another reference to Draco and the NEP may appear in the Latin inscription just below the Hebrew "*Ruach Elohim,*" which translates to "A dragon guards the entrance to the magical Garden of the Hesperides, and without Hercules, Jason would not have tasted the delights of Colchis."

There is yet another level from which these various symbols can be interpreted. According to the Hermetic dictum "as above, so below" the individual can be seen as a microcosm of the universe as projected from the terrestrial locus out onto the celestial sphere. In this case, the central axis of the NEP, which reaches up to the circumpolar regions – known in the Hindu

---

[100] Alternatively, the two "horizontal" taus can be read as designating the two equinoctial loci; i.e., Taurus and Scorpio, whereas the middle (vertical) tau would indicate the summer solstice (i.e., Leo, during the Age of Taurus).

cosmology as Mount Meru – corresponds to the spinal cord, spanning from the lower generative region to the central solar plexus nerve ganglia, to the upper diencephalon and cortical regions of the brain.

According to Garstin:

> In the first place the body may be considered as divided into two main portions, namely the head and trunk together, and the legs. The centre between them is at the base of the spine, and *running throughout the whole trunk is this spinal cord, the axis of the body as Mount Meru is that of the earth*.
>
> In the trunk are the seven Lokas or regions to which the seven Chakras, Centres or Lotuses correspond. These, working from the base up, are Muladhara, Svadisthana, Manipura, Anahata, Vishuddha and Ajna, the highest being Sahasrara. This latter is the highest centre of the manifestation of consciousness in the body, and is thus the abode of the supreme Shiva-Shakti.[101]

The mystery of alchemy at the human scale is fundamentally concerned not with the literal transmutation of lead into gold, but with the refinement of the latent generative energies in such a way that otherwise dormant regions of the brain are activated, new chemicals (some of which are psychoactive) are produced, and a higher state of consciousness is achieved; one reaches Nirvana in a sustainable fashion, and begins to resonate with the astral and planetary influences constantly impinging upon us that heretofore have rung only upon deaf ears. In short, one achieves a real communion with the divine.

The Porta Alchemica can thus be seen as an encoded reference to the relationship between the human microcosm and celestial macrocosm, which shows its creators to have been intimately familiar with the structure of the celestial sphere. The quotation at the portal's base, which references the hollowing out of Amenta – as well as the statement "If you make the earth fly upside down with its wings, you may convert torrential waters to stone" found under the symbol for Venus – can be read as cryptic references to a knowledge of the diurnal axial rotation of the earth. It is almost a certainty that the portal's architect was aware of the heliocentricity of the solar system – knowledge that the Church may not have seen fit to publicly disseminate at the time. It was not until the latter part of the 17th century that the Copernican-derived model was officially accepted by the Church, and we must remember that Galileo was threatened with torture in 1633 by the Roman Inquisition if he would not recant his views on the heliocentric model – ultimately being placed under house arrest for the last 8 years of his life (1634-1642).

---

[101] Garstin. *The Secret Fire*. Chapter V. Italics added.

I The Magician

IN *THE PICTORIAL Key to the Tarot*, Waite describes this trump as "having the countenance of divine Apollo," referring to the lemniscate placed above the figure's head as "the mysterious sign of the Holy Spirit." He then goes on to suggest that this symbol is connected to Christ, and the "Jerusalem above."[102]

The lemniscate may be seen to represent the analemma, or the apparent annual path of the Sun as seen from Earth. If one were to record the position of the Sun from the same vantage point – and the same time of day – for an equally spaced number of intervals throughout the course of the year, the resultant shape would be just such a "figure 8." Placement of the lemniscate above the head of the figure suggests an identification with the ecliptic axis; the angular positioning of the figure's arms corresponding to the celestial axis. Further corroborating such a reading of this trump is the ouroboric belt around the figure's waist, which stands in the same relation to the central axis of the figure as the ecliptic does to the ecliptic axis. Additionally, the angle formed by the central axis of the magician's body and extended arms is 23.5°. This is the exact angular relationship between the ecliptic and celestial axes [Figure 1].[103]

These considerations suggest a solar connotation to this card, and although such may be the case, a more complete reading would include the relationship between the ecliptic and celestial axes vis-à-vis the phenomenon of precession.

The Magician can thus be seen to embody the mysterious forces[104] at play which are responsible for equinoctial precession, representing the mediating,

---

[102] Waite. *The Pictorial Key to the Tarot*. Part II; Section 2; The Magician.

[103] An alternative reading reverses this relationship, such that the NCP is considered as the central "fixed" axis, with the NEP appearing to rotate about it throughout the course of the year. At this timescale – the "mundane" year of 365 days, as opposed to the "Great Year" of 25,920 years – the Magician's ouroboric belt represents the annual journey of the Sun as witnessed from a geocentric perspective.

[104] These "mysterious forces" are of course nothing more than gravitational interactions, from the scientific perspective. Whether one views precession as the result of the gravitational effects of the Sun and Moon on the earth, or the rotation of our Sun around a companion star,

or binding force between the stable ecliptic axis centered in the constellation Draco, and the mutable celestial axis which circumambulates the former endlessly – in a sense, declaring an eternal obeisance.

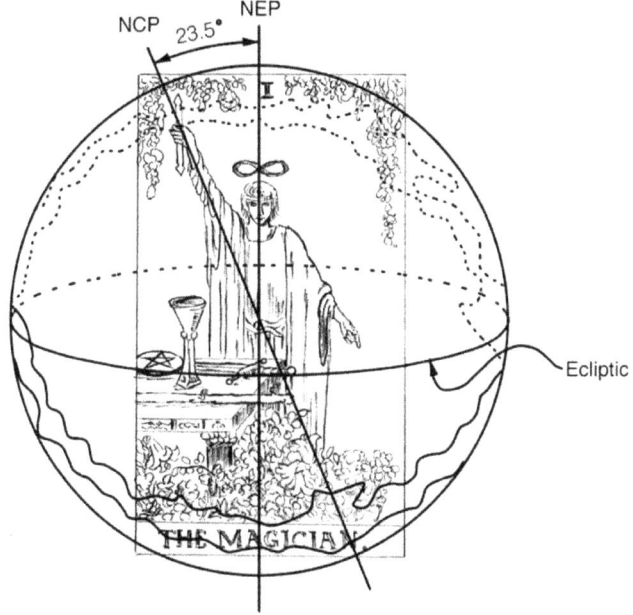

[Figure 1: Author's rendering of the Rider-Waite Magician trump, interpreted with respect to the celestial sphere.]

Massey refers to the precessional relationship between the northern circumpolar extents of these two axes in his discussion of the following passage from the Egyptian Book of the Dead:

> "Let me make head for thy staircase. Let me have charge of those who convey thee, who are attached to thee, and who are of the stars that never set." These are the seven that pull at the rope, or as we should say, that keep the law of gravitation and equipoise; the seven arms of the balance, or the seven bonds of the universe; the seven tow-ers that became later the seven rowers, sailors, or Kabiri. These are sometimes called the seven spirits of Annu, that is at the pole, the mount of glory in the stellar mythos.[105]

The circumpolar concept of the mount of glory was a function of the stellar phase of Egyptian chronometry. Later, when the mode of timekeeping had moved into the solar phase, the mount became a figure of the equinoctial

---

the common element is gravity. Perhaps it is not too far off to suggest that the gravity of science is in many ways analogous to the Holy Spirit of theology.

[105] Massey. *Ancient Egypt, the Light of the World.* Book 9. Page 551.

I THE MAGICIAN

Sun on the horizon.[106] This dualistic representation of the heavenly mountain may in fact explain why the lemniscate in the Rider-Waite trump is depicted horizontally, rather than vertically, for the horizontal lemniscate is the symbol par excellence of the Sun as seen facing east (or west) at the equator, whereas the vertically oriented glyph indicates a N-S orientation.[107]

According to Massey:

> The Egyptian rule of perspective is positively based upon the right hand being the upper, and the left the lower. In the scenes of Hades the blessed on the right hand are represented as those above, whilst the damned, those on the left hand, are down below. So that in facing the east, their upper land of the south was on the right hand, the lower on the left ... The Egyptians had a Sabean orientation still more ancient ... Time was when the right hand was also the east, ab, the right hand and the east side. On the other hand, the west (sem) is the left side, semhi is the left hand, Assyrian samili, Hebrew samali. Now for the east to be on the right and the west on the left, the namers must face the north, and that this the region of the Great Bear was looked to as the great quarter, the birthplace of all beginning, is demonstrable. Those, then, who looked northward with the east on their right and the west on their left hand were naming with their backs to the south and not their faces; nor were their faces to the east. This mode of orienting was likewise the earliest with the Akkadians, who looked to the north as the front, the favourable quarter, the birthplace. They, like the dwellers in the equatorial regions, had seen the north was the starry turning point, and the quarter whence came the breath of life to the parched people of the southern lands.[108]

We thus see both orientations described succinctly within a single image in the Rider-Waite Magician, and this motif of the interchangability of both perspectives (equatorial, and circumpolar) will be seen to be used in several other trumps as well.

Waite says the following, regarding the Magician's lemniscate:

> With further reference to what I have called the sign of life and its connexion with the number 8, it may be remembered that Christian Gnosticism speaks of rebirth in Christ as a change "unto the Ogdoad."[109]

The gnostic concept of the ogdoad refers to the ancient geocentrically conceived notion of an eighth celestial sphere, existing outside the Saturnian "7th heaven." Each sphere was arranged concentrically according to the

---

[106] Ibid. Book 6. Page 348.
[107] The vertical lemniscate is achieved by recording the Sun's position *at its zenith* while facing due south (in the Northern Hemisphere) or north (in the Southern Hemisphere) throughout the course of the year.
[108] Massey. *A Book of the Beginnings*. Section 1. Page 15. Ellipses added.
[109] Waite. *The Pictorial Key to the Tarot*. Part II. Section 2. The Magician.

Chaldean order, which posited an increasing radial distance of each planet from the central location of Earth in proportion to the apparent speed of the particular luminary in question [Figure 2].

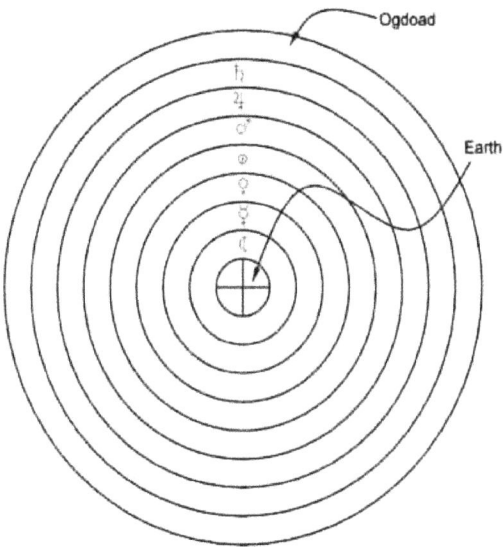

[Figure 2: Classical Chaldean order of planets, and the Ogdoad]

The ultimate result of the striving for a permanent state of higher consciousness is achievement of the gnostic "divine Anthropos," and a relative immortality: this is the attainment of the Ogdoad.[110] The following engraving by Rosicrucian apologist Robert Fludd depicts a divine figure reaching from Heaven to Earth, spanning each of the celestial spheres between. This is reminiscent of the divine Androgyne, divine Anthropos, and Adam Kadmon, and can be seen to represent an anthropomorphization of the cosmic forces connecting the mortal and divine via the great chain of being. The figure grasps a chain in each hand, one connecting it to the terrestrial based "ape of Thoth," representing the celestial pole;[111] the other reaches up past the highest

---

[110] This is based on Gurdjieff's division of "man" into 7 categories, as expounded by Ouspensky (See *In Search of the Miraculous*, Chapter 2). One can infer that it is "man number 7" who possesses the "causal body," and is "immortal within the limits of the solar system" (*In Search of the Miraculous*. Chapter 5. Page 94); whereas "man number eight" has fully crystallized that body, and is working to develop an even more rarefied metaphysical constitution. Gurdjieff equated Christ with "man number eight" (*In Search of the Miraculous*, page 319).

[111] The ape was at one time a figure of the pole in the Egyptian mythology (see Massey. *Ancient Egypt, the Light of the World*. Book 11. Page 721). This is not to preclude the association of Taht (Thoth) with the Moon, as he was – presumably during the lunar or luni-solar phase of chronometry, which followed the earlier Sabean mode. But Taht was also ruler of *Smen*, the circumpolar heaven associated with Ursa Major (see Massey. *A Book of the Beginnings*. Book 13. Page 83) – station of the pole during the Age of Gemini/Taurus.

I THE MAGICIAN

celestial regions to the cloud of Jehovah, the ineffable realm associated with the ecliptic pole, and *Keter*, which is centered in the Cat's Eye Nebula [Figure 3].

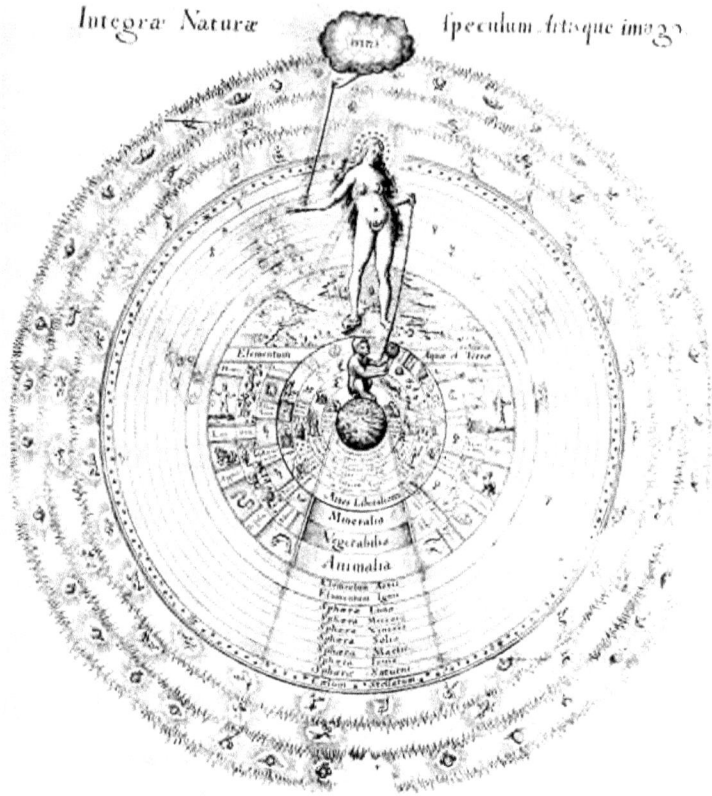

[Figure 3: Designed by Robert Fludd. Frontispiece to *Utriusque Cosmi, Maioris Scilicet et Minoris, Metaphysica, Physica, Atque Technica Historia* (*The Metaphysical, Physical, and Technical History of the Two Worlds, Namely the Greater and the Lesser*) by Robert Fludd, published in Germany circa 1617-1621 CE. Image source: Wikimedia Commons.]

According to Case, the lemniscate symbolism of this trump refers to a cross-section of the spinal column, the two lobes of the figure signifying the Ida and Pingala nerve currents which must be harmonized to pull the Kundalini up through the center of the spinal column, or Sushumna, from the lower to the higher chakras. This is achieved through a mastery of the Solar Force via the transmutation of the generative energies, mediated by the Magician (objective mind).[112]

Since the overall import of this trump extends far beyond the limits of any single planetary sphere, it is not ascribed to any of the Hebraic letters with

---

[112] Case. *The Secret Doctrine of the Tarot.* Chapter 2.

planetary values. Instead, it is suggested that *Alef* is a more fitting candidate, as it is the mediating and aetheric force of the supernal triad, אמש.[113]

The Magician as a form of Apollo (per Levi, Waite, and Crowley) or the Sun can also be read as a reference to Horus, whose Greek equivalent is Hercules.[114] This circumpolar constellation is the locus of the NCP during the Age of Leo, which is ruled (in the Ptolemaic system) by the Sun. Hercules and Leo can be considered the starting point of the Great Year, as they correspond to the circumpolar and ecliptic loci of the polestar and vernal Sun, respectively, during the Golden Age – of which the Great Sphinx is said to be a monument.

We thus see the Magician as primarily connected with the mediating or binding force which ties the celestial and ecliptic axes together in their dialectic relationship throughout the course of the Great Year. Grafted upon this central meaning – as a subordinate branch, of sorts – is the notion of this trump representing the primal locus of the NCP in the course of the Great Year, corresponding to the Golden Age of Leo. This second layer of meaning will take on additional significance as we move through the next three Trumps, as it will be seen that the sequence from Magician (I), to High Priestess (II), to Empress (III), relates primarily to the three circumpolar loci of the Masseian heptanomis which correspond temporally with the ancient Ages during which the reckoning of time was Sabean (i.e. nocturnally reckoned by the stars) and ostensibly matriarchal; whereas the Emperor inaugurates the Age of Aries, in which a luni-solar chronology predominates, concomitant with the inception of a global patriarchal regime.

---

[113] Kaplan. *Sefer Yetzirah*
[114] Massey. *Ancient Egypt, the Light of the World.* Book 5. Page 320.

II The High Priestess

THE POMEGRANATE MOTIF in both this and the Empress trump recalls the myth of Persephone.[115] For reasons that will be elucidated in the next chapter, it seems more fitting to ascribe Persephone to the Empress. The pomegranate was also associated with the Greek mother goddess, Rhea/Cybele[116] ("Mother of the Mountain"), who was depicted as flanked by two lions, or riding a chariot pulled by them [Figure 1].

[Figure 1: Rhea (Cybele) in chariot drawn by 2 lions – from Ai Khanoum, Bactria (Afghanistan), 2nd century BCE. Image source: Wikipedia.]

---

[115] The reference here is to the Rider-Waite versions of these trumps.

[116] Massey, in *The Natural Genesis, Book 10* mentions the following with respect to this Goddess: "Rhea, or Kubele [Cybele], was worshipped in two characters. She was adored in Phrygia as Idaia-Mater, the mother of knowledge, and held in her hand the pomegranate sign of her own full-wombed fruitfulness; a type also of the seed within herself." Bracketed comment added.

This double lion imagery may originate from the ancient Egyptian concept of the dualistic "mount of the equinox." The use of the lion as zootype for the depiction of the Sun on the eastern and western horizons during the equinox derives – according to the "alternative Egyptological" perspective – from some 13,000 years ago, when the vernal point was in Leo. According to Massey:

> In the eschatology it is said of the house on high, "Tum buildeth thy dwelling, the Lion-faced God (Tum or Atum) layeth the foundation of thy house, as he goeth his round" in fulfilling the solar circle, which was completed with the twelve thrones, twelve stars, twelve gates, or twelve foundations of the final zodiac. This foundation, as the imagery shows, was extant at the time when the solar lion-god first rose up in the strength of the double lions, and the mount of the vernal equinox was in the sign of Leo.[117]

The lion was subsequently (in the Age of Taurus) associated with the beginning of the inundation in Egypt, when the Sun had entered Leo during the summer. The Mother of the Mountain (Rhea) was associated with Ursa Major during this period,[118] and it seems most likely that the depiction of this goddess in connection with the lions in [Figure 1] derives more immediately from this later phase.[119] It is postulated that the Mother of the Mountain was primordially associated with the NEP, as the unchanging source of the waters throughout the course of precession. This concept will be discussed further in the chapter on the Devil trump.

Reference to the heptanomis (see [Figure 2]) shows that the circumpolar astronome adjacent to Hercules is Corona Borealis, which Massey considered the precessional "keystone."[120] The method of dividing the polar precessional cycle into 7 units based on the same number of discrete constellations naturally produces some degree of overlap in relation to the corresponding zodiacal Ages. Thus, the polar station of Corona Borealis is associated with both the Ages of Leo and Cancer.

As Clavis Corona was "primary polestar of the seven which formed the circle of the crown,"[121] the proposed connection of this trump with Corona

---

[117] Massey. *Ancient Egypt, the Light of the World*. Book 11. Page 716.
[118] Massey. *A Book of the Beginnings*. Section 14. Page 148.
[119] The bas-relief dates from the latter part of the Age of Aries, which suggests the double force of the lions to have been associated with the summer heat. It is possible that the crowned figure at the top represents Sirius. The crescent Moon and Sun-like symbol to the right may refer to the New Moon either coming out of or entering into conjunction with the Sun. Overall, the imagery appears to signify the heliacal rising of Sirius and the Moon during the period of inundation.
[120] Massey. *Ancient Egypt, the Light of the World*. Book 9. Page 602.
[121] Ibid.

II THE HIGH PRIESTESS

Borealis may be considered in light of Waite's statement that the High Priestess is in some respects the "highest and holiest of the Greater Arcana."[122] During the Age of Cancer – the sign ruled by the Moon – the polestar was localized in the region of Corona Borealis. Furthermore, in Celtic mythology, this constellation is associated with the mother/fertility/Moon goddess, Arianrhod.[123,124] Finally, Waite describes this trump as representing the "Moon nourished by the supernal light of the Mother."[125] In this statement, an illuminating relationship is suggested: the "supernal light" in this case may signify the continual circumpolar influx of stellar energies from Cair Arianrhod (Corona Borealis) during the "lunar" Age of Cancer.

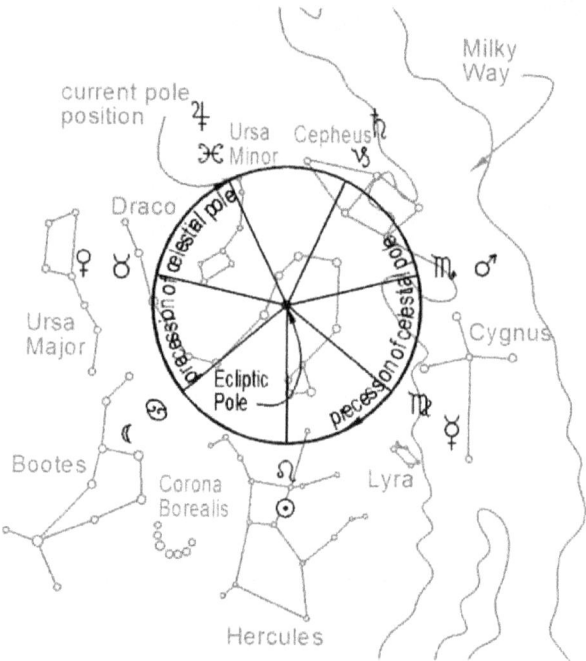

[Figure 2: The heptanomis, depicted with Aeonic planetary rulerships.]

The Greek cross upon the figure's breast[126] can be interpreted in this context to reference the vernal equinox during the Cancerian Age. Also suggestive is the Moon at the figure's feet, which echoes the shape of Corona Borealis. The peculiar diadem gracing the female avatar's head can be seen as a lunar symbol, which references the phases of the Moon [Figure 3].

---

[122] Waite. *The Pictorial Key to the Tarot*. Part II. Section 2. The High Priestess.
[123] Mackillop. *Oxford Dictionary of Celtic Mythology*. Page 24.
[124] Graves, Robert. *The White Goddess*. Kindle edition. Page 95, 99.
[125] Waite. *The Pictorial Key to the Tarot*. Part II. Section 2. The High Priestess.
[126] Cf. the Rider-Waite version of this trump.

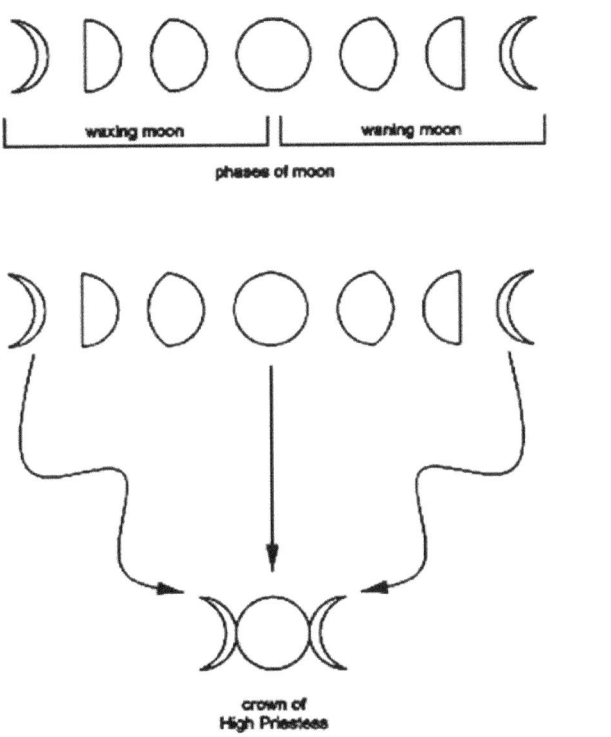

[Figure 3: The crown of the High Priestess as illustrative of the phases of the Moon]

Additionally, it may double as an indication of a fixed point of rotation, signifying the NEP, as the spatial relationship between the crescent Moon at Arianrhod's feet and her lunar crown are analogous to that of the NEP and Corona Borealis, with the Greek Cross between the two signifying the actual location of the NCP during the Age of Cancer [Figure 4]. Waite indicates that the High Priestess was associated with Isis.[127] This is congruent with the present interpretation, as Arianrhod is connected with Isis,[128] and Isis with the Moon.[129,130]

This orientation suggests the pillars of Jachin and Boaz to be indicative of east and west. Typically, these so-called "pillars of Solomon's Temple" are considered to represent various dialectic pairings, such as the north and south,

---

[127] Waite. *The Pictorial Key to the Tarot*. Part II. Section 2. The High Priestess.

[128] In *The White Goddess*, Graves associates Arianrhod with the "Triple Goddess" (page 95), who is subsequently identified with a form of the Virgin Mary (page 138). The Virgin Mary is demonstrated by Massey to be cognate with Isis (see *Ancient Egypt, the Light of the World*; Book 3 – page 136).

[129] Graves, Robert. *The White Goddess*. Kindle edition. Page 98.

[130] Massey. *Ancient Egypt, the Light of the World*. Book 1. Page 27.

II THE HIGH PRIESTESS

or the summer and winter solstices.[131] Levi – from whom Waite was certainly drawing – indicates in his *Dogme et Rituel de la Haute Magie* that such pillars, and associated *Urim* and *Thummim* sphinxes represent "above/below," as well as "east/west."[132] Whether this "above/below" could be interpreted as "north/south" is not made clear, but it seems implicit, as examination of a celestial sphere will show [Figure 5].

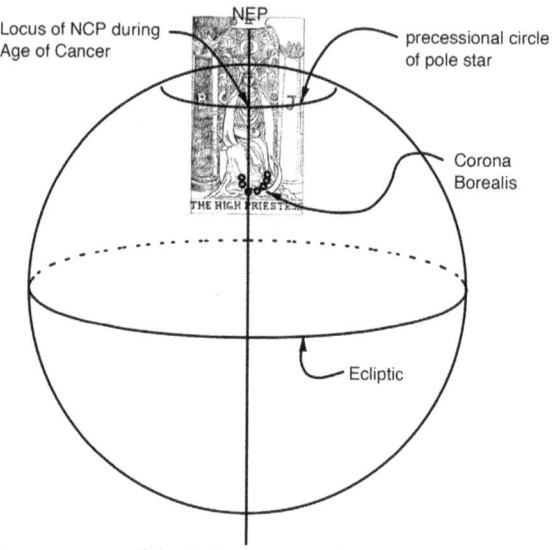

[Figure 4: An interpretation of the Rider-Waite High Priestess trump, with respect to the celestial sphere. Author's rendering.]

Case ascribes *Gimel* to this trump, rationalizing that *Gimel's* shape may have been derived from the ideograph of a bow, thereby connecting this card to the Moon.[133] However, using that logic, it would appear that *Beit* more closely resembles a bow than *Gimel*. Furthermore, according to Massey, *Beit* represents one half of the circle of heaven,[134] i.e., a bow. The fact is, one can imagine anything from the shapes of letters, but does so doing necessarily imbue them innately with the qualities arising in the mind's eye?

Waite refers to the High Priestess as the "Secret Church" – perhaps an allusion to the so-called "Invisible College" of Rosicrucian import, which is depicted in the frontispiece to the *Speculum Sophicum Rhodotauroticum* ("The Mirror of Wisdom of the Rosicrucians") by Theophilus Schweighardt [Figure 6].

---

[131] Hall, Manly P. *The Secret Teachings of All Ages*. Page 308.
[132] Levi, Eliphas. *Dogme et Rituel de la Haute Magie*. Part II: The Ritual of Transcendental Magic. Chapter XXII, The Book of Hermes. Page 129. Translated by A.E. Waite.
[133] Case. *The Secret Doctrine of the Tarot*. Chapter 4.
[134] Massey. *A Book of the Beginnings*. Section 13. Page 120.

CELESTIAL ARCANA

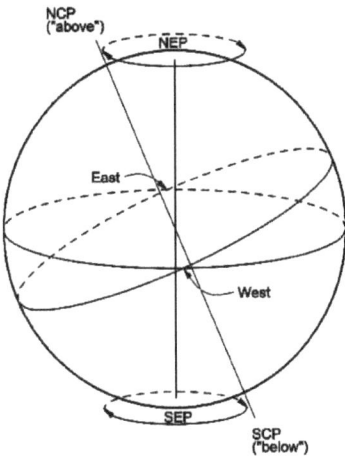

[Figure 5: "Above/Below" as "North/South"]

[Figure 6: Frontispiece to the *Speculum Sophicum Rhodotauroticum* ("The Mirror of Wisdom of the Rosicrucians") by Theophilus Schweighardt. Author's Rendering.]

II THE HIGH PRIESTESS

Although there are undoubtedly several layers of meaning ascribable to the various symbols employed in this illustration, at one level it is in fact a veiled representation of the celestial sphere, oriented such that the vertical extents of the ecliptic axis can be seen to bisect the composition; centrally justified with respect to the space between Corona Borealis and Hercules [Figure 7].

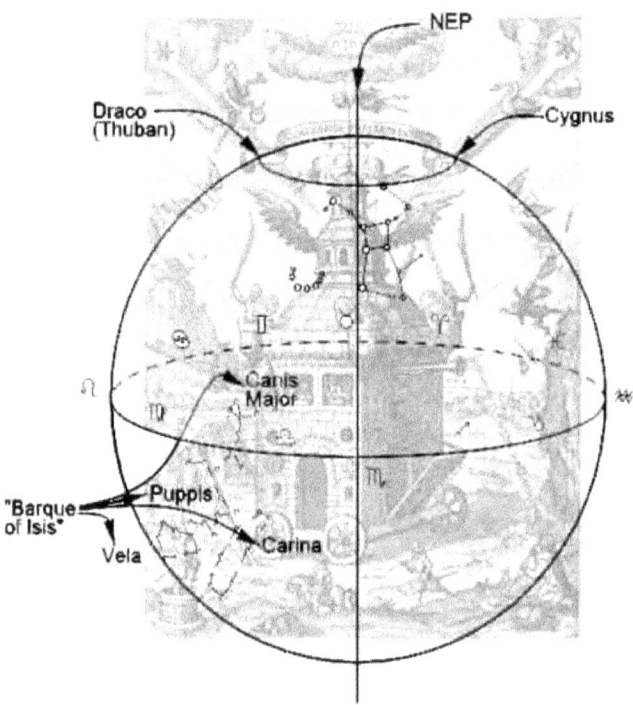

[Figure 7: Schweighardt's "Invisible College" as representative of a specific orientation of the Celestial Sphere. Author's rendering.]

Here again, we see the arm of Jehovah reaching down from the cloud we have previously identified with the Cat's Eye Nebula of Draco. To the right, the Cygnian swan heralds the supernova of 1604 CE, and below this, the figure depicted falling or jumping from the cliff can be read as a dual reference: both to the mythical suicide attempt of Cygnus, as well as to the concept of the changing of Ages; which as we shall discover in our analysis of the Fool, is symbolized by analogous imagery of "stepping off into the unknown." Likewise, the trumpet protruding from the circular window of the tower may refer to the heralding of the incipient Age of Aquarius, as it points directly to the interstitial region between that constellation and Pisces, roughly corresponding to the present vernal point.

To the left of the vertical spire/ecliptic axis, a man holding a serpent floats in the air, contrapuntal to the swan of Cygnus. Typically, this figure is associated with yet another supernova observed in 1604 – this one by Johannes Kepler, in the constellation Ophiuchus (the serpent bearer). Schweighardt himself says as much in his *Mirror of Wisdom*, but there is another undisclosed meaning that may have been intentionally obfuscated by Schweighardt's supernova reference.

Referring to our celestial sphere, it becomes readily apparent that the circumpolar asterism diametrical to Cygnus is none other than Draco; or more properly, the tail of Draco, in which the star Alpha Draconis (Thuban) is located. This star was referred to as Dayan Esiru ("the crown of heaven")[135] by the ancient Sumerians, and served as polestar roughly from 3,400 to 200 BCE, thereby corresponding both to the Ages of Taurus and Aries. Inclusion of the crown atop the head of the serpent is suggestive of this connotation.

Below this crowned serpent, Noah's Ark or the barque of Isis rests atop what is presumably the western equinoctial mount, corresponding to the region of Virgo, Leo, and Cancer on the celestial sphere. These were the three signs associated with Isis that corresponded to the summer flood season during the Age of Taurus, as they rose heliacally with the Sun and Sirius during the months of inundation. During our present era, they have shifted to occupy the autumnal season; hence their association with the mount of the equinox in the west.

On the celestial sphere, the location of this Isiac region corresponds roughly to the hand holding the sword with the admonition of *CAVETE!* ("beware!") inscribed above. We may do well to take heed of this admonition, considering the fact that what we have tentatively ascribed to the western equinoctial mount is also to be associated with the north, as evidenced by the designation of the left edge of the illustration as *Septentrio* (the North Wind).

The diminutive *crux immissa*/Celtic cross hybrid protruding from the cottage below the sword may offer a clue to the solution of this dilemma. The *crux immissa* ("Latin cross") suggests a predominately solstitial connotation – as the crossbar is off-center, corresponding to the inequality of day and night associated with the solstices – whereas the equal arms of the Greek cross represent the equal length of day and night associated with the equinoxes. The hybridization of the two, along with the vertical inscription of "*PER MULTA DISCRIMINA RERUM*" ("through much dividing of the facts") leading down to the so-called "well of opinion" ("*PUTEUS OPINIONUM*"), can be interpreted to suggest that the determination of solstitial and

---

[135] Allen, Richard Hinkley. *Star Names – Their Lore and Meaning.* Page 207.

II THE HIGH PRIESTESS

equinoctial reference points is contingent upon the perspective we choose (i.e. whether that perspective is tropical or sidereal).

Resolution of this apparent discrepancy is not as difficult as it may at first appear, for the cardinal directions written at each of the four edges of the illustration were at one time individually associated with a particular "fixed" star:[136] the North to Regulus ("Watcher/Guardian of the North"), ascribed to the summer solstice; the West to Antares ("Watcher/ Guardian of the West"), given to the autumnal equinox; South to Fomalhaut ("Watcher/Guardian of the South"), associated with the winter solstice; and the East to Aldebaran ("Watcher of the East"), ruler of the vernal equinox.[137]

The cardinal directions associated with these "fixed" stars are clearly nothing more than relictual echoes of a distant past, when the vernal point was actually to be found in the constellation Taurus. This would place the initial codification of the above referenced fiducial system somewhere in the 4300-2100 BCE range. At present, the vernal point can be found in the region of Fomalhaut, some 90° from its "original" Taurean locus.

What emerges from these considerations is the inevitable conclusion that – at one level – this mobile tower was designed to hermetically illustrate the phenomenon of precession. Moreover, the central justification of the circumpolar *"Spiritus Sanctus"* ("Holy Spirit")[138] region we have identified with Corona Borealis and Hercules, suggests prominence was ascribed to this dyad.

This circumpolar juncture between the Sun (Hercules/Leo) and Moon (Corona Borealis/Cancer) marks the end of the Hindu Golden Age, and in a sense, also the "Fall" of humanity, with its associated lapse into lower states of consciousness. If the entire Yuga cycle were divided into two units of day and night, the diurnal Ages of enlightenment would stretch from the Ages of Sagittarius to Cancer, with the nocturnal Ages of relative privation corresponding to Gemini through Capricorn [Figure 8].

The Leo/Cancer juncture which marks the exit from the Satya Yuga may be considered to be centered in Clavis Corona (Alphecca), the key jewel of Ariadne's crown, which is also the Celtic Kaer-Bediwyd – "Ark of the World" – the heptanomal astronome Massey ascribes to the Sun and Moon.[139] According to the present model, it is suggested that this is the station of the

---

[136] These are the so-called "royal stars" of the ancient Persian Magi, who Massey suggests may have been instructed by the Druids. See Massey's *A Book of the Beginnings*, Book 6, page 220.

[137] George A. Davis, Jr. "The So-Called Royal Stars of Persia". From *Popular Astronomy Magazine*, Volume LIII, No. 4. April 1945.

[138] See banner lebeled "SS" on spire in [Figure 6].

[139] Massey. *The Natural Genesis*. Book 9. Page 55.

pole corresponding to the Age wherein the transcendental gnosis was implanted in the womb of collective consciousness before humanity's forgetfulness became too great, and separation from the divine source of the central Brahmanic Sun became the dominant condition. Here the Shekinah (NCP) and Blessed Holy One (NEP) conceived the foundations of a gnosis that might endure the vicissitudes of the Piscean spiritual Winter, in which the ultimate level of separation would occur. [140]

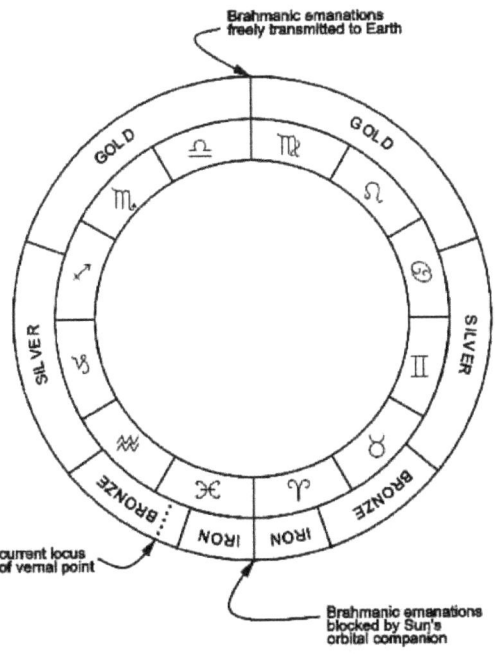

[Figure 8. Author's rendering of Yuga cycles, as per Sri Yukteswar in *The Holy Science*.]

This lunar/solar pair can further be correlated to the primordial Lilith and Adam of Hebraic lore. The temptation of Lilith by the serpent can be interpreted as the listing of the celestial pole from the Alpheccan axis towards the tail of Draco, station of the pole during the "lower" Ages of Taurus and Aries. Massey correlates Lilith with Ta-urt and Ursa Major,[141] corresponding to the circumpolar region midway between Alphecca and Polaris. This locus

---

[140] See Massey's comments concerning "the Fall" in *The Natural Genesis* (Book 12. Page 320), where he suggests the eating of the forbidden fruit to have occurred during the 1st of the "seven days of precession." This corresponds to the first, or primary, astronome of the heptanomis; i.e. Corona Borealis (Massey. *Ancient Egypt, the Light of the World*. Book 9. Page 602) whereas the last (7th) day is figured in the adjacent circumpolar asterism, Hercules (Ibid. Book 9. Page 619).

[141] Massey. *A Book of the Beginnings*. Section 18. Page 383.

II THE HIGH PRIESTESS

of the heptanomis corresponds to the Age of Gemini, which marks the "official" expulsion from the paradise of the Golden Age, as the Autumn or Fall of the Great Year, read also as the "Fall" of humanity – Adam and Eve as the gemini, or original twins, cast forth from the Garden at the end of the Cancerian Age.

Schweighardt provides an image in chapter 2 of his *Mirror of Wisdom* that appears to reference a geocentric view of this lunar/solar axis, a perspective which can essentially be read as if from within the tower of the Invisible College [Figure 9]. Here the solar and lunar circumpolar elements are reversed, and to the right of the central ecliptic axis we see an alchemist with various accouterments in a cavern of sorts. To the left, a man wades in a body of water, working towards some incomprehensible ends with a spoon and bowl or cup. The book before him is opened to display the word "*labore*" ("work").

[Figure 9: Image from Schweighardt's *Mirror of Wisdom*. Author's rendering.]

If we consider the horizon to generally depict the ecliptic, it will be noted that the rising Sun corresponds roughly to the location of the Galactic Center in the region of Scorpio/Sagittarius. The presence of the scales in the cave to the right is suggestive of Libra, and so here again we can surmise the orientation of the implied celestial sphere to be such that the ecliptic axis bisects the zodiacal Scorpio and Libra, as well as the circumpolar Hercules (cognate with Leo, and the Sun) and adjacent Crown of Immortality (cognate with Cancer, and the Moon) [Figure 10].

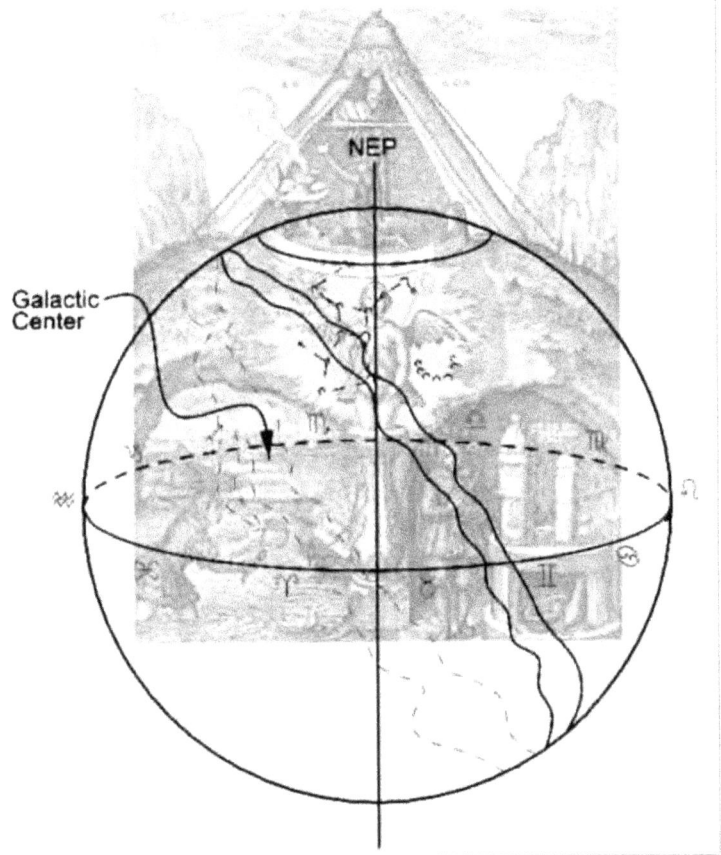

[Figure 10: Image from Schweighardt's *Mirror of Wisdom*, interpreted as representing a view from "within" the celestial sphere. Author's rendering.]

What are these men doing? Perhaps they are to be seen as the workers antecedent to the Fall; those who go before, in the "Golden Ages" of Scorpio and Libra, preparing the way in anticipation of the eventual expulsion from Eden.

This preparation is culminated with the conception of the parergon at the Leo/Cancer juncture which is the equipoise between light and darkness, and the forerunner of the Autumnal Equinox of the Great Year located at Cancer/Gemini, from whence the Fall of humanity proceeds.

The ultimate price for separation from the divine is paid on the vernal Piscean cross: the Sut-Typhonian pole of Ursa Minoris. This marks the

II THE HIGH PRIESTESS

circumpolar nadir, or Winter of the Great Year, and it is interesting to note in this connection that Sut was characterized as a lunar unicorn by Massey:

> The Unicorn was a type of Sut, and the Lion of Horus; and their conflict is described in our legend –
>
> > "The Lion and the Unicorn
> > Were fighting for a farthing,
> > The Lion beat the Unicorn
> > Up and down the garden!
> > The Lion and the Unicorn
> > Were fighting for a crown,
> > The Lion beat the Unicorn
> > Up and down the town!"
>
> The farthing is a fourth; and they fought for a fourthing, or a quarter of the moon; equal to the seven days during which darkness was put to flight; and the crown is the full, round disk of the moon.[142]

In this instance, Massey references a monthly periodicity; but the concept applies equally on the precessional time scale, in which case the "crown" would correspond to Ariadne's Wreath; keystone of the precessional "arch," or mount. One of the earliest zootypes of Sut was the hippopotamus,[143] or unicorn,[144] which Massey considered to be representative of the pole-station corresponding to the region of Corona Borealis/Bootes.[145] The farthing each was vying for would then connote two opposed eighths of the Dark Ages at the precessional nadir – one descending, and the other ascending. The lion beats the unicorn on the ascent from the Typhonian solstice, claiming the wreath of immortality in the Age of Leo; but this crown is then cast away in the Ages to follow, when Lilith is "tempted" by the serpent, and it is here that the unicorn of Monoceros reigns supreme on the vernal mount of the equinox (see [Figure 11]).

The point of equipoise between the lion and unicorn can thus be said to reside at the primary circumpolar astronome of Clavis Corona, where the influences of the Sun and Moon are intermingled at the Leo/Cancer vernal equinoctial mount, collected in the chalice of the Northern Crown.

The remaining 5 astronomes of the heptanomis can be arranged according to the order in which each of the ancient Egyptian circumpolar zootypes serve as the locus for the NCP throughout the course of the Great Year: hippopotamus/crocodile (Ursa Major/Draco), jackal (Ursa Minor), ape

---

[142] Massey. *Lecture #3*. Paragraphs 24-25.
[143] Massey. *Ancient Egypt, the Light of the World*. Book 8. Page 496.
[144] Massey. *A Book of the Beginnings*. Section 13. Page 105.
[145] Massey. *Ancient Egypt, the Light of the World*. Book 9. Page 613.

(Cepheus), ibis (Cygnus), and tortoise (Lyra).[146] These NCP loci roughly correspond to the following ecliptic-based vernal points: Taurus, Pisces, Capricorn, Scorpio, and Virgo, respectively – which in turn give the following order of Ptolemaic domicile rulers: Venus, Jupiter, Saturn, Mars, and Mercury [Figure 12].

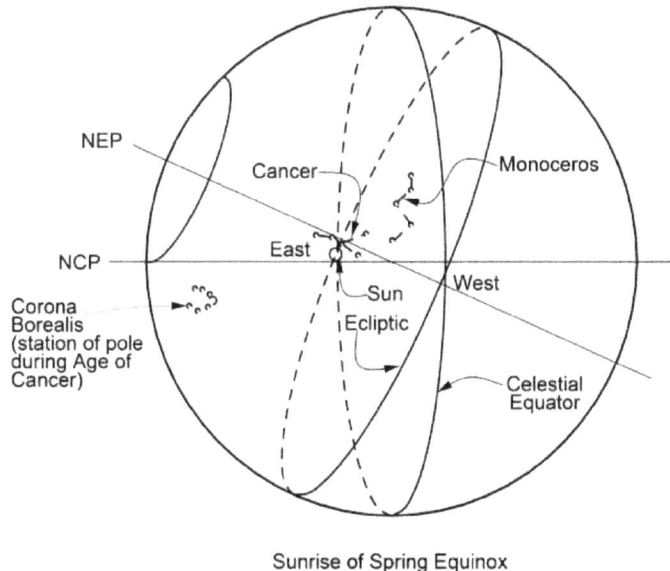

Sunrise of Spring Equinox
circa 7,000 BC

[Figure 11]

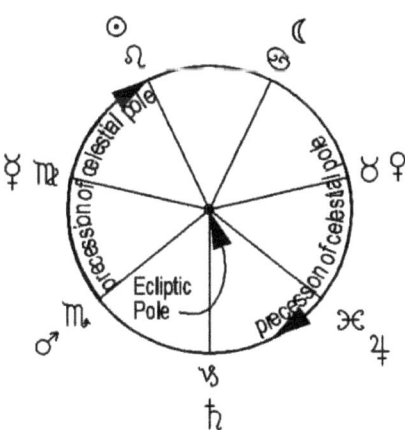

[Figure 12]

---

[146] Ibid. Pages 601-602.

II THE HIGH PRIESTESS

This arrangement of 5 circumpolar astronomes, with Hercules and Clavis Corona completing the heptanomis as the Solar and Lunar keystones of Mount Meru is reminiscent of the V.I.T.R.I.O.L. diagram from *Geheime Figuren der Rosenkreuzer* ("Secret Symbols of the Rosicrucians"), an 18th century commentary on the Emerald Tablet of Hermes Trismegistus [Figure 13]. Here we have the solar and lunar influences of the Sut-Horus nexus being poured into the chalice of Clavis Corona, in this case with Mercury serving as a uniting element to form a kind of supernal triad. This can be seen as reference to the fact that the solar and lunar stations of the pole (Hercules, and Corona Borealis, respectively) are bounded on either side by a Mercurial Age – one at the height of the Golden Age (Virgo), the other at the expulsion from paradise (Gemini) [Figure 14].

[Figure 13: V.I.T.R.I.O.L. diagram from *Geheime Figuren der Rosenkreuzer, aus dem 16ten und 17ten Jahrhundert* (*Secret Symbols of the 16th and 17th Century Rosicrucians*). Author's rendering.]

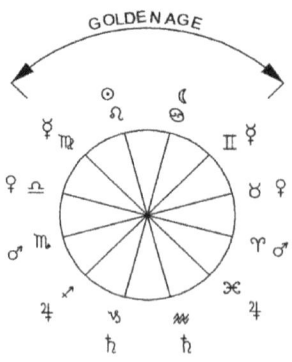

[Figure 14]

Mercury thus acts as messenger, transmitting the sacred light of Brahma via the birth of the Leonean Sun from the Virginal Mother; i.e. through the transition from the Age of Virgo to the Age of Leo. Here Atum/Adam (Hercules) is wed to Lilith (Corona Borealis), who wears the septenary crown of heaven; their union being represented by the mingling of essences into the grail which is carved from Clavis Corona. The seed thus planted, the holy dyad leave their circumpolar mount, as the world axis lists towards the serpent, who appears first as the mercurial serpent of Enki/Ea[147] in the Garden during the Age of Gemini. It was here that the jewel of the crown of Lilith (*Clavis Corona*, or *Gemma*) was "cast down" (i.e., "fell away" from its central locus as NCP), as the *Lapis Exilli* – the Stone of Exile.

Another way of interpreting Schweighardt's view from within the Invisible College is as a representation of the inner physiology of the later stages of the Great Work [Figure 9]. Here we see the head in a coronal cross-section, our point of view presumably facing anteriorly. The two hemispheres are divided sagittaly at the Scorpio/Libra juncture, with the *HINC SAPIENTIA* ("from this, Wisdom") pedestal representing the medulla, pons, and diencephalon regions of the brain, and the embryo of the Parergon corresponding to the pineal gland.

The Sun and Moon in this interpretation refer to the marriage of the higher "masculine" and "feminine" functions of the brain, variously described as Higher Mental and Higher Emotional functions; Purusha and Prakriti; or simply the action of the left and right hemispheres of the diencephalon upon the surrounding regions, after the refined generative impetus has made its way up through the spinal column. The dark cave in which the alchemist performs his secret work corresponds to the right hemisphere of the brain – the intuitive and "lunar" aspect of consciousness. In contrast, the sunlit outdoor activity of the laborer equates to the left hemisphere – the locus of rational and "solar" thinking.

Schweighardt describes the Parergon as follows in *The Mirror of Wisdom*:

> Is this Parergon general or special? The more widely it is practised the better is its effect, and so it shall be spoken of here next. Thou shalt see its theory in the figure on the page: His father Sun (which Trismegistus says), Mother Moon; he bore the wind in his belly, his nurse is the earth. This is the matter and subject of our philosophy *or of our general physiology*, which are provided by time and occasion, not by money. For this thou needest no wishing-cap or bag of fortune, nor special art or athletic speed, but only time and place. Contemplate my figure properly and well, the most important thing is hidden

---

[147] Ea was associated with the "serpent of wisdom" (Massey. Lecture #5. Page 114). This serpent was a zootype of Sut (Ibid. Page 125), who was the first form of Mercury/Hermes (Massey. *A Book of the Beginnings*. Section 1. Page 9).

II THE HIGH PRIESTESS

therein and it is impossible to indicate it more clearly. No father would place it more clearly before the eyes of his son than I have done before thee, wherefore I beg and enjoin thee (lest thou desire aught more useful and profitable to find in this): let this figure be highly and well recommended to thee, observe it, contemplate it, examine it not once but often, for there is nothing included in it in vain, but can be seen with our open eyes, that thou mayest boldly believe, for I am not here as a deceiver but as a brother and friend, wherefore I have not minced my words, but spoken everything freely, openly, and roundly, against the will and good opinion of many.

Schweighardt's Parergon is a version of the "One Thing" mentioned in the Emerald Tablet of Hermes Trismegistus:

True it is, without falsehood, certain and most true. That which is above is like to that which is below, and that which is below is like to that which is above, to accomplish the miracles of [the] *one thing*. And as all things were [created] by contemplation of one, so all things arose from this *one thing* by a single act of adaptation. The father thereof is the Sun, the mother the Moon. The wind carried it in its womb; the earth is the nurse thereof. It is the father of all works of wonder throughout the whole world. The power thereof is perfect. If it be cast on to earth, it will separate the element of earth from that of fire, the subtle from the gross. With great sagacity it doth ascend gently from earth to heaven. Again it doth descend to earth, and uniteth in itself the force from things superior and things inferior. Thus thou wilt possess the glory of the brightness of the whole world, and all obscurity will fly far from thee. This thing is the strong fortitude of all strength, for it overcometh every subtle thing and doth penetrate every solid substance. Thus was this world created. Hence will there be marvelous adaptations achieved, of which the manner is this. For this reason I am called Hermes Trismegistus, because I hold three parts of the wisdom of the whole world. That which I had to say about the operation of Sol is completed.[148]

There are many interpretations possible for this text; the following one is suggested for the present context: The One Thing is representative of the catalyzing and crystallization of higher states of consciousness, in which the pineal gland is awakened from its otherwise dormant state, through the controlled administration of the Solar Force. This One Thing's Father is the Sun and Mother is the Moon. The Sun also represents the Pingala, and the Moon the Ida, two of the central triad of Nadis, or subtle energy currents, which are said to become merged and resonate with the central Sushumna Nadi that travels up the spinal column in the raising of Kundalini energy (i.e., the Azoth).

---

[148] Steele and Singer. *The Emerald Table*. Page 486. Italics and bracketed comments added.

This results in the awakening of the 6th chakra through a merging of the generative and solar forces after a process of refining the Universal Substance (i.e., *spiritus animates*). But this is merely the "parergon," that is, the subsidiary to the main work, just as the NCP is in a subsidiary relationship to the NEP throughout the precessional cycles. The Ergon, or "master work," is the opening of the 7th chakra, above the head, corresponding to the 4th or causal body, the development of which opens many degrees of freedom not accessible to the lower 3 bodies, one of which is immortality within the limits of the solar system.[149]

We thus see two levels of interpretation at work here. The explanation just offered refers to the microcosmic level of the individual. At the macrocosmic level, the Parergon is associated with any discrete locus of the heptanomis (symbolized by the NCP), and thus a specific Age in which an individual may work out their own spiritual ascent. The Ergon, associated with the NEP, symbolizes the heart of the entire precessional cycle, and thus the spiritual evolution of humanity as a whole. This is the Great Work.[150]

---

[149] Author's interpretation. Also see Ouspensky's discussion of the 4th, or causal, body (*In Search of the Miraculous*, page 319).

[150] Cf. Case. *The Tarot: A Key to the Wisdom of the Ages*. Chapter 14: "The [...] march of culture, civilization, and ameloriation ... is called the Great Work." Ellipses added.

III The Empress

THE RIDER-WAITE VERSION of this trump presents us with a number of suggestive alterations from the preceding *Tarot de Marseille* (TdM) and tarrochi/minchiate analogs. Examination of the Minchiate Fiorentine deck reveals a close similarity between its Eastern and Western Emperors [Figure 1] to the Empress and Emperor of the TdM [Figure 2]. In both versions, a complementary relationship between the imperial pair is suggested.

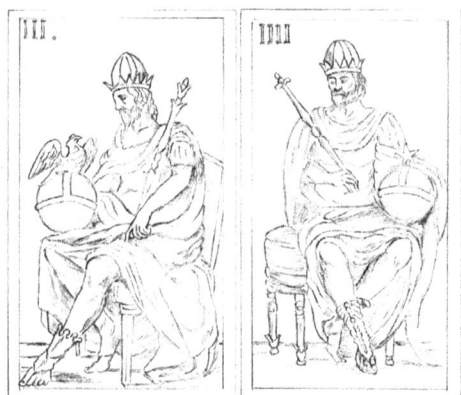

[Figure 1: Western Emperor (III) and Eastern Emperor (IIII) from Minchiate Fiorentine deck, circa 1820 (reproduction of earlier deck). Author's rendering.]

In the TdM and tarrochi decks, the Empress is shown holding a *globus cruciger*. In the Rider-Waite card, we find this symbol echoed obversly as the sign of Venus. These "cross-bearing orbs" appear to be Christianized versions of earlier pagan representations of the celestial sphere, in which the equinoctial loci were indicated by simple X's to denote the intersection of the ecliptic and celestial equator.[151] In the case of the *globus cruciger*, the equinoctial connotation still obtains, for this symbol represents the intersection of the ecliptic with its polar meridian at the equinox [Figure 3].

---

[151] Molnar, Michael. "Symbolism of the Sphere". *The Celator.* June 1998. Pages 1-2.

CELESTIAL ARCANA

[Figure 2: Empress and Emperor of the *Tarot de Marseille* by Jean Dodal, circa 1701. Image source: Wikipedia.]

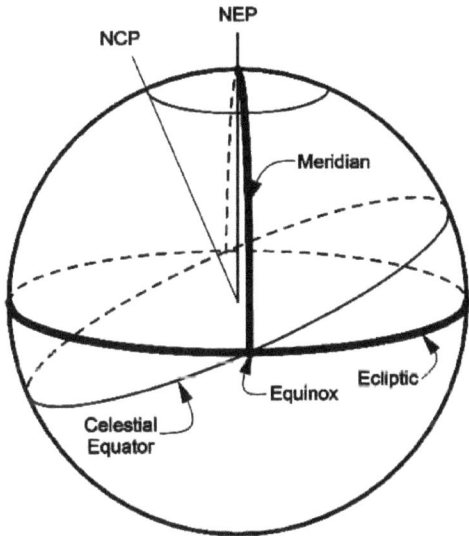

[Figure 3: The *globus cruciger* interpreted as an equinoctial symbol. Author's rendering.]

In the TdM, the two *globus crucigers* are complementarily tinted red and green – calling to mind the corresponding associations of the green shoots of spring, and the full ripe redness of fall, etc. Within the timeframe that the tarrochi was created (1300-1400 CE), the sidereal equinoctial loci were roughly equivalent to Virgo (fall) and Pisces/Aries (spring).

Regarding our hypothesis that the sigil of Venus represents (at one level) an inversion of the *globus cruciger*, it is interesting to note that the symbol for

III THE EMPRESS

antimony is also the reverse of this Venusian glyph. In medieval alchemy, this "metal" represented one of the intermediate steps in the production of the Philosopher's Stone, which was purported to not only facilitate the transmutation of lead into gold, but was also associated with the so-called "elixir of immortality." The specific alchemical operation involving antimony was referred to as "refining the star regulus," an intermediate step involving the admixture of traces of iron to stibnite (an ore of antimony), the result being the creation of a star-shaped crystalline substance from which the product derived its name.

It is undoubtedly no coincidence that the chief star of the constellation Leo is none other than Regulus, the "heart of the lion," and the placement of the inverted antimony symbol within the heart-shaped shield of the Rider-Waite Empress expands her putative celestial dominion to include not only the autumnal equinoctial locus of Virgo, but Leo as well. The stellar crown she wears can be interpreted to signify the 12 stars of Coma Berenice, which mark the position of the North Galactic Pole (NGP) [Figure 4]. Additionally, it should be noted that one of the Chinese appellations for Alpha Leonis (Regulus) is "The Empress."[152]

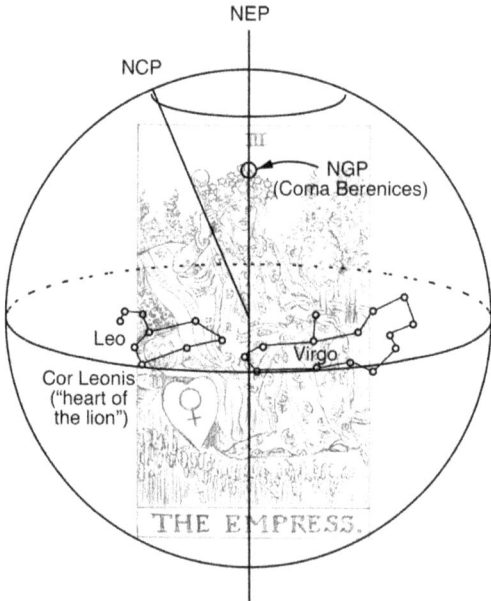

[Figure 4: The Empress as embodiment of the Leo/Virgo/Coma Berenices region of the celestial sphere. Author's rendering.]

---

[152] This is according to Diana K. Rosenberg: "The Chinese called Regulus 'The Empress' (interesting that both the eastern and western traditions considered this star 'royal') and indeed, there are many prominent female leaders here, manifestations of Egyptian Lion-goddess Sekhmet and Ishtar/Venus in her warrior mode."

One of the earliest forms of this trump can perhaps be found in the Locri Persephone pinax of 450 BCE, which depicts the goddess opening a cista from which the solar hero Adonis emerges [Figure 5]. This bas-relief appears to reference the Sun (depicted by the 8-pointed star[153]) in the constellation Crater, antiscion of Leo. The mirror-like device floating in mid-air to the left may symbolize the "mirror of Aphrodite," another name for the glyph of Venus.[154] Taking a view from outside the celestial sphere, the following correlations may be ascertained: the globus/pomegranate[155] on the desk doubles as the symbol for antimony, and by extension the star Regulus in Leo; the Greek krater represents the constellation Crater, and Persephone signifies the constellation Virgo [Figure 6].

[Figure 5: Locri Persephone Pinax, circa 450 BCE. Image source: Wikipedia.]

---

[153] The 8-pointed star has variously been associated with the Sun, Venus, and Sirius. The eight-rayed star is a symbol of the genetrix Ishtar, and Ishtar is cognate with Venus. The connection of the eight-rayed star with Sirius is inferred through the fact that Ishtar was an Akkado-Assyrian form of Hathor, and Sirius is the star of Hathor. It is possible that this is a reference to some form of the phoenix cycle. [See Massey, *Lecture #9*, Page 234; *Lecture #1*, Page 12; *The Natural Genesis*, Book 9, Page 80; *Lecture #9*, Page 135; *Ancient Egypt, the Light of the World*, Book 2, Page 729.]

[154] Stearn, William T. "The Origin of the Male and Female Symbols of Biology". *Taxon*. Volume 11, Issue 4. May 1962: 109-113.

[155] The pomegranate can also be read as a symbol of the celestial sphere, in which the multitude of seeds represent the starry heavens, with the calyx at the top signifying the circle of precessional polestars which circumscribe the north ecliptic pole. This calyx is also said to be the original design influence of the traditional "king's crown," or Solomon's crown [Schram, Peninnah]. These considerations suggest an association of the pomegranate with the *globus cruciger*; in this case, the cross at the top would likely represent the ecliptic axis, which is functionally equatable to the Sun, and by extension, Christ.

III THE EMPRESS

Persephone was anciently associated with Isis,[156] and Isis with Virgo;[157] she was considered the goddess of Death and Resurrection.[158] Circa 450 BCE, Crater and Virgo rose heliacally with the Sun at the autumnal equinox, ushering in the annual period of Nature's senescence. As the myth goes, each year Persephone was obliged to spend the autumn and winter months in the Underworld with Hades, because she had eaten from a pomegranate he had offered to her.[159] The ripe pomegranate was an ancient symbol of late autumn and the death of the year.[160] As the Sun passed into the autumn and winter constellations, Virgo (Persephone) was no longer visible at daybreak, becoming associated instead with the nocturnal heavens, i.e. the Underworld. It was not until the Sun had entered Pisces – some 6 months later, at the vernal equinox – that she began her emergence from the Underworld. It was at that point that Virgo became visible for increasingly shorter nocturnal periods, and by the time the Sun had entered Taurus, the balance had tipped to favor her diurnal tenure.

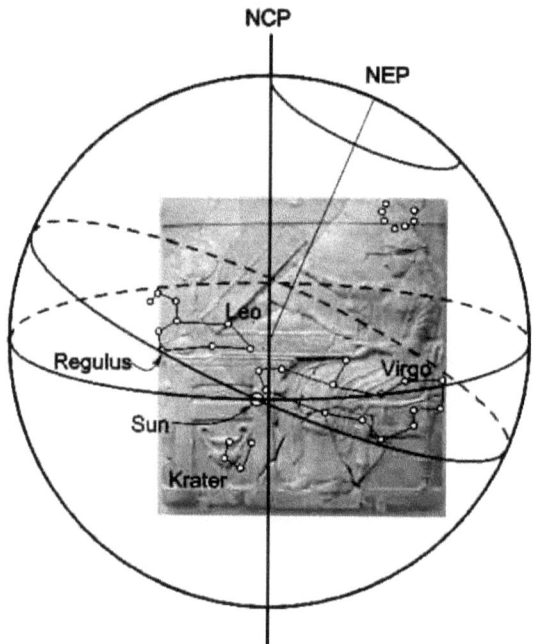

[Figure 6: Autumnal point of Sun circa 450 BCE.]

---

[156] Apuleius. *The Golden Asse*. Book XI, Chapter 47.
[157] Massey. *Ancient Egypt, the Light of the World*. Book 5. Pages 295-296.
[158] Graves. *The White Goddess*. Kindle. Page 250.
[159] Gantz, Timothy. *Early Greek Myth: A Guide to Literary and Artistic Sources*. Johns Hopkins University Press. 1996. Volume 1. Pages 65-67.
[160] Graves. *The Greek Myths*. Kindle. Page 28.

CELESTIAL ARCANA

Waite indicates one of the connotations of this trump to be the gate of entrance of a soul into this world.[161] This would correspond to the so-called Silver Gate, or the Gate of Humanity, located in the region of Taurus, where the ecliptic and Galactic Equator intersect. The diametrically opposed crossing of the same two circles occurs again in the region of Scorpio, and is known as the Golden Gate, or Gate of the Gods – where souls journey out from the Earthly planes, further inwards towards the Galactic Center. These Gates will be discussed in more detail in our analysis of both the Hierophant and Death trumps.

We thus seem to have a strange double reference to the Virgo/Leo/Coma Berenice region of the celestial sphere, as well as the area of the ecliptic/galactic nexus localized in Taurus. The solution is evident, however, upon examination of the relationship of the autumnal point to the vertically oriented ecliptic axis, and the circumpolar heptanomis. If we choose the NEP as our vertical axis (as opposed to the NCP), and trace an arc from the NEP through the region of Coma Berenices (corresponding to the NGP), the resultant arcsegment intersects the heptanomal circle through the tail of Draco, near Thuban, polestar during the Age of Taurus [Figure 7]. During that period, this region – which includes Ursa Major – was associated with Ishtar, "Lady of the Mountain";[162] i.e. goddess of the Celestial Pole.

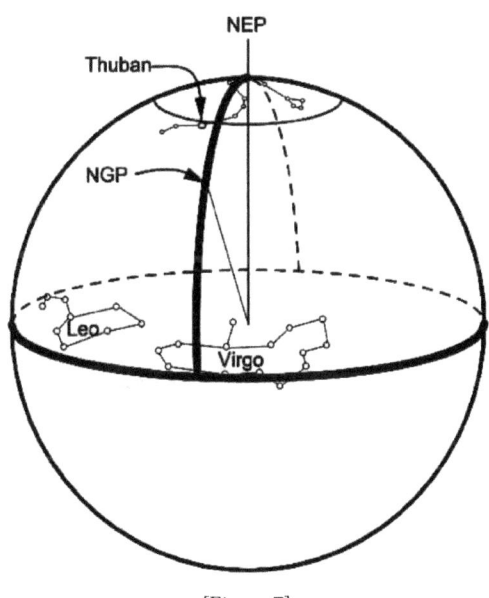

[Figure 7]

---

[161] Waite. *The Pictorial Key to the Tarot*. Part II; Section 2; The Empress.
[162] Massey. *A Book of the Beginnings*. Section 14. Page 171.

III THE EMPRESS

Here, then, the contradiction is resolved: Venus (ruler of Taurus) is attributed to the Empress, and she takes her place upon the mount of the ecliptic pole at the tail of Draco.

This association of Venus with the circumpolar mount and the constellation Draco appears to be referenced in the depiction of Venus Urania[163] ("heavenly Venus") by Christian Griepenkeri (1878) [Figure 8]. Here we see the goddess portrayed as a type of solar avatar, her centrally elevated placement and nimbus indicative of the NEP and its connection to the Sun. To her right, a cherub holds a statuette around which is coiled a serpent; another to her left plays the lyre. In this case, we have a reference to both circumpolar loci associated with Venus: the tail of Draco, corresponding to the Age of Taurus; and Lyra, corresponding to the Age of Libra [Figure 9]. A relationship is here suggested, where the ineffable godhead is associated with the unchanging NEP, while making itself manifest through a variety of discrete forms. These material representations of the spiritual can be seen as the avatars which are said to incarnate throughout the various precessional Ages, demarcated at certain stations of the NCP throughout the cycle of the Great Year. Possible examples of this concept may include the various avatars of the Hindu Vishnu; the solar Christos as Piscean Ichthus; and as we shall see in our discussion of the Hanged Man, Allfather Odin taking on the form of Mercury.

[Figure 8: Venus-Urania by Christian Griepenkeri (1878). Image source: Wikipedia.]

---

[163] According to Levi [*Transcendental Magic, Its Doctrine and Ritual.* Page 134], this trump is indeed attributed to "the Venus-Urania of the Greeks."

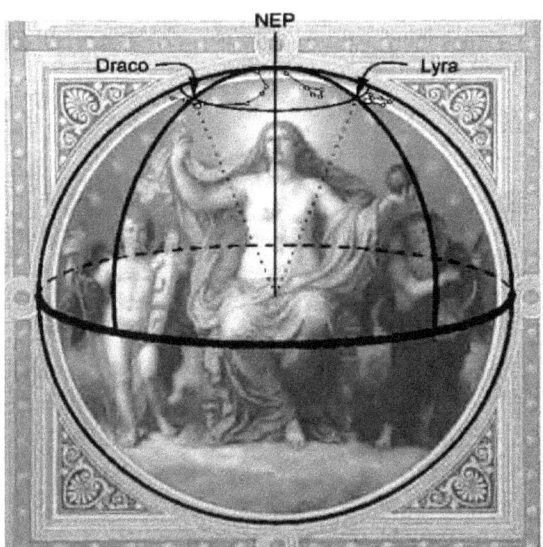

[Figure 9: Venus-Urania as a figure of the Solar and Venusian aspects of the Mount of Heaven.]

IV The Emperor

IN OUR DISCUSSION of the Empress, we considered the possibility that the Venusian symbol represented an inverted form of the *globus cruciger*, referencing the fact that both Eastern and Western Emperors of the Minchiate deck wield such accouterments. In the case of the Rider-Waite Emperor, we find them replaced by a *crux ansata*. Like the *globus cruciger*, this symbol can be seen to reference the equinox, in this case with the circular component representing the Sun rising above the horizon at the intersection of the ecliptic and celestial equator [Figure 1].

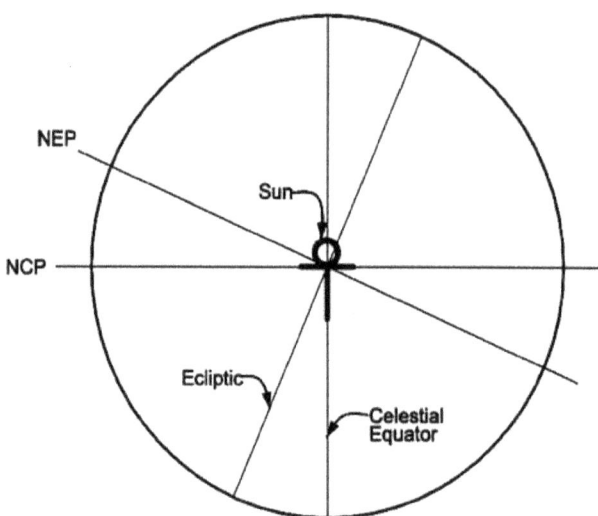

[Figure 1: The *crux ansata* as symbol of the equinox.]

According to Waite, the Emperor is the husband of the Empress, by imputation.[164] Our current model wherein the dragon's tail/Thuban/Ursa Major region of the heptanomis is attributed to the Empress resonates to such an interpretation, since that circumpolar astronome is the polar station

---

[164] Waite. *The Pictorial Key to the Tarot.* Part I. Section 2; The Emperor.

corresponding to both the Taurean and Aryan Ages. Thus, during the Age of Aries, the Emperor represents the masculine and diurnal aspect of the Sun's vernal location, whereas his wife embodies the nocturnal (and feminine) counterpart: the aeonically equivalent stellar region of the north celestial pole, visible only at night.

The Emperor represents the first sequential card of the Major Arcana in which the symbolism has shifted from a primarily circumpolar (and therefore stellar/nocturnal) character during the Ages in which "the woman was dominant,"[165,166] to an ecliptic-based solar and masculine character. Although the situation is undoubtedly more nuanced and complex, the transition from the Age of Taurus to the Age of Aries can be seen as marking the turning point from a matriarchal culture with pantheistic and Goddess-based religious overtones to a patriarchal one in which monotheism was fiercely embraced in the Middle East. Instead of the ancient Sabean system of reckoning the immensity of precessional time via observation of the slowly changing orientation of the circumpolar heavens, a heliacal system was established which gave preeminence to the vernal equinoctial locus of the Sun as the measure of precessional drift.

Crowley suggests this card to be the first in sequence below the Abyss.[167] In the present case, the Abyss is interpreted to refer to the circumpolar regions of the heavens, as they were anciently conceived of as a watery abyss. Since the ecliptic – where Aries, or the Emperor, is located – is found below the celestial pole, it is indeed below the Abyss.

Case also distinguishes this trump from the previous 4, noting they represent the forces of involution and movement towards a central locus (compare with the circumpolar nature of those trumps according to the present system), whereas the Emperor symbolizes the evolutionary "outflow of intelligence" from subjective to objective states.[168]

The fact that Aries is still widely associated with the equinox, even though the vernal point is now ostensibly to be found in Aquarius is testament to the dissociation of Western culture from the actual observation and active participation in the cycles of Nature characterized by the ancient Pagan cultures. In this connection, Crowley's inversion of the *Tzadik* and *Hei* attributions as noted previously is quite possibly an indirect reference to the precessional shifting of the vernal point, thus explaining why in his table he has ascribed Aquarius to the Emperor, and Aries to the Star – for tropically,

---

[165] Massey. *Ancient Egypt, the Light of the World.* Book 7. Page 440.
[166] i.e., the Ages of Virgo, Leo, Cancer, and Gemini.
[167] Crowley. *The Book of Thoth.* Page 78.
[168] Case. *The Secret Doctrine of the Tarot.* Chapter 6.

IV THE EMPEROR

the vernal point is considered to be Aries (the Emperor), but sidereally, it is Aquarius (the Star).

During the Age of Aries, the NCP was localized in the region between the tail of Draco and Ursa Minor, and was associated with the god Sevekh (Sebek/Sobek).[169,170] The identification of this god with Draco and Aries is evident in the Egyptian depictions of him with a crocodile's head (Draco), and ram's horns (Aries) [Figure 2]. Here we have potential evidence of an acknowledgement of the relationship between the stellar and solar chronometries, as both the circumpolar NCP locus and vernal point references are encapsulated within the same iconography. Perhaps this god is exemplary of a transition period from the stellar to solar mythical paradigm, in which both perspectives were given a similar weight.

[Figure 2: Sebek (Sevekh). Author's rendering]

According to Massey, Sevekh is the 7-headed dragon in the Revelation of St. John the Divine, constellated as Ursa Minor. He is villified along with the Great Mother (i.e., Ta-urt/Draco) because their earlier preeminence was associated with the Egyptian stellar cult of Sut-Typhon, who kept their time by the circumpolar orientation of those two stellar "dragons," in conjunction with Ursa Major and Sirius.[171] Once the mode of chronometry had shifted

---

[169] Massey. *Lecture #5*. Page 126.
[170] Massey. *Ancient Egypt, the Light of the World*. Book 11. Pages 707-708.
[171] Massey. *Lecture #5*. Also see *A Book of the* Beginnings. Section 17. Pages 282-283.

from stellar to luni-solar, the former methods were anathematized in the mythos. In Revelation 17, we read:

> [1] And there came one of the seven angels which had the seven vials, and talked with me, saying unto me, Come hither; I will shew unto thee the judgment of the great whore that sitteth upon many waters:
>
> [2] With whom the kings of the earth have committed fornication, and the inhabitants of the earth have been made drunk with the wine of her fornication.
>
> [3] So he carried me away in the spirit into the wilderness: and I saw a woman sit upon a scarlet coloured beast, full of names of blasphemy, having seven heads and ten horns.
>
> [4] And the woman was arrayed in purple and scarlet colour, and decked with gold and precious stones and pearls, having a golden cup in her hand full of abominations and filthiness of her fornication:
>
> [5] And upon her forehead *was* a name written, MYSTERY, BABYLON THE GREAT, THE MOTHER OF HARLOTS AND ABOMINATIONS OF THE EARTH.
>
> [6] And I saw the woman drunken with the blood of the saints, and with the blood of the martyrs of Jesus: and when I saw her, I wondered with great admiration.
>
> [7] And the angel said unto me, Wherefore didst thou marvel? I will tell thee the mystery of the woman, and of the beast that carrieth her, which hath the seven heads and ten horns.
>
> [8] The beast that thou sawest was, and is not; and shall ascend out of the bottomless pit, and go into perdition: and they that dwell on the earth shall wonder, whose names were not written in the book of life from the foundation of the world, when they behold the beast that was, and is not, and yet is.
>
> [9] And here *is* the mind which hath wisdom. The seven heads are seven mountains, on which the woman sitteth.
>
> [10] And there are seven kings: five are fallen, and one is, *and* the other is not yet come; and when he cometh, he must continue a short space.
>
> [11] And the beast that was, and is not, even he is the eighth, and is of the seven, and goeth into perdition.
>
> [12] And the ten horns which thou sawest are ten kings, which have received no kingdom as yet; but receive power as kings one hour with the beast.
>
> [13] These have one mind, and shall give their power and strength unto the beast.

IV THE EMPEROR

[14] These shall make war with the Lamb, and the Lamb shall overcome them: for he is Lord of lords, and King of kings: and they that are with him *are* called, and chosen, and faithful.

[15] And he saith unto me, The waters which thou sawest, where the whore sitteth, are peoples, and multitudes, and nations, and tongues.

[16] And the ten horns which thou sawest upon the beast, these shall hate the whore, and shall make her desolate and naked, and shall eat her flesh, and burn her with fire.

[17] For God hath put in their hearts to fulfil his will, and to agree, and give their kingdom unto the beast, until the words of God shall be fulfilled.

[18] And the woman which thou sawest is that great city, which reigneth over the kings of the earth.

There are countless possible interpretations of this highly symbolic language, but we will confine ourselves to two explanations, both from an astromythological perspective.

The Great Harlot who sitteth upon the beast has been identified with the dragon-horse Ta-urt, or Draco,[172] and the 7-headed dragon with Sevekh-Kronus, or Ursa Minor.[173] Verse 7 thus makes it evident that the temporal reference point of this vision would correspond to a time in which the polestar was localized in Ursa Minor, since that constellation would correspond to the "carrier" of Draco in its revolutions.

Of course, that temporal reference point applies to the present day, since in fact the station of the pole is currently in the last star of Ursa Minor (Alpha Ursa Minoris). During the timeframe this text was supposedly written (81-96 CE),[174] the station of the pole was nominally in Epsilon Ursa Minoris, the 5th star of this constellation in the precessional sequence.

The 7 mountains in the present interpretation refer to the 7 stars of Ursa Minor. These correspond to the 7 kings of whom it is said: "five are fallen, and one is, and the other is not yet come." This dates the nominal timeframe for the prophecy to around 1000 CE; since it was at this point that the station of the pole had moved from the 5th to the 6th star of Ursa Minor in the precessional sequence. This answers the question as to why John of Patmos may have been prophesying in 100 CE about some future condition in 1000 CE: he was referring to the changing of the pole on a micro-level – from one star to the next within the single constellation of Ursa Minor, not the motion through the 7 astronomes of the heptanomis *in toto*. Significantly, The Holy

---

[172] Massey. *Ancient Egypt, the Light of the World*. Book 10. Page 686.
[173] Massey. *The Natural Genesis*. Book 6. Page 349.
[174] Stuckenbruck. "Revelation". (Article in *Eerdman's Commentary on the Bible*, 2003. Pages 1535-1536.)

Roman Empire of central Europe started its millennial tenure during this period (962-1806 CE), and as has previously been mentioned, this timeframe coincides with the precessional locus in which the possible solar "lensing" effects of Sirius are at their maximum in the Northern Hemisphere.[175] This leads us up to the question regarding the identity of the 7th star, or king, who "is not yet come": in this context the answer is Alpha Ursa Minoris, the polestar of our current Age.

Given this interpretation, it is difficult not to read into the aforementioned verses a prophetic condemnation of the zeitgeist of the Piscean Age. The misogynist overtones of the text may simply be a byproduct of a literalization of the anthropomorphic genders ascribed to the constellations, but they could also be explained as the echoes of sentiment passed on through the centuries of doctrinal transmission by the priestly caste in the transition from a matriarchal to a patriarchal eschatology.

The question as to why Aries (the Lamb) should overcome the circumpolar acronychal reguli of the summer solstice (Ursa Minor and Draco) is answered by the fact that the chronometry had shifted from stellar to solar. Thus, during the Age of Taurus, the Harlot (Draco), the Dragon (Ursa Minor), and Aquarius were the acronychal circumpolar counterparts of the Sun's passage through the underworld while in Leo. As we shifted to a Solar and patriarchal regime in the Age of Aries, the predominance became associated with the vernal point. Thus, the Lamb (Aries) overcomes the Harlot (Draco) and Beast (Ursa Minor).

Massey offers a slightly different interpretation of the Revelation – one based on a larger scale of precession, which views the passage from star to star in the 7-fold sequence as referring to the cycle of precession *in toto*. Thus, the 7 stars or kings are seen to represent each of the circumpolar astronomes of the heptanomis. Working from the assumption that the primordial first astronome is the "mount of Man" (i.e., Hercules), corresponding to the Age of Leo, he goes on to date the temporal frame of reference for verse 10 to be the Age in which the station of the pole was localized in the 6th astronome, i.e., the constellation Lyra, which corresponds to the Age of Libra/Virgo.[176]

The 10 horns that are kings are said to represent the 10 zodiacal constellations before the ecliptic had been divided into 12 signs.[177] The transition between the two divisioning systems is said to be traceable as far back as the end of the Age of Virgo, some 13,000 years ago.[178]

---

[175] See chapter on The Great Seal.
[176] Massey. *Ancient Egypt, the Light of the World*. Book 11. Page 702.
[177] Ibid. Pages 715-716.
[178] Ibid.

IV THE EMPEROR

From these considerations, it is evident that Massey hypothesized the Revelation to have originated from a very ancient Egyptian eschatological mythos that was reworked by the emerging patriarchate to encode the transition from the Age of Taurus to the Age of Aries. This interpretation sheds a revealing light on Crowley's statement that the Emperor represents "a generalization of the paternal power."[179]

Finally, it should be noted that the insignia of the Holy Roman Empire was the very 2-headed eagle Crowley depicts on his version of this trump. The Emperor bears a *globus cruciger* – yet another symbol of the Holy Roman Empire. According to Crowley, this symbolism indicates that the Emperor's "energy has reached a successful issue," and that "his government has been established."[180]

---

[179] Crowley. *The Book of Thoth*. Page 78.
[180] Ibid.

V The Hierophant

ACCORDING TO WAITE, the pillars between which the Hierophant is seated are not to be considered equivalent to the Boaz and Jachin pillars of the High Priestess.[181] If, as has been postulated, the latter refer to the east and west, it is reasonable to infer the Hierophant's pillars to signify the north and south.

Waite goes on to say that "Grand Orient states truly that the Hierophant is the power of the keys, exoteric orthodox doctrine, and the outer side of the life which leads to the doctrine..."[182, 183] These keys, which lie at the Hierophant's feet, between the twin monk figures, are the same as those depicted in the Vatican coat of arms [Figure 1]; they are the keys to the gates of heaven given to Peter by Christ.

[Figure 1: The Keys of the Vatican Coat of Arms.]

As is the case with most sacred texts, there are multiple layers of meaning that can be ascribed to the symbols used in the New Testament. From an astrotheological standpoint, these keys can be seen to represent the means of entry into the two gates of heaven: the silver key unlocks the Silver Gate, or "Gate of Humanity"; the gold key unlocks the Golden Gate, or "Gate of the Gods."

---

[181] Waite. *The Pictorial Key to the Tarot*. Part II. Section 2; The Hierophant.
[182] "Grand Orient" is a pseudonym Waite used.
[183] Waite. *The Pictorial Key to the Tarot*. Part II. Section 2; The Hierophant.

In his commentary on Cicero's "Dream of Scipio," Macrobius identifies these gates as the diametrically opposed heavenly crossroads represented by the two points of intersection of the ecliptic and Galactic Equator, as projected upon the celestial sphere [Figure 2]. The Silver Gate is located in the arm of Orion, between the constellations of Taurus and Gemini.[184] It is said to represent the entry point of souls incarnating upon the terrestrial plane. Individuals who have attained the requisite degree of spiritual evolution upon Earth are said possess the means of exiting this life through the Golden Gate, located in the Sagittarius/Scorpio region of the zodiac.[185] The evolved souls are thus capable of migrating towards the Galactic Center, ever nearer to the ineffable source of light and consciousness.

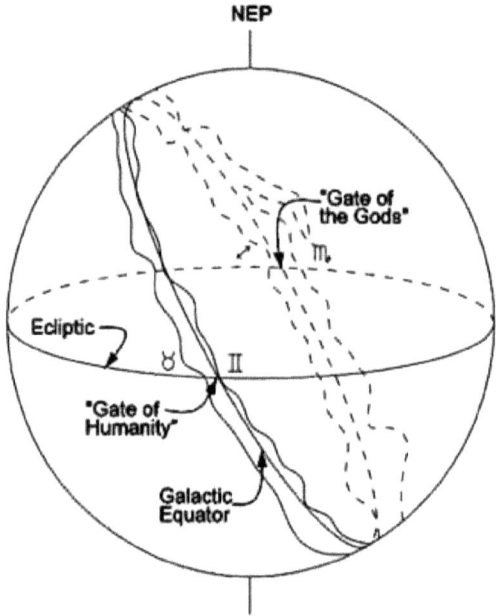

[Figure 2: The Gate of the Gods, and the Gate of Humanity]

In a sense, the Silver Gate can be considered "exoteric," as it pertains to those incarnational lessons endured further out from the Galactic Center. The suggested zodiacal attribution to the Hierophant would therefore be the ecliptic/galactic nexus in the Orion region.

---

[184] Pike, Albert. *Morals and Dogma*. Chapter XXV – Knight of the Brazen Serpent. Although Pike associates Cancer and Capricorn with these two gates, he points out the fact that what is being referred to are the tropical signs (which siderally equate to Taurus/Gemini and Sagittarius/Scorpio).

[185] Ibid.

V THE HIEROPHANT

If we orient ourselves facing eastward at the equatorial mount, with the left pillar representing north, and right the south, the encoded celestial purview of the Hierophant will become apparent [Figures 3 & 4]. With the keys themselves representing the crossing of the ecliptic and Galactic Equator, it is evident that the figure of the Hierophant corresponds to the constellations Taurus and Orion, and the twin monks to Gemini. The papal crown can be read as the inverted "v" portion of Taurus, with the three vertically arranged crosses corresponding to the belt stars of Orion. These crosses suggest the triple crucifixion on Calvary, and the three Marys said to be present at this event. In fact, these very belt stars are known as "The Three Marys."[186]

[Figure 3: The Rider-Waite Hierophant. Author's rendering.]

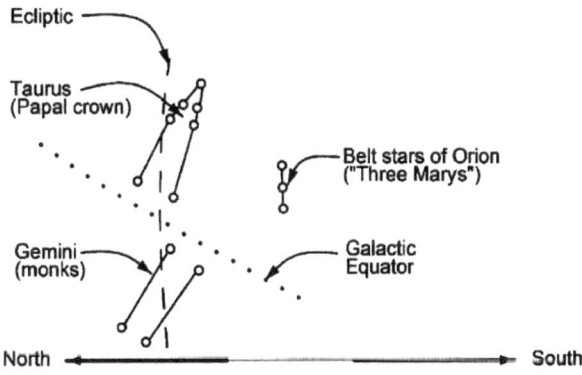

[Figure 4: A proposed celestial correspondence to the Rider-Waite Hierophant]

---

[186] Allen, Richard Hinckley. *Star names – Their Lore and Meaning*. Page 316.

Crowley attributes Taurus to this card, and indicates that its main reference is to the establishment of a connection between macrocosm and microcosm;[187] in other words, the activation of higher centers of consciousness such that in essence, the individual is capable of communicating to a certain degree with the divine.

The Harris-Crowley version of this trump depicts a "woman girt with a sword" standing before the Hierophant. She is alternatively referred to as the Scarlet Woman – indicating Isis, and hence one of her aspects, Sirius – as well as "Venus as she now is in this new aeon."[188] The subtext here may refer to the phoenix cycle, which we have suggested to be characterized by the heliacal rising of Venus and Sirius. In the Age of Taurus, this phenomenon occurred when the Sun was in Leo. Today, as we transition from vernal point Pisces to Aquarius,[189] the height of the Egyptian flood season[190] – if considered as the summer solstice – would occur when the Sun is held aloft by Orion's arm, between Gemini and Taurus. The Hierophant may be considered as Osiris/Orion proper, as opposed to Taurus per se, as Osiris is the "original" patriarch, i.e. the Father of the Egyptian trinity (Isis/Osiris/Horus).

Crowley's statement that "the rhythm of the Hierophant is such that he moves only at intervals of 2,000 years"[191] is clearly a reference to the approximate time-span for a single precessional Age. Within this context, the Hierophant as Initiator is to be seen as the source of wisdom pertaining to the Mysteries; the major dispensation of which occurs at the beginning of each Age, presumably through some globally visible Avatar who ushers in a new mode of relation between the human and divine.

This "new mode" may in fact be the re-emergence of an old one, adjusted for the particular circumstances of the impending Age. According to Manly P. Hall:

> When the Aquarian Age is thoroughly established, the Sun will be in Leo, as will be noted from the explanation previously given in this chapter regarding

---

[187] Crowley. *The Book of Thoth*. Page 78
[188] Ibid. Page 79.
[189] As the current vernal point is approximately located at 5° Pisces, and given the standard value of 72 years per 1° of precessional motion, it is technically some 360 years, or 1-1/2 more phoenix cycles (here a phoenix cycle is considered to be 243 years) until this transition is fully effectuated. Moreover, if we are to gauge the transition into the Age of Aquarius by the phoenix cycle with consideration given to additional layers of celestial cycles, such as those of eclipse and conjunction for the various other luminaries, the dawning of this solar "Aeon of Horus" would seem to be most properly located somewhere within the interval of 2322 – 2565 CE, which is punctuated by a phoenix cycle at each end.
[190] As the Nile has been dammed since the 1970's, the annual flooding no longer takes place.
[191] Crowley. *The Book of Thoth*. Page 80.

V THE HIEROPHANT

the distinction between geocentric and heliocentric astrology. Then, indeed, will the secret religions of the world include once more the raising to initiation by the Grip of the Lion's Paw (Lazarus will come forth).[192]

The "explanation previously given" is stated as follows:

> The important point to be remembered is that when the sun was said to be in a certain sign of the zodiac, the ancients really meant that the sun occupied the opposite sign and cast its long ray into the house in which they enthroned it. Therefore, when it is said that the sun is in Taurus, it means (astronomically) that the sun is in the sign opposite to Taurus, which is Scorpio. This resulted in two distinct schools of philosophy: one geocentric and exoteric, the other heliocentric and esoteric. While the ignorant multitudes worshiped the house of the sun's reflection, which in the case described would be the Bull, the wise revered the house of the sun's actual dwelling, which would be the Scorpion, or the Serpent, the symbol of the concealed spiritual mystery.[193]

This appears to suggest that equinoctial eclipse cycles are esoterically used to demarcate specific vernal points for any particular Age. For example, when the vernal Sun enters the sign of Aquarius, the antipodal locus of the full Moon – and more particularly, the truly eclipsed Moon – will be found in Leo. The Sun in Aquarius has thus "cast its long ray" across the body of Earth, to be fully reflected in its celestial "twin" and counterpart which is the "Sun of the nocturnal waters" – the full Moon.

As will be discussed further in the chapter on the Devil trump, this "nocturnal Sun" is – according to Massey – a form of Neptune or Poseidon, a telling fact which reveals in part the dense astrological encoding embedded within the mythological language of the ancients, for Poseidon is to be seen to play a key role in the Kretan legend from which this trump is said to derive.

According to Crowley:

> It is impossible at the present time to explain this card thoroughly, for only the course of events can show how the new current of initiation will work out.

> It is the aeon of Horus, of the Child. Though the face of the hierophant appears benignant and smiling, and the child himself seems glad with wanton innocence, it is hard to deny that in the expression of the initiator is something mysterious, even sinister. He seems to be enjoying a very secret joke at somebody's expense. There is a distinctly sadistic aspect to this card; not unnaturally, since it derives from the legend of Pasiphae, the prototype of all the legends of Bull-gods.[194]

---

[192] Hall, Manly P. *The Secret Teachings of All Ages.* The Zodiac and its Signs. Page 157.
[193] Ibid. Page 156.
[194] Crowley. *The Book of Thoth.* Page 79.

The legend of Pasiphae is given as follows by the "Pseudo-Apollodorus":[195]

> Minos aspired to the throne [of Krete], but was rebuffed. He claimed, however, that he had received the sovereignty from the gods, and to prove it he said that whatever he prayed for would come about. So while sacrificing to Poseidon, he prayed for a bull to appear from the depths of the sea, and promised to sacrifice it upon its appearance. And Poseidon did send up to him a splendid bull. Thus Minos received the rule, but he sent the bull to his herds and sacrificed another... Poseidon was angry that the bull was not sacrificed, and turned it wild. He also devised that Pasiphae should develop a lust for it. In her passion for the bull she took on as her accomplice an architect named Daidalos... He built a wooden cow on wheels, ... skinned a real cow, and sewed the contraption into the skin, and then, after placing Pasiphae inside, set it in a meadow where the bull normally grazed. The bull came up and had intercourse with it, as if with a real cow. Pasiphae gave birth to Asterios, who was called Minotauros. He had the face of a bull, but was otherwise human. Minos, following certain oracular instructions, kept him confined and under guard in the labyrinth. This labyrinth, which Daidalos built, was a cage with convoluted flextions that disorders debouchment.[196]

King Minos in this legend is a figure of the vernal Sun during the Age of Taurus,[197] and Pasiphae represents the Moon[198] – or more particularly, the Moon opposite the Sun.

According to Massey:

> The dove is a bird of breath or spirit, and the Pleiades are six doves named from peleia, the dove. The scorpion is a figure of 6, as well as the sign of Serk, to breathe, the genus Scorpio being determined by the six eyes. These two signs of breath and number 6 are vis-à-vis in the zodiac, where each succeeds the passage of the waters. Thus when the Sun entered the sign of Scorpio in the [lunar] month of Hathor, and the body of Osiris was shut up in the ark of the six lower signs by the evil Typhon, it was the time at which the Pleiades arose with their sixfold symbol of breath above, and of the regeneration of Osiris *in the ark, or cow, of the Moon*, which they rose to accompany in the sign of the Bull.[199]

Here is the "love and love" of Crowley[200] – the dove and the serpent, or the Pleiades and Scorpio; the two diametrically opposed signs which mark the

---

[195] An unknown mythographer of the 1st–2nd centuries CE, at one time mistakenly identified as Apollodorus of Athens.
[196] Pseudo-Apollodorus. *Bibliotheca*. 3. 8 – 11 (trans. Aldrich). Web. <theoi.com/Ther/Minotauros.html >
[197] Massey. *A Book of the Beginnings*. Section 1. Page 39.
[198] Pausanias. *Description of Greece*. 3.26.1. Web. <theoi.com/Text/Pausanias3B.html>
[199] Massey. *The Natural Genesis*. Book 12. Page 287. Bracketed comment and italics added.
[200] Crowley. *The Book of Thoth*. Page 79.

V THE HIEROPHANT

crossing points of the ecliptic and Galactic Equator. Here we see a key to the significance of the hollow cow created for Pasiphae: it is a symbolic type of the ark, or barque, of Isis – the boat in which Osiris as the Sun sails in safe passage through the underworld; it is none other than Neptune (Poseidon), the nocturnal Sun, or the Moon.

The fact that the Moon is debilitated in Scorpio may go a long way in explaining the "sadistic" overtones indicated by Crowley. According to Fagan, Taurus – not Scorpio – rules the sexual organs, and a debilitated Moon opposite this sign would certainly cast a disharmonious overtone upon any cycle of time for which such a configuration was considered a prime influence. The discomfiture of the Moon in Scorpio can be explained by the fact that Scorpio is ruled by Mars, and this bellicose "active-negative" planetary influence may bear a natural antipathy and antagonism towards the "passive-negative" lunar vibrations. Indeed, Fagan states in his Solunars Handbook[201] that the transit of Mars to the Moon is "A pain-bearing transit..." and further goes on to state that "Should the angles of the quotidians be directed to the Moon and Mars simultaneously, it portends trouble with the opposite sex, often accompanied with violence and loss of temper." It should also be noted that King Minos (Sun in Taurus) was said to ejaculate serpents and scorpions (both references to Scorpio) into his lovers; a curse placed upon him by Pasiphae (the Moon).

---

[201] Fagan. *Solunars Handbook*. Page 102.

VI The Lovers

THE TWO FIGURES flanking the central angelic mediator in this trump have been compared to Adam and Eve, the overall connotation sometimes being formulated as "marriage."[202] In the Rider-Waite version of this card, two trees are depicted on either side of a mountain; one associated with each human figure. In the words of Waite: "Behind the man is the tree of life, bearing twelve fruits, and the tree of the knowledge of good and evil is behind the woman; the serpent is twining around it."[203]

[Figure 1: The Rider-Waite Lovers trump interpreted with respect to the NEP, NCP, and Mount of Equinox/Pole. Author's rendering.]

---

[202] Waite. *The Pictorial Key to the Tarot.* Part I. Section 2; The Lovers.
[203] Ibid. Part II. Section 2; The Lovers.

CELESTIAL ARCANA

Building on our previously developed hypotheses, we can identify the recurring motifs of the polar/equatorial mount, the eternal tree of the ecliptic axis, the mutable tree of the celestial axis, and the mysterious binding force between the two. The "twelve fruits on the tree of life" relate to the 12 zodiacal constellations associated with the ecliptic and its pole, whereas the serpent twining around the second tree represents the celestial pole, enveloped by the constellation Draco in its circadian revolution atop the circumpolar mount [Figures 1 & 2].

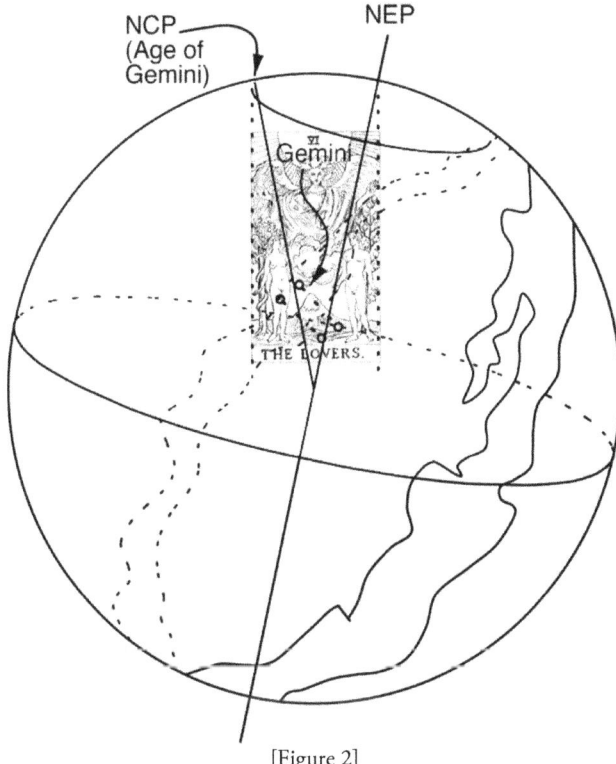

[Figure 2]

The stigmatization of the Woman as cause of the Fall of humanity and expulsion from the Garden of Eden appears to have its roots in the usurpation of the stellar and gynocentric cultus of ancient Egypt – in which the Great Mother was preeminent – by the solar and patriarchal regime in which Jehovah was the central deity. According to Massey:

> The change in Israel from the worship of El-Shaddai to the worship of Ihuh, from the Elohistic to the Jehovistic god, corresponds to the change from the stellar to the solar worship in the astronomical mythology.[204]

---

[204] Massey. *Ancient Egypt, the Light of the World.* Book 8. Page 499.

VI THE LOVERS

This trump has been attributed to *Zayin* by Crowley and Case.[205] In Hebraic tradition, *Zayin* is associated with a sword, which connotes division, and therefore also complementarity.[206] It represents the mutual facilitation of the woman and man – one towards the other – the typical characterizations of which, as we draw closer to the Age of Aquarius, begin to seem ever more anachronistic. Indeed, much of the commentary about this trump has been dressed up in antiquated notions about the roles of the sexes, which illustrate the ongoing reverberations of the rejection of feminine power by the patriarchal cult several thousand years ago. The following passages from Case's dissertation on this card will prove the point:

> The woman in God is the basis of His creative work. She is the Substance – that which stands under, as the foundation. She is the source of the urge for self-expression, even as the subjective mind of man is the seat of the emotions and desires ... When he [(the trained occultist)] entrusts a piece of work to the subjective mind he feels as certain that it will be accomplished as does the husband who asks his wife to prepare a certain dish for dinner, or to sew on a button – only more so ... Just as a loving wife delights in serving her husband, and just as a loving mother gives her son all that she has, even to her very life, so does Prakriti in both aspects, work joyously for Purusha,[207] who is, when manifested as the Ego in the heart of man, both her Son and her Lord.[208]

The mythic expulsion of the primordial *gemini* from the Garden of Paradise contains multiple layers of meaning. On one level, it can be seen as a reference to the loss of humanity's innate awareness of the higher states of consciousness, which are by rights its sovereign possession. According to the present model, this falling away from the "divine presence" is a natural consequence of the decrease in declination of the emanations from the Galactic Center as focused by the Sun – where the most potent lensing effect of the Brahmanic Logos is associated with the summer solstice angle of incidence for these emanations during the Ages of Virgo, Leo, and Cancer.

The Summer of the Great Year gave way to the Fall some 6,000-8,000 years ago, during the Age of Gemini. This was observed by the ancients and preserved in their astronomical eschatology:

> We learn from Plutarch that ... Ta-Urt ... deserted and came over to the side of Horus, and was pursued by a serpent. Ta-Urt was the Great Mother in the constellation of the Great Bear, the old harlot of the heptanomis who deserted

---
[205] See Crowley's *Book of Thoth*, and Case's *Secret Doctrine of the Tarot*.
[206] Ginsburgh. *The Hebrew Letters*. Page 113.
[207] Case equates Purusha with the "Supreme Spirit," and "Originating Principle" of all manifestation, and Prakriti with the agency through which Purusha manifests on the material plane. A hierarchy is implied in this relationship.
[208] Case. *The Secret Doctrine of the Tarot*. Chapter 8. Ellipses added.

Sut [i.e., Bootes] and joined herself to the solar Sebek-Horus [i.e., Ursa Minor] as "the great mother of him who was married to his mother."[209]

This refers to the precessional drifting of the NCP from the circumpolar regions of Bootes and Corona Borealis during the end of the Cancerian Age, into the regions of Draco and Ursa Minor during the Ages of Taurus and Aries. The Great Mother as Ursa Major used to be "allied" with the primordial Sut as the constellation Bootes, in that the two revolved more or less around a common center during the Age of Cancer. Her "going over" to the side of Sebek-Horus, and subsequently being "followed by a serpent" references the shift of the NCP from the region of Bootes to the region between Ursa Major and Draco during the Taurean Age, where the Genetrix and Serpent shared a common center.

The timeframe in which this shift or "betrayal" occurred corresponds to the Age of Gemini – the first portion of which may have been characterized by a period of innocence and abundance, where the primordial pair "walked with the Lord"; the latter half being demarcated by the expulsion from Paradise, and the subsequent vilification of the Genetrix.

Adam and Eve can thus be seen as the "original" *gemini*, or "twins" – not in the historical sense, as this motif far predates the specific Hebraic rendition; but from a typological perspective. Earlier forms of this primordial pair can be found in the Egyptian anthropomorphic dyads of Sut/Horus, and Shu/Tefnut, who share a somewhat complex astromythological relationship. They are variously depicted as twins, brother and sister, or consorts.

The origins of the twin symbolism can supposedly be traced at least as far back as 6,000-5,000 BCE, when Shu and Tefnut were acronychal figures of the equinox, constellated in Sagittarius. Their diurnal form, represented by the twins Sut and Horus – or Castor and Pollux, in the later Greek rendition – corresponded to the actual constellation Gemini. Anciently, Sut ruled the South and Horus the North.[210] The war continually raged between the elemental powers of inundation and drought; water and fire; summer and winter; Horus and Sut – but at the equinox, these forces were reconciled, and the reconcilers were Shu and Tefnut. They are depicted as a double-headed Sagittarius – whose body is notably leonine, rather than equine – in the rectangular zodiac of Denderah [Figure 3].[211]

In the Age of Taurus, Tefnut became associated with Cassiopeia, and Shu with Cepheus. They were depicted as lions, as they were the nocturnal heralds

---

[209] Massey. *Ancient Egypt, the Light of the World*. Book. 10. Page 686. Bracketed comments and ellipses added.
[210] Ibid. Book 6. Page 327.
[211] Ibid. Book 5. Page 297; Book 6. Page 326.

VI THE LOVERS

of the entrance of the summer Sun into the constellation Leo.[212] This iconography is also relevant to the preceding Age of Gemini, as during that period Cepheus and Cassiopeia were the circumpolar acronychal reguli when the summer Sun was in Virgo. Sut and Horus were at one time – presumably in the Age of Gemini – depicted as twin lions,[213] probably in reference to the solstices, as it was during the summer solstice that Leo could be seen rising heliacally when the Sun was in Virgo. During the winter solstice, the lion embarked upon its journey into the Underworld in the west, as the winter Sun entered Pisces in the east.

[Figure 3: Double-headed Shu (front) and Tefnut (rear) as depicted in the rectangular ceiling zodiac in the Temple of Hathor – Denderah, Egypt. Author's rendering.]

Horus represented the force that tamed the summer heat via the waters of inundation and life. Sut embodied the winter solstice, when the Sun was in the "abyss," which was constellated as Cetus – the ancient leviathan, dragon of the deep – whose acronychal counterpart is the constellation Hydra that finally ascends from the pit in Revelation.[214] This is the very same "abyss of source" which later became constellated as the Aquarian waterbearer, revealing a connection between the Great Beast, and the Great Mother – yet another form of the primordial dyad, that was even more anciently depicted as a feminine pair that constituted a biune Genetrix before the complementary male figure in the form of a Twin, Son, and finally Father had been added to the astronomical mythos.[215,216]

---

[212] Massey. *A Book of the Beginnings*. Section 16. Page 245.
[213] Massey. *Ancient Egypt, the Light of the World.*. Book 5. Page 253; Book 6. Pages 325-328.
[214] Ibid. Book 5. Pages 280-284; Book 6. Page 325. In one interpretation, the beast released from the pit may represent the constellation Hydra, or the 7-headed "Apap-dragon," that can be seen to rise above the western horizon as we transition from the Age of Pisces to Aquarius.
[215] Ibid. Book 5. Pages 279-80.
[216] Massey. *The Natural Genesis*. Section 8. Pages 456-457.

VII The Chariot

A WINGED DISC surmounting an emblem consisting of a vertical axial element orthogonally piercing a horizontal circular form (the Hindu lingum and yoni) is rendered on the front of the chariot in the Rider-Waite version of this trump. Amongst the various celestial connotations implied by this glyph, perhaps the most relevant is that of the "triumph" of the Sun at the summer solstice: it is here, at the apex of the analemma, that Sól[217] appears stationary for 3 days, and exerts a maximum influence before descending towards her autumnal abode.

The Druids of ancient Britain had stones that represented the lingum and yoni in a similar fashion as that depicted on the escutcheon of the stone chariot in this trump. The lingum stone was called the Said-Stone, Seven-Stone, *Syth*-Stone (Erect Stone), or *Maen-Llüd* (Many-sided Stone). The yoni stone was known as the Rocking-Stone, Ark-Stone, and Stone of Keridwen. The combination of both stones, as illustrated in the Rider-Waite and Levi versions [Figures 1 & 2], yielded the *Pelydr*, or Beam (i.e., "Ray"), which was said to represent the rising Sun on both equinoxes and solstices. In the words of the 19th century Arch-Druid of Wales, Myfyr Morganwy:

> The Seven-stone in connection with the Ark-stone was with our ancestors the Beam (*Pelydr*), /|\ being the rays of the rising Sun on the equinoxes, and solstices, converging to a focus – an eye of Light – in the centre of the Ark-stone. As the seven or the *beam* coming from the Sun into contact with the earth, caused the goddess Keridwen to conceive and bring forth living beings, so also the beams which were represented by the other three stones (*tri maen gorsaf*) coming into contact with the Ark-stone, representing the womb of the goddess, gave being to the Throned-poets or Bards, who were to be the sons of Nature to teach the people the language of God in their own language. In another aspect the *beam* represented the "cyfriu," name of the Trinity; in another sense it represented the "thrice-functioned" Hu, the Interpreter, Viceroy of the Eternal (*Celi*).[218]

---

[217] Sól is the Old Norse goddess of the Sun.
[218] Massey. *A Book of the Beginnings*. Section 1. Page 675.

The apparent stasis of the Sun at the summer solstice is indicated by the implied immobility of the chariot hewn of stone, and the nature in which the driver is depicted as embedded or trapped within. The current sidereal location of the Sun at the summer solstice is longitudinally aligned with Orion, which corresponds tropically to Cancer, and it is to the latter that the symbols of this trump appear to refer, although the reality is not so simple.

The overall triangular structure of the elements of the card echoes the shape of this constellation when viewed from a vertical orientation upon the celestial sphere. From this perspective, the stellar diadem worn by the driver can be seen to represent the star Iota Cancri, and the square breastplate that of Gamma Cancri. The zodiacal belt worn by the figure corresponds to the ecliptic, where the star Delta Cancri (Asellus Australis) is to be found, and the sphinxes correlate to Alpha and Beta Cancri (Acubens and Altarf, respectively) [Figure 1]. Within this context, the lunar crescents present upon the driver's shoulders denote the rulership of this sign by the Moon.

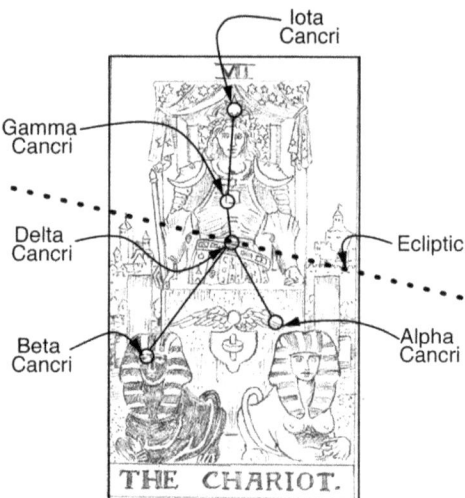

[Figure 1: The Rider-Waite Chariot trump as representative of the constellation Cancer. Author's rendering.]

Eliphas Levi reproduces this trump from an undisclosed source in *Transcendental Magic, Its Doctrine and Ritual* as "...a victor crowned with a circle adorned with three radiant golden pentagrams"[219] [Figure 2]. In the present interpretation, these pentagrams are symbolic of Sirius, which is depicted in ancient Egyptian hieroglyphs and Senegalese symbolism as a 5-pointed star [Figures 3 & 4]. The triad of stars thus refer to the three diurnal

---

[219] Levi. *Transcendental Magic, Its Doctrine and Ritual.* Part II: The Ritual of Transcendental Magic. Chapter XXII – The Book of Hermes.

VII THE CHARIOT

"water signs" and the associated heliacal rising of Sirius during the three months of inundation. During the Age of Taurus, these signs were Virgo, Leo, and Cancer, but in the Age of Aries (wherein the signs became mistakenly "fixed," with Aries forever representing the vernal point), they had shifted to Leo, Cancer, and Gemini; the central pentagram in this case represents Cancer [Figures 5, 6, & 7].

In addition to the Sirian connotations of the pentagram, we also have the association of this glyph with the Sun/Venus/Earth conjunction every 8 years. The dual reference can be read as an allusion to the simultaneous heliacal rising of Venus and Sirius – that 243-year cycle previously identified with the phoenix cycle [Figure 8].

[Figure 2: The Chariot, from Levi's *Transcendental Magic, Its Doctrine and Ritual*.]

[Figure 3: Egyptian hieroglyph for Sirius ("Sopdet")]

[Figure 4: Senegalese (West Africa) representation of Sirius ("Yoonir")]

131

CELESTIAL ARCANA

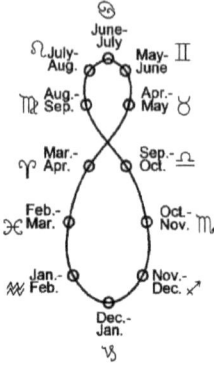

[Figure 5: Ptolemaic fixed (tropical) zodiac and its relation to the analemma.]

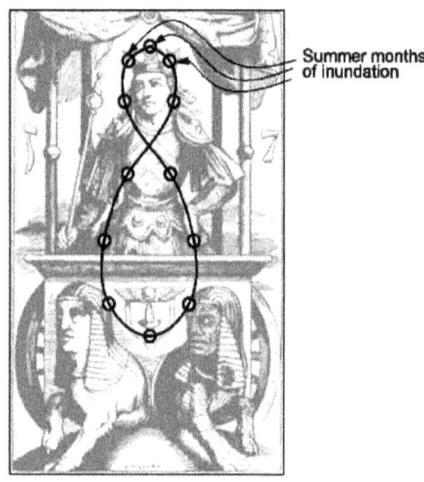

[Figure 6: The 3 pentagrams of the Solar Hero's crown as representative of the summer months of inundation.]

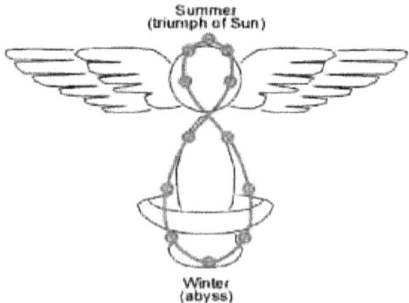

[Figure 7: An interpretation of the winged sphere and lingam/yoni symbols from the Levi Chariot trump.]

VII THE CHARIOT

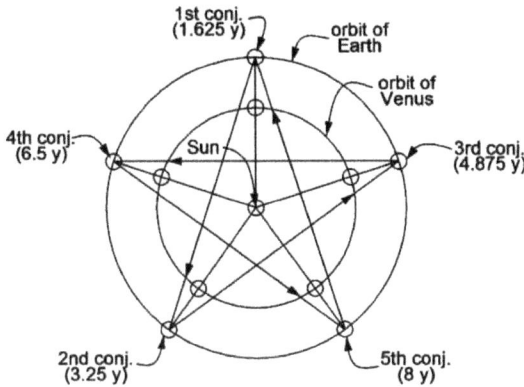

[Figure 8: Conjunctions of Earth, Venus, and Sun over an 8-year period]

According to our precessional model, the Chariot should be considered as organically related to the High Priestess, analogous to the relationship between the vernal point and NCP during any particular Age. Interestingly, Waite says the following regarding the driver of the chariot: "…if he came to the pillars of that Temple between which the High Priestess is seated, he could not open the scroll called tora, nor if she questioned him could he answer."[220] Waite's exclusive reference to the High Priestess in his discussion of this card suggests that he considered them to be closely related, whereas his assertion of the inability of the latter to fathom the laws or mechanisms of the former may signify the distinct and separate natures of the two functionaries.

In the currently proposed system, although the two are inexorably bound through the mysterious dynamics of precession, observation of the diurnal province of the Chariot of the Sun at the Cancerian vernal point can never be performed simultaneously with the corresponding nocturnal observation of the NCP in Corona Borealis (which we have ascribed to the High Priestess). Following this rationale further, if the Charioteer were to come to the temple of the High Priestess, he could only do so at night; since as we have previously established, her domain is fundamentally circumpolar and therefore only visible nocturnally. But as night falls, and we observe the *aselli* rearing above the horizon, the only visible circumpolar asterisms within the precessional circle are the jackal (Ursa Minor) the ape (Cepheus), and the dragon (Draco); Corona Borealis lies some 90° below the horizon.

Anciently, this constellation was purportedly connected to an Egyptian god of some 13,000 years ago, Iu or Iu-sa, the "original" Wandering Jew, who is also the eternally-incarnating Jesus.[221] At the precessional level, then, one

---

[220] Waite. *The Pictorial Key to the Tarot.* Part II. Section 2; The Chariot.
[221] Massey. *Ancient Egypt, the Light of the World.* Book 12. Page 751.

reason for the close association of this trump and the High Priestess is explained through the relationship of the Virgin and Child. One form of the primordial Virgin was the circumpolar genetrix Rerit, the sow or hippopotamus, which has been proposed to have originally corresponded to the polar station of Corona Borealis. The diurnal male equivalent to the polar nocturnal mother is the vernal equinoctial locus of the Sun in Cancer, which was anciently the crib, or manger.[222] Iu is depicted as ass-headed, or riding an ass, which has been interpreted as an originating astromythological motif for the later entrance of Christ into Jerusalem, riding upon an ass.[223,224]

Case attributes Cancer to this trump, and like Waite also makes a connection between the Chariot and the High Priestess. Citing the rulership of Cancer by the Moon, he connects the High Priestess with "Diana, the many breasted."[225] He describes her as a moon goddess, but more anciently, she was a naturally derived symbol of the inundation that represented the many-breasted circumpolar Genetrix, Rerit.[226] Later (during the Ages of Taurus and Aries), she became associated with the nocturnal acronychal counterpart of the Sun (i.e., Aquarius) during the summer solstice.[227] It seems likely that in the lunar phase of chronometry that followed the primordial stellar one, Diana (Aquarius/Capricorn) became associated with the Moon by virtue of the Full Moon marking the antipodal locus of the Sun during the summer flood season.

This card has been ascribed the Hebraic letter *Chet*.[228] Notably, *Chet* is said to be composed of the two preceding letters, *Zayin* and *Vav*, which are then united with a *chatoteret*, or *chupah*.[229] The *chupah* is a rectangular canopy used in Jewish wedding ceremonies, and it is – among other things – a representation of the vault of the heavens.[230] This is clearly depicted in most versions of this trump as a star-spangled drapery adorning the *chatoteret*.

In the context of the present trump, the wedding *chupah* may be seen to symbolize the return of the heavenly Bride, as given in the Hebraic apocryphal text 2 Esdras:

> Behold, the time shall come, that these tokens which I have told thee shall come to pass, and the bride shall appear, and she coming forth shall be seen,

---

[222] Ibid. Page 732.
[223] Ibid. Book 8. Page 506.
[224] Matthew 21.
[225] Case. *The Secret Doctrine of the Tarot*. Chapter 9.
[226] Massey. *A Book of the Beginnings*. Section 14. Page 160.
[227] Massey. *Ancient Egypt, the Light of the World*. Book 5. Page 302.
[228] Cf. Crowley's *Book of Thoth*, and Case's *Secret Doctrine*.
[229] Ginsburgh. *The Hebrew Letters*. Pages 122-136.
[230] Ibid.

VII THE CHARIOT

that now is withdrawn from the earth. And whosoever is delivered from the foresaid evils shall see my wonders. For my son Jesus shall be revealed with those that be with him, and they that remain shall rejoice within four hundred years.[231]

Massey suggests the Bride to symbolize Sirius, and her return to be indicative of the completion of a Sothic cycle of 1,460 years.[232] The story of the marriage of the Bride and the Lamb in Revelation is said to have a similar root meaning, ultimately deriving from the ancient Egyptian chronometry marking the return of Sirius in the civil calendar to match its heliacal rising point in the Sothic calendar.[233]

2 Esdras is thought to have been written somewhere between 90-96 CE.[234] Considering this timeframe, it is possible that the text may correspond to a foretelling of the "return" of Sirius circa 141 CE. Presumably, the content had been compiled from sources preceding this date, and Massey suggests the fundamental elements of the myth to have been developed at least as early as 2,410 BCE, as the vernal colure was traversing from Taurus to Aries.[235]

Around 2,781 BCE, the Sun was well into the transition between these 2 signs, and a Sothic Cycle had just been completed. According to Massey, the Sun had entered Aries by 2,410 BCE. If indeed the marriage of the Bride and the Lamb represents a further development of the prior marriage of the Bride (Hathor/Sirius) and Calf (Sun in Taurus) of the Taurean era (and why shouldn't it?), this may explain the stated 400 year interval within which "they that remain shall rejoice," since the period from 2,781-2,410 BCE is 371 years.

This, then, is one explanation of the marriage of the Bride and the Lamb, wherein the constellation Cancer (i.e., the Chariot) is the sign in which the solstitial Sun resides at the heliacal rising of Sirius: *it is the representation in mythic form of the entrance of the solar Horus into the sign of the Lamb (Aries) at the fulfillment of a Sothic cycle, measured by the heliacal rising of Sirius (the Bride) at the commencement of the summer solstice.*

But there is another mode in which to view this myth that underscores the difficulty of accurate interpretation of any astronomical mythology, due to the multiple layers of symbolic accretion and language barriers developing over thousands of years. According to Massey, there was a shift from a lunar to luni-solar chronometry that occurred when the summer solstice had shifted

---

[231] 2 Esdras. Verses 26-28. KJV.
[232] Massey. *Ancient Egypt, the Light of the World*. Book 11. Pages 712-722.
[233] See chapter entitled "The Domicile System and World Ages."
[234] Gottheil, Littmann, Kohler. From "Esdras" entry of *The Jewish Encyclopedia*. 1906. <web.jewishencyclopedia.com>
[235] Massey. *Ancient Egypt, the Light of the World*. Book 11. Pages 712-722.

sidereally into Cancer,[236] i.e., during the Age of Aries. In this mode of reckoning time, Isis took on the additional characteristics of the full Moon, which marked the sign acronychal to the Sun. Here Khunsu was representative of the newly emergent Solar Force of spring, and he was depicted with a full moon atop his head in reference to the newly established luni-solar form of reckoning. This new phase of chronometry corresponds to the building of Solomon's temple – not a temple made of hands, but a celestial temple constructed or revealed on the basis of luni-solar time.[237]

We may even go a step further, and suggest that the transition from Aries to Pisces has been encoded in a similar luni-solar form, as the birth of the Christ (Sun in Pisces) from the Virgin (full Moon in Virgo), which ostensibly occurred around 141 CE, if timed strictly according to the 1st Sothic cycle to occur in the Age of Pisces. Notably, Massey dates Jehoshua Ben-Pandira's birth to circa 154 BCE, which is 295 years prior to the Sothic cycle's culmination – well within the 400 year interval described above. Pandira is supposed by some to represent the most plausible candidate for a "historically accurate" version of Jesus. He was known to have been an initiate into the Egyptian mysteries, upon which he expounded to those who "had ears to hear." He was stoned, and then hung on a tree for sorcery in accordance with Jewish law.[238]

It is proposed that the combination of the observance of the Sothic cycles overlaid with the inauguration of each new Age at the entrance of the vernal colure into the successive zodiacal signs makes for a complex dynamic of shifting mythological narratives, in which an anthropomorph associated with one particular season or sign may eventually become associated with an altogether different one, yet still maintain its fundamental identity.

Thus far, all of the symbolism discussed in relation to this trump has led us back to the constellation Cancer, sign of the summer solstice in the Age of Aries. However, de Gébelin indicates the triumph of Osiris depicted in this card [Figure 9] to represent the vernal equinox, not the summer solstice:

> Osiris advances; he comes in the form of a king triumphing, his scepter in hand, his crown on his head: he is in the chariot of a warrior, drawn by two white horses. Nobody is unaware that Osiris was the primary god of the Egyptians, the same one as that of all the Sabaean people, or that he is the physical sun symbol of a supreme invisible divinity, but who appears in this masterpiece of Nature. He was lost during the winter: he reappeared in springtime with a new radiance, having triumphed over all against whom he made war.[239]

---

[236] Massey. *A Book of the Beginnings*. Section 13. Page 105.
[237] Ibid. Pages 105-106.
[238] Massey. *The Natural Genesis*. Book 13. Pages 491-492.
[239] DeGébelin. *Monde Primitif*. Volume 8, Book 1: The Game of Tarots.

VII THE CHARIOT

Similarly, Case proposes this trump to be another form of the Emperor, which he associates with the spring equinox.[240]

There are numerous ways to explain this, but perhaps one of the simplest is through the movement of the Sun's vernal point through Taurus, Aries, and Pisces, over the course of some 6,300 years. In the Age of Taurus, Osiris (as the Sun in Taurus/Orion) was considered triumphant at the vernal equinox, as he had become resurrected from the depths of the underworld of winter.[241] Today, the colure of the Sun is to be found in Orion at the summer solstice; thus, Osiris triumphed over the winter season in the Age of Taurus, and today he triumphs at the height of his power as anthropomorph of the summer solstice.

[Figure 9: De Gébelin's version of the Chariot. Author's rendering.]

According to Case, the scepter of de Gébelin's charioteer represents the lingam and yoni; also the Purusha and Prakriti, united with the lemniscate of Solar Force [Figure 10]. Here again we can see the motif of marriage, via the unification of Purusha and Prakriti via Solar Force. This relates to the balancing of the objective and subjective states of consciousness, whereby the one equally informs the other. Intuition is conjoined with the complementary force of rationality, the result being a synthesis of the two. Prophecies are made, oracles uttered... but the balance is key. At the root of it all is the generative force, which must become integrated into the higher mental and emotional functions of the brain. In so doing, the hemispheres become balanced, and the pineal gland activates.

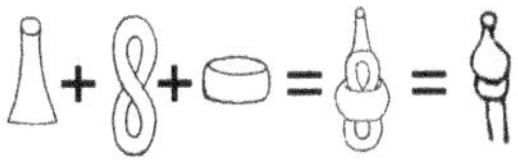

[Figure 10. Mediation of the Purusha (Linga Stone) and Prakriti (Yoni Stone) via the Solar Force (Lemniscate). Author's rendering.]

---

[240] Case. *The Secret Doctrine of the Tarot*. Chapter 6.
[241] Massey. Lecture #4: "Gnostic and Historic Christianity".

But how is this achieved? There may be any number of ways, but we will briefly confine ourselves to one, which may be particularly relevant to this trump. There appears to have been an ancient Egyptian technique practiced and promoted by the "heretic king" Akhenaten that involved the controlled administration of the Solar Force directly to the brain by looking into the Sun when it is low on the horizon, although most records of it were apparently expunged by a later regime.[242] It is postulated that this is one of the alchemical methods of directing the Solar Force in order to awaken the latent higher functionality of the brain. Akhenaten's wife, Nefertiti says the following: "Thou disk of the Sun, thou living god! There is none other beside thee. Thou givest health to the eyes through thy beams."[243]

Interestingly, Akhenaten reigned during the culmination of a Sothic cycle, circa 1,320 BCE. Akhenaten translates as "the adorer of Aten," where Aten represents the Sun disk. He changed the city of his namesake, Akhetaten, into Pa-Aten-Haru, which translates as "the city of delight for the solar disk," where the solar disk is personified as the youthful sun-god, Aten.[244] Inscribed on one of the stelae [Figure 11] discovered in the ruins of Pa-Aten-Haru is the following:

> On this day was the king in Khu-aten, in a tent of byssus. And the king – life, prosperity, and health to him! – changed Khu-aten [Akhenaten], which was its name, into Pa-aten-ham [Pa-Aten-Haru] (that is, the city of the delight of the disk of the Sun.) *And the king appeared riding on the golden court-chariot*, like the disk of the Sun, when it rises and sheds over the land its pleasant gifts, and he directed his course where the beautiful road has its beginning. ... *The beams of the sun's disk shone over him with a pure light, so as to make young his body daily.*[245]

It is clear from these statements that the royal couple believed the Sun's radiations to possess regenerative properties.

Kabbalistically, *Chet* represents the descent of the soul through the 7 heavens to Earth.[246] This is similar to the Masonic doctrine regarding the so-called Gate of Man (or the Gate of Humanity) – the gate of entrance for souls becoming incarnate on Earth.[247] It is connected to Cancer "exoterically" (i.e., tropically), but in reality (sidereally) it is the crossing point of the ecliptic with

---

[242] Cf. Wayne Purdon. "Secret Mysteries of the Sun Revealed". *New Dawn Magazine*. 2014. <web. newdawnmagazine.com>.

[243] Henry Brusch-Bey. *A History of Egypt under the Pharaohs*. Volume 1. Page 450.

[244] Massey. *A Book of the Beginnings*. Section 8. Page 341.

[245] Brusch-Bey. *A History of Egypt under the Pharaohs*. Volume 1. Page 451. Bracketed comments, ellipses, and italics added.

[246] Ginsburgh. *The Hebrew Letters*. Page 124.

[247] Albert Pike. *Morals and Dogma*. Chapter XXV – Knight of the Brazen Serpent.

the Galactic Equator – situated between Gemini and Taurus, and longitudinally aligned with Orion (Osiris), Sirius (Isis) and Columba (the Dove). Here again we have another explanation for the seemingly contradictory attributions of this trump to Osiris by some sources, and Cancer by others.

[Figure 11: Stela from the ruins of Pa-Aten-Haru, depicting Akhenaten and Nefertiti receiving emanations from the Aten, or Sun disk. Author's rendering.]

VIII Strength

AS PREVIOUSLY NOTED, Waite's juxtaposition of the ordering of Strength and Justice implies that he intended at some level to signify a connection between Strength and the constellation Leo. In his discussion of this trump, he references "A woman, over whose head there broods the same symbol of life which we have seen in the card of the Magician…closing the jaws of a lion."[248] This "symbol of life" is equivalent to the lemniscate and analemma – solar symbols par excellence. The reference here may be to the Sun as ruler of Leo.

[Figure 1: The Rider-Waite Strength trump. Author's rendering.]

In addition to her lemniscate halo, the figure is also seen to wear a wreath of some sort upon her head, possibly an allusion to the wreath or crown of immortality conferred to Ariadne, wife of Dionysus.[249] One of the forms that Dionysus was said to have assumed was that of a lion. According to Massey:

---

[248] Waite. *The Pictorial Key to the Tarot*. Part II. Section 2; Strength.
[249] According to Richard Hinckley Allen, Corona Borealis was anciently referred to as Corona Ariadnes, and was associated with Ariadne, daughter of Minos in the Kretan legend. See *Star Names: Their Lore and Meaning*.

The solar Dionysius was known by the name of "the roarer, and he was also portrayed as a lion-headed god. In the Bacchae of Euripides, he is invoked by the chorus to manifest in his might, and appear as a flaming lion.[250,251]

The Visconti-Sforza Strength trump depicts either Hercules slaying the Nemean Lion, or Samson poised to overcome the Hebraic version of that same beast [Figure 2]. Both mythic figures can be seen as representative of the same Solar Hero, the 12 Labors of Hercules being an ancient celestial allegory that poetically depicted the nature of the Sun's journey through the zodiac. But no key or formula will easily unlock the multiple layers of meaning contained within these myths; we are met at every instance of generalization with an exception to the rule we seek to impose. Much of this is undoubtedly due to the various transitions from the Sabean to Lunar to Solar modes of chronometry, and the subsequent layering of mythic accretions upon the original astronomical and astrological conceptualizations.

[Figure 2: The Visconti-Sforza Strength trump. Author's rendering.]

Thus, before Hercules was a solar hero, he was a stellar one. His conquering of the lion symbolized the overcoming of the intense heat of summer by the arrival of Sirius and Orion on the pre-dawn horizon when the Sun was in Leo. During the Age of Taurus, this signified the coming of the Nile floods, bringing the much-anticipated waters to slake the parched earth:

---

[250] Massey. *Ancient Egypt, the Light of the World*. Book 8. Page 504.
[251] In ancient Greece, the constellation Ara was known as the Altar of Dionysus [Allen. *Star Names: Their Lore and Meaning*]. This may be a continuation of the earlier Egyptian conception of Ara as Altar of the Equinox, during the Taurean Age [Massey. *The Natural Genesis*. Book 11. Pages 201-204]. When the vernal Sun disappeared under the western horizon, Ara rose in the southeast, along with its longitudinal counterparts to the east and northeast: Scorpio, Corona Borealis, and Bootes.

VIII STRENGTH

The first celestial hero was not the Sun, but the conqueror of the sun and solar heat. He was represented by the Dog-Star [Sirius] not only as the fire-god, but a god over fire; and at the season when the sun was in the sign of the lion and the heat in Africa was intolerable, then Sut, as Dog-Star, or Orion, was hailed as the conqueror of this cause of torment. The lion, as is apparent from its place in the zodiac, was the type of the furious summer fire, hence Samson, like the later Hercules, slays the lion as his first feat of strength, and out of the slain lion comes the honey.[252]

One of the most widely recognized leonine symbols of the ancient Egyptians is the Great Sphinx. The alternative Egyptological view of the Sphinx sees it as having been built during the Age of Leo, some 10,400-12,500 years ago. It was to serve as a monument to the Golden Age, as well as a chronograph of sorts to measure the cyclical progression of the Sun and celestial pole throughout the course of the Great Year.

The connection of the Sphinx with Atum (and by extension, Hercules) is suggested by Massey in the following passage from *Ancient Egypt, the Light of the World*:

Thus, the sphinx is a monument that commemorates the founding of the equinox in the double horizon, and as this was assigned to Atum Harmachis, it may account for the Hebrew traditions which associated Adam with the equinox, Adam being a Jewish form of the Egyptian Atum … we may date the sphinx as a monument which was reared by these great builders and thinkers, who lived so largely out of themselves, some thirteen thousand years ago.[253]

This period – during which the vernal point was transiting from the sign of the Virgin to that of the Lion – was referred to as *Zep Tepi*, or "the first time," by the ancient Egyptians.[254] This suggests they considered that particular locus of the precessional cycle to be the primary "alpha and omega" point. In this connection, we should note that the preeminence anciently ascribed to the constellation Hercules is borne out by the fact that it is the only station of the heptanomis to be depicted anthrotypically.

If in our present model Corona Borealis is considered as the first or primary station of the pole, Hercules corresponds to the seventh, and last. The seven circumpolar astronomes were imagined as seven mounds, mountains, or islands, and as the NCP passed from one to the next, the

---

[252] Massey. *A Book of the Beginnings*. Section 14. Pages 145-146.
[253] Massey. *Ancient Egypt Light of the World*. Book 6. Pages 338-339. Ellipses added.
[254] Ibid. Page 377: "The traditions lead one to think that profound secrets were buried in the building of the sphinx, as was the way with these builders, who put all they knew into all they did. We gather from the *Stele of Tahtmes* that the monument was built to commemorate the sacred place of creation, or, literally, 'of the first time,' an Egyptian expression generally used for the creation or 'in the beginning.'"

preceding island was said to become submerged. Here we have an astronomical allegory to the sinking of the seven islands of Atlantis, and as the pole progressed through the final astronome of the Man (Hercules), it was recorded allegorically as the "destruction of Man," and at that point, all seven Atlantean "islands" became submerged.[255] In this interpretation, the myth of the sinking of Atlantis and the supposed connection between this ancient civilization to the Sphinx becomes evident, as the so-called "destruction of Man" was witnessed astronomically and recorded monumentally during the Age of Leo.

Waite associates Strength, or Fortitude, with "...*innocentia inviolata*, and with the strength which resides in contemplation."[256] These ideas are said to be symbolized by "...the chain of flowers, which signifies, among many other things, the sweet yoke and the light burden of Divine Law when it has been taken in to the heart of hearts."[257] Here one may consider the *innocentia inviolata* to refer to the "spiritual essence" which is carried from life to life through sequential incarnations, symbolized by the chain of flowers; an essence which remains unspoiled by the various trials and "evils" through which it must pass in order to become further tempered, refined, and evolved.

Crowley appears to touch upon this rendering of *innocentia inviolata* as an essential spiritual quality that external influences are unable to diminish in the following passage from *The Book of Thoth* in which he gives his own interpretation of Fortitude, as seen through the lens of Lust:

> "Come forth, o children, under the stars, & take your fill of love! I am above you and in you. My ecstasy is in yours. My joy is to see your joy."
>
> "Beauty and strength, leaping laughter and delicious languor, force and fire are of us."
>
> "I am the Snake that giveth knowledge & delight and bring glory, and stir the hearts of men with drunkenness. To worship me, take wine and strange drugs whereof I will tell my prophet, & be drunk thereof! They shall not harm ye at all. It is a lie, this folly against self. The exposure of innocence is a lie. Be strong, o man! Lust, enjoy all things of sense and rapture: fear not that any God shall deny thee for this."[258]

Further on, he states:

The main subject of this card refers to the most ancient collection of legends or fables. It is necessary to here go into the magical doctrine of the succession

---

[255] Massey. *Ancient Egypt, the Light of the World*. Book 9. Pages 607-608.
[256] Waite. *The Pictorial Key to the Tarot*. Part II. Section 2; Strength.
[257] Ibid.
[258] Crowley. *The Book of Thoth*. Page 92.

VIII STRENGTH

of the Aeons, which is connected with the procession of the zodiac. Thus, the last Aeon, that of Osiris, is referred to Aries and Libra, as the previous Aeon, that of Isis, was especially connected with the signs of Pisces and Virgo, while the present, that of Horus, is linked with Aquarius and Leo.[259,260]

This refers to the precession of the equinoxes, and the "magical doctrine" associated with this phenomenon parallels the concept of the changing nature of the dominant astral radiations permeating Earth during any particular Age.

Crowley mentions that the "seers in the early days of the Aeon of Osiris" (i.e. Age of Aries/Taurus) were terrified at the thought of the arrival of our present Age, as they supposedly regarded the changing of the polestar as a catastrophe. He associates this fear with the repudiation of the Beast and the Scarlet Woman in the 13th, 17th, and 18th chapters of Revelation.[261]

---

[259] Ibid. Page 93.

[260] But Crowley appears to have erred in his Aeonic sequence. He equates the "current Aeon" with Aquarius and Leo. This relates to our present Age, which is nominally that of Aquarius, and the diametrically opposed sign of Leo is the locus of the Full Moon as the "Sun of the nocturnal waters." He states that the last Aeon (i.e. the one prior to the current Aquarian Aeon) was associated with Aries and Libra; the Aeon prior to *that* being associated with Pisces and Virgo, whereas in fact it is just the opposite: the last Age corresponded to Pisces and Virgo, and the one prior to that was associated with Aries and Libra [Figures 3 & 4].

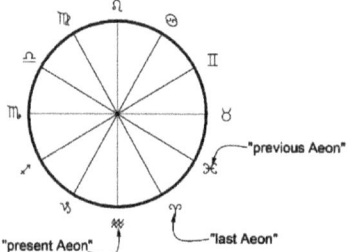

[Figure 3: Wrong order of Ages, with Pisces and Aries inverted.]

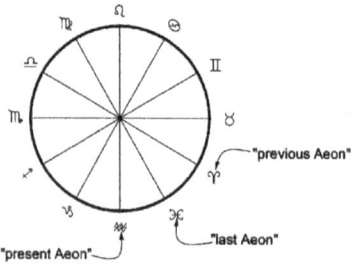

[Figure 4: Correct order of Ages]

[261] Crowley. *The Book of Thoth*. Page 94.

The Beast and Scarlet Woman represent Sut or Sevekh (the Beast, or 7-headed Dragon) and Ta-urt, the Genetrix – zootypes of Ursa Minor and Ursa Major, respectively. Prior to the vilification of the matriarchate, this "red whore of Babylon" appeared as the red hippopotamus of Ta-urt (or Apt), constellated in the signs of Bootes and Corona Borealis.[262]

The color red was connected with the setting Sun, and thus the entrance into the underworld, or the underworld itself.[263] Additionally, menstruation – as the earliest mode of blood sacrifice – became attached to the concept of the "bloody" Sun's passage through the abyss of the nocturnal waters as the Osiris Tesh-Tesh.[264] During the summer months of inundation, this invisible passage was figured as the 3 acronychal culminations of the lunar cycle with the Moon reaching its fullness thrice over in the signs opposite the Sun.[265]

"Tesh-Tesh" was also used as an epithet for the Nile during the inundation, as at the beginning of the flood season the "waters of salvation" were seen to run red, perhaps as the result of seasonal algal blooms or the so-called "red tides." The red ascribed to the Harlot is thus seen to derive from the seasonal and celestial rhythms demarcated by the ancient Egyptians at the level of the day, month, and year, as the daily setting or "dying" Sun, the monthly "blood sacrifice," and the yearly "red tide" of the Nile.

Crowley says that the figure astride the lion in his Atu is

> a form of the Moon, very fully illuminated by the Sun, and intimately united with him in such wise as to produce, incarnate in human form, the representative or representatives of the Lord of the Aeon.[266]

This appears to be an allusion to the Full Moon in the sign opposite the Sun, and a reference to the changing of the vernal point from Pisces to Aquarius.

The "lion-serpent" of this Atu refers to the constellations of Leo and Hydra, which were depicted by the ancient Egyptians as a single constellation [Figure 5]. Hydra was seen as the evil "Apap serpent" or dragon which rose acronychally each season during the periods of subsidence, and was thus seen to threaten the land with drought before being overcome by the solar Horus as he passed into the sign of Leo during the summer flood season.

The 7-headed beast upon which the Whore of Babylon rides can be seen to represent both the constellation Hydra where the Full Moon is to be found at the vernal equinox in this incipient Age of Aquarius, as well as the

---

[262] Massey. *Ancient Egypt, the Light of the World.* Book 11. Page 698.
[263] Massey. *A Book of the Beginnings.* Section 22. Pages 583-584.
[264] Massey. Lecture 1: *The Historical Jesus and Mythical Christ.* Page 19.
[265] See the chapter entitled "The Enneagram, Zodiac, and Celestial Sphere" for a discussion of the relationship between the 3 abysmal winter months, and the 90 "taboo days" of menstruation.
[266] Crowley. *The Book of Thoth.* Page 94.

VIII STRENGTH

circumpolar constellation Ursa Minor – the current station of the pole – which has been identified with the seven heads of the dragon Sevekh-Kronus. One of the ancient meanings of "Babylon" is "the seat in the circle," or "the circular seat," also rendered as the *Bab* of *El*, i.e. the circle of the gods, which is associated with Kheft, goddess of the north,[267] the original "Ancient of Days."[268] In this context, the "seat in the circle" refers to the station of the pole, which in the earlier stellar mythos corresponded to Corona Borealis. The heptagonal sigil [Figure 7] that Crowley illustrates in his discussion of this Atu can thus be seen to represent the seven stars of Corona Borealis, Ursa Major, or Ursa Minor, as they superseded one another as primary "seat in the circle." Given these considerations, the Holy Grail which the Scarlet Woman holds aloft is seen to symbolize both the constellation Crater; sacramental goblet of the inundation,[269] as well as Corona Borealis; keystone of the circumpolar heptanomis.

[Figure 5: Leo and Hydra in the Denderah Zodiac. Author's rendering.]

Another interpretation of this sigil involves the plotting of conjunctions of the Earth, Sun, and center of gravity of the asteroid belt.[270] The concept is the

---
[267] Massey. *A Book of the Beginnings*. Section 20. Page 513.
[268] Ibid. Section 12. Page 40.
[269] Massey. *Ancient Egypt, the Light of the World*. Book 5. Page 296.
[270] Collin. *The Theory of Celestial Influence*. Page 79.

same one that gives rise to the pentagonal representation of the 8-year conjunction cycle of Earth, Sun, and Venus. In the case of the asteroids, the conjunction takes place every 468 days, and by connecting the points of conjunction as they occur sequentially, a figure analogous to Crowley's heptagon is obtained [Figure 6].

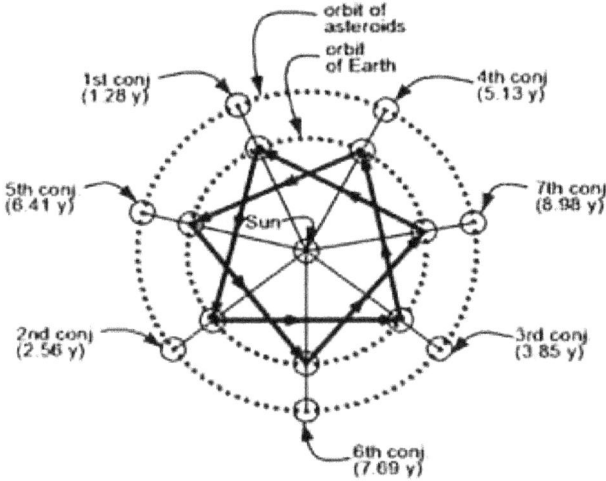

[Figure 6: 9 year conjunction cycle of asteroid belt center of gravity, Earth, and Sun. Based upon average orbital period of the asteroids' "chief concentration" as given by Rodney Collin in *The Theory of Celestial Influence*.]

In this connection, it is interesting to note that Zechariah Sitchin[271] has postulated that the asteroid belt is the result of a collision between the moon of a rogue planet ("Nibiru") which was somehow ejected from its original orbit, and a primordial super-planet ("Tiamat") of our solar system, which was once located between Jupiter and Mars. Nibiru is said to enter our solar system every 3,600 years, potentially wreaking cataclysmic havoc upon Earth as a result of its gravitational and electromagnetic influences.

According to Sitchin, the ancient Sumerians were well aware of this rogue planet, referring to it as "the planet of the crossing," or the planet whose orbit periodically crossed the plane of the ecliptic. Sitchin asserts that the Anunnaki, or "those who from Heaven to Earth came" were a highly evolved species that catalyzed the evolution of humanity from primates through genetic engineering. The reason for this "evolutionary jump-start" is said to have involved the creation of a slave-species of sorts, which provided the Anunnaki with a means of mining the gold of Earth in order to atomize it into the failing atmosphere of their home planet.

---

[271] Sitchin. *The 12th Planet*.

VIII STRENGTH

Whether Crowley was alluding to any of these concepts in his sigil of Babalon is unclear, but it is noteworthy that the complete version of this symbol, the Sigillum Sanctum Fraternitatis A∴A∴ (the holy Sigil of the Brotherhood of the Silver Star) included in the frontispiece of *The Book of Thoth* bears as "X" within a circle which could be read as an allusion to the "planet of the crossing" [Figure 7]. This interpretation is a tentative proposition, as Crowley had obviously never read Sitchin, although this does not preclude the possibility of an occult tradition involving similar concepts which both men drew from – intentionally, or otherwise.

[Figure 7: Crowley's Astrum Argentium sigil. Author's rendering.]

The notion of either a comet or rouge planet invading our solar system is an ancient one, as demonstrated in the writings of Plato (the *Timaeus*), which refer to such an event which altered the declination of Earth, as given in the myth of Phaeton. What connection this celestial invader may have to the Silver Star (Sirius) is not clear. A speculative possibility is that Nibiru hailed originally from the Sirian system, and was ejected from it millions of years ago through a supernova event, eventually making its way into our solar system where it was gravitationally captured in an elongated 3,600 year orbit.

IX The Hermit

THIS TRUMP HAS been described as a "blending of the ideas of the Ancient of Days, and the Light of the World."[272] In the Zohar, a "mysterious Ancient One" is mentioned, which is an apparent reference to the primordial Ancient of Days. The Ancient One desired to manifest himself, and so created the *Aleh*, or "hosts of heaven." These he conjoined with *Mi*: the "highest pole" and "summit of the heavens above." The merging of *Aleh* and *Mi* resulted in the so-called *Alhim*, or *Elohim*.[273] These were the seven guardians or "watchers, who turn[ed] round together"; in other words, the seven circumpolar constellations of the precessional circle, which were seen to rotate about the Great Bear during the Age of Taurus. Later, these seven came to represent the seven stars of Ursa Major itself.[274]

The conjoining of "the hosts of heaven" (the stars) and the "highest pole" by the Ancient of Days can be seen as a reference to the celestial tethering of the NCP (which binds the hosts of heaven) to the NEP (the highest pole in the eternal precessional cycle). The "mysterious Ancient One" thus makes himself manifest to the astronomer priests and priestesses after thousands of years of stellar observations which reveal the crown of crowns as the sevenfold circumpolar heptanomis, the central jewel of which is figured in the stars of Corona Borealis.

According to Massey, the Hebraic Ancient of Days superseded the earlier Egyptian Genetrix, the four quarters of her circumpolar paradise becoming appropriated by the emerging patriarchate to form the name of the new god, JHVH.[275]

This assimilation of the original Egyptian astronomical mythos continued into the Christian era, as is evidenced in the New Testament's reference to

---

[272] Waite. *The Pictorial Key to the Tarot.* Part II. Section 2; The Hermit.
[273] Nurho de Manhar. *The Zohar: Bereshith to Lekh Lekha.* Chapter entitled "The Occult Origin of Alhim". Web. <sacred-texts.com>
[274] Massey. Lecture #5. Pages 124-126. Bracketed note added.
[275] Massey. *A Book of the Beginnings.* Section 14. Pages 152-153.

Christ as the Light of the World (John 8:12). This epithet was originally applied to the Egyptian Ra, spirit of the Sun. We read in chapter XV of *The Book of the Dead*:

> Hail to thee who risest up from the horizon as Ra...Adoration to thee, who arises out of the Golden, and givest light to the earth on thy day of birth. Thy mother bringeth thee forth upon her hands, that thou mayest give light to the whole circumference which the Solar Orb enlighteneth.[276]

Here we have a reference to the rising of Sirius (the mother who "bringeth forth") with the Sun (Ra) during the season of inundation.

As the primary mode of chronometry shifted from stellar to solar, the conception of the Ancient of Days was transformed to connote the mountain of the equinox, rather than the circumpolar one. In his later form, the aged nature of the god was representative of the "old" Sun after the autumnal equinox, as he descended from the western mount into the wintry underworld. This became commemorated in the celebration of All Souls Day and All Hallows Eve (Halloween), when "the ghosts of those who have died during the year assemble together, and prepare to follow the Sun through the underworld as their leader into light."[277] Presently, the Sun's position during this period of the year is in the constellation Virgo.

We thus see that this trump is in a sense connected with the "descent into Hell" (i.e. into the underworld). Crowley seems to be referring to this concept in his discussion of the Hermit when he states that Virgo "forms the crust over Hades," and that the trump "recalls the legend of Persephone."[278] Here the crust over Hades is to be understood as the earthy Virgo poised upon the western horizon of the equinox, below which the nether regions of the abyss await [Figure 1].

Such is the role of Virgo in our present Age, some 12-13,000 years after the vernal Sun used to be found in that sign. In that era, the Virgin and Sun rose together to usher in the beginning of the ancient Golden Age, when the station of the pole was to be found in the constellation Lyra. In this connection, note the vertical staff or pole the hermit wields, and the star within his lamp: a pole and a star... Although the star itself is partially occluded by the lantern, it is clearly intended to represent the six-pointed "star of David" [Figure 2]. Perhaps it isn't too inconceivable to posit a correlation here between Lyra, station of the pole during the Age of Virgo, and King David's Harp, another name for this circumpolar constellation.[279] In

---

[276] *The Egyptian Book of the Dead.* Translated by P. Le Page Renouf.
[277] Massey. *A Book of the Beginnings.* Section 7. Page 293.
[278] Crowley. *The Book of Thoth.* Page 89.
[279] Allen. *Star Names: Their Lore and Meaning.*

IX THE HERMIT

this case, rather than King David's Harp (the entire constellation of Lyra), the reference is to King David's Star – the brightest star in the constellation, alpha Lyra, or Vega.

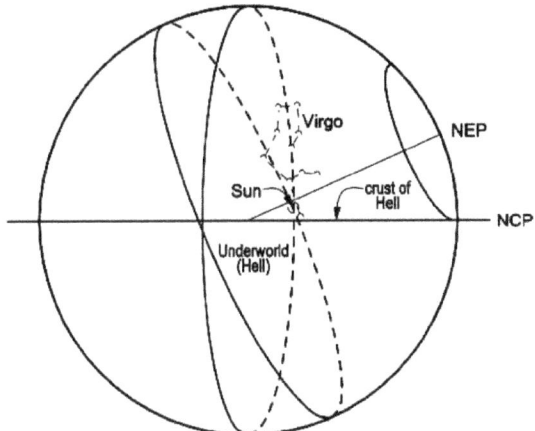

[Figure 1: Current locus of the Sun at the Atumnal Equinox (September 21).]

[Figure 2: The Rider-Waite Hermit trump. Author's rendering.]

According to Case, the 4th and 7th ordinal Hebraic characters (*Dalet*: 4, and *Zayin*: 7) should be related to the Hermit, since it is the 10th trump in sequence, and 10 is the extension of 4 (1+2+3+4 = 10), as well as the reduction of the extension of 7 (1+2+3+4+5+6+7 = 28, and 2+8 = 10).[280] It is admittedly difficult to seriously entertain this numerical "puzzle-game" as a way of elucidating fundamental truths about the nature of any of the trumps. One could just as easily argue that the Hermit has an affinity for all cards 1-9,

---

[280] Case. *The Secret Doctrine of the Tarot*. Chapter 11.

since it is the first to reach the culmination of the "base 10" series, etc., etc. But upon closer inspection, one would see that both conditions are true, and in one sense the Hebraic alphabet is a "base 4" numerical system that after 3 iterations yields 10, and after 7 iterations yields 22.

Case hints at a similar discomfiture regarding this mode of numerology, nonetheless accepting it as a historical precedent worthy of further study:

> Perhaps these correspondences cannot be justified by ordinary rules of logic; but, logical or not, they form a recognized part of occult doctrine. As such they are important clues to the meaning of the Tarot.[281]

At the root of the correspondences is the Tetragrammaton, or "name of God," composed of the four letters *Yud*, *Hei*, *Vav*, and *Hei* (final). *Yud*, being the first letter of the divine name is said to represent unity, and to be present as an active principle in all phenomena. The first *Hei* is a manifestation of the universal *Yud* as dualism; *Vav* represents the union of *Yud* and the first *Hei*, and the final *Hei* signifies the fourth element as a completion of the first 3, as well as the beginning of a new sequence of 4 elements, as a "*Yud* in germ."[282] Thus:

Yud	Hei	Vav	Hei (final)
1	2	3	4
4	5	6	7
7	8	9	10

In fact, the entire Hebraic alphabet is said to be organized according to this principle, in the following way:

Yud	Hei	Vav	Hei (final)
1-Alef	2-Beit	3-Gimel	4-Dalet
4-Dalet	5-Hei	6-Vav	7-Zayin
7-Zayin	8-Chet	9-Tet	10-Yud
10-Yud	11-Kuf	12-Lamed	13-Mem
13-Mem	14-Nun	15-Samech	16-Ayin
16-Ayin	17-Pei	18-Tzadik	19-Kuf
19-Kuf	20-Reish	21-Shin	22-Tav

Furthermore, as each trump may be assigned a unique Hebraic letter (22 trumps = 22 letters), every trump should be expected to have strong symbolic connections to those within its own "tetragrammaton value class" (i.e., trumps 1, 4, 7, 10, 13, 16, and 19 of the tetragrammaton value class "*Yud*" all have affinities for each other, etc.).[283]

---

[281] Case. *An Introduction to the Study of the Tarot.* Chapter 2.
[282] Ibid.
[283] Ibid.

IX THE HERMIT

It must be pointed out, however, that whereas such relationships may be valid for the Kabbalistic doctrine attached to the Hebraic alphabet, the same cannot be said for their application to the Major Arcana.

In the first place, many different numbering schemes have been proposed throughout the course of the Tarot's history, and the one Case uses employs Waite's Strength/Justice inversion of the previous *Tarot de Marseille* order. This new numbering scheme clearly has the advantage of massaging the order of zodiacal trumps into conformity with those of the "12 simples," but to follow any kind of extensive path of correspondences between Hebraic letter and Major Arcana ascriptions such as those suggested in the above table is potentially to invest much time in error. Here again, the occult blind comes into play: what is the proper order of the planetary trumps? Or yet another possibility: the planetary trumps may deny fixed categorization, as the planets are "the wanderers," par excellence.

Case says the following regarding his position on the Kabbalistic connections imagined to be the basis for the structure and iconography of the Tarot:

> My object is neither to prove the accuracy of the Qabalistic interpretation of the Tetragrammaton, nor to defend the doctrines that have been deduced therefrom. I merely seek to show how the inventors of the Tarot used Qabalistic ideas as the basis for their alphabet of symbols.[284]

The obvious question is since it has already been established that there is no conclusive evidence regarding the origin of the Tarot, how can one so readily accept the very specific notion that its inventors were Kabbalists? The main reason seems to center around the equivalency between number of trump cards and letters in the Hebrew alphabet. One may just as "reasonably" conclude that the Tarot was originally conceived by a group of geneticists, since there are 22 autosome pairs in the human organism. The more one studies the different theories regarding the origin and meaning of the Tarot, the more it becomes evident that their common thread is the ability (and perhaps even the "imperative") of humans to find patterns and meaning in phenomena which might otherwise be characterized as chaotic and meaningless. In the words of Case:

> The knitting together of apparently unrelated ideas which results from diligent search for, and prolonged meditation upon, these hidden connections, will be found to be one of the principle benefits of this study.[285]

---

[284] Ibid.
[285] Ibid.

Notwithstanding a reticence to venture too deeply into the murky waters of derivative Kabbalistic correspondences, it is here noted that Case's central conclusions regarding this trump and its close symbolic relationship with the other 2 trumps represented by *Dalet* and *Zayin* still maintain for the present system, although for altogether different reasons.[286] In our case, both *Dalet* and *Zayin* are associated with Mercury, and thus fulfill the required affinity for the trump represented by *Yud* – i.e., the Hermit, since the Hermit is representative of Virgo, and Virgo is ruled by Mercury.

---

[286] See chapter on The Hanged Man for further elucidation.

X Wheel of Fortune

IN THE RIDER-WAITE version of this trump, the four apocalyptic beasts square the circle of a centrally placed wheel, the spokes of which alternately bear the inscriptions of Taro/Rota/Tora, and the Tetragrammaton [Figure 1]. Reading the beasts as the fixed zodiacal signs – Aquarius (angel), Scorpio (eagle), Leo (lion), and Taurus (bull) – it becomes evident that our reference point is "outside" the celestial sphere, with the central axis of the wheel corresponding to either the NCP, or NEP.

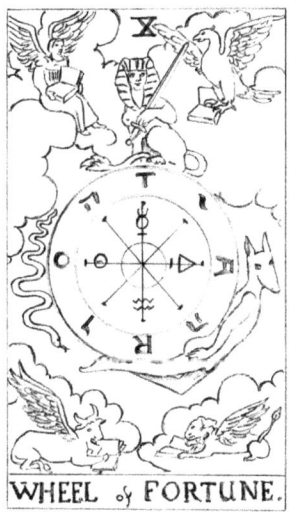

[Figure 1: The Rider-Waite Wheel of Fortune trump. Author's rendering.]

The three figures arranged about the wheel's circumference are described by Waite as a sphinx, hermanubis, and typhon.[287] If we equate the central hub with the NEP, we should expect to find three circumpolar constellations that bear spatial and symbolic equivalencies with these three zootypes. Orienting ourselves from a bird's eye view "above" the Cat's Eye Nebula and

---

[287] Waite. *The Pictorial Key to the Tarot*. Part I. Section 2. Class I. The Trumps Major, otherwise Greater Arcana.

maintaining the relative positions of the four fixed signs as depicted in the card, we see that the sphinx corresponds to the circumpolar region of Cygnus, the serpent or typhon to Ursa Minor, and the hermanubis to the area demarcated by Ursa Major, Bootes, and Corona Borealis [Figure 2].[288]

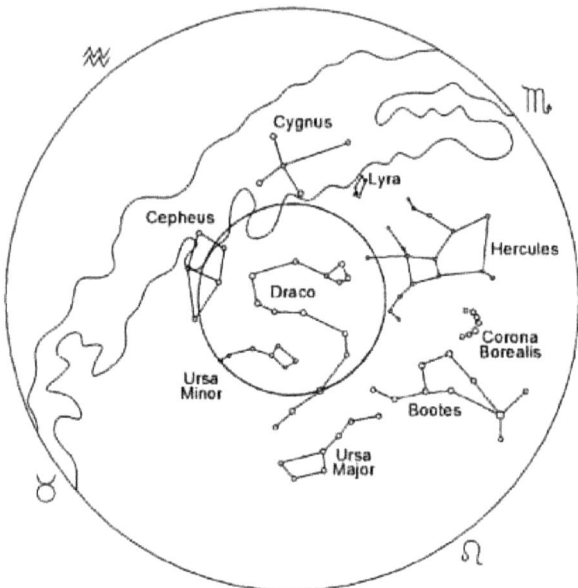

[Figure 2: The circumpolar constellations as they correspond to the Rider-Waite Wheel of Fortune]

The sphinx (i.e., lion) as ideograph of the Cygnian astronome is explained when viewed from a stellar acronychal perspective, during the Age of Taurus. When the summer inundation had reached its peak with the Sun in Leo, Aquarius – and its paranatellons Cepheus and Cygnus – rose as the Sun set [Figure 3].[289]

---

[288] Note the spatial correlation between the sword of the sphinx directed towards the eagle/Scorpio, and the cross of Cygnus pointing to the same constellation, down through the path of the Milky Way [Figures 1 & 2].

[289] According to Massey, "when the Sun was in the sign of the lion, Kepheus was visible very low down in the Northern Hemisphere, at the same time Leo was hidden in the solar radiance. Thus Kepheus took the place of Leo as guide of the Sun, or indicator or lawgiver, the Regulus in person, as the paranatellon of Regulus the star (Cor-Leonis) in the lion, and, being so low down in the Northern Hemisphere, he may be described as seen under the feet of Judah or the lion. In one Egyptian planisphere reproduced by Kircher, the figure of Shu-Anhar, as Cepheus, fills all three decans of the Waterman. He wears upon his head the two ostrich feathers, which read Ma-Shu; in his stretched-out right hand he holds the sceptre or rod, and in his left he grasps the arrow. He is portrayed in his marching martial attitude, and, as a paranatellon of the lion, is literally the lawgiver between the feet of the Lion of Judah. Cepheus,

X WHEEL OF FORTUNE

As the serpent's location along the wheel corresponds to Ursa Minor, Waite's allusion to "typhon in his serpent form"[290] is interpreted as a reference to Sut-Typhon, otherwise known as Sut-Anup, or Sut-An (later, Satan),[291] whose seven heads[292] were constellated in the seven stars of the Lesser Bear – son of the Greater Bear, who was the "original" typhonian Genetrix.[293]

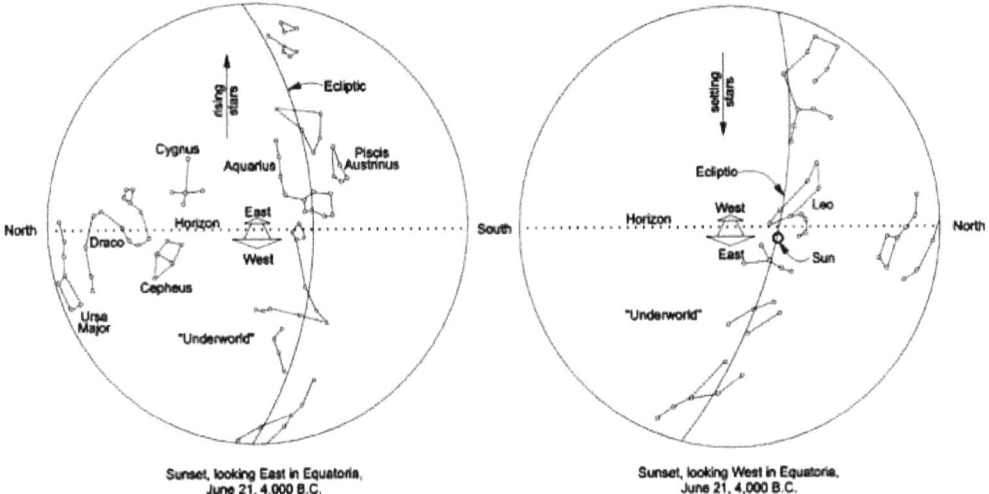

[Figure 3: Sunset looking east and west in Equatoria on the summer solstice, circa 4000 BCE]

The hermanubis is depicted as jackal-headed in this trump, but another form of it is the ibis-headed zootype, which is associated with the constellation Cancer.[294] Significantly, the circumpolar region occupied by the hermanubis in the Rider-Waite trump is inclusive of Corona Borealis, station of the pole during the Age of Cancer.

The four letters of the Tetragrammaton radiating from the central hub of the wheel recall a similar compositional device as utilized in Schweighardt's depiction of the Invisible College [Figure 4],[295] except in the case of the Fortune trump, one's viewpoint is from above the NEP, whereas in Schweighardt's illustration, it is orthogonal to it [Figure 5]. In both cases, the

---

the lawgiver, be it understood, has two stars. One is Regulus, the heart of the lion, the other, the northern constellation." [Massey. *A Book of the Beginnings*. Section 16. Pages 244-245.]

[290] Waite. *The Pictorial Key to the Tarot*. Part II. Section 2; Wheel of Fortune.
[291] Massey. *A Book of the Beginnings*. Section 23. Page 663.
[292] These are the 7 heads of the serpent/dragon Sevekh Kronus.
[293] Massey. *A Book of the Beginnings*. Section 9. Page 426.
[294] Drummond. *Oedipus Judaeica*. Verse 11, Chapter 15.
[295] Schweighardt, Theophilus. *Speculum Sophicum Rhodotauroticum* (*The Mirror of Wisdom of the Rosicrucians*).

four letters JHVH can be read to signify the four corners or divisions of the heavens based upon the solstices and equinoxes according to the ancient Egyptian mode in which these quarters were mapped out in the circumpolar heavens during the stellar phase of chronometry. The Tetragrammaton may have ultimately been derived from this matriarchal and stellar source.[296]

[Figure 4: Frontispiece to the *Speculum Sophicum Rhodotauroticum* ("The Mirror of Wisdom of the Rosicrucians") by Theophilus Schweighardt. Author's rendering.]

---

[296] Massey. *A Book of the Beginnings.* Section 14. Pages 152-153: "It was on account of the feminine origin of Jehovah that it was considered blasphemy to pronounce the name. […]In the gnostic account of the beginning attributed by Irenaeus to Marcus, it is said the deity uttered the first word of four letters. This word was Arke (ἀρχή)—the Greek form of the famous Tetragrammaton, which with the Hebrews was the name of four letters. … Ark (Eg.), as before said, means to encircle, encirclings, enclosings, settings, endings, weavings; arkai is to appoint a limit, fix a decree, and signifies *finis.* The first circle or arc observed in heaven as a measure of time was that of Ἄρκτος (Arktos), the Bear whose revolution made the first (Arctic) circle round the pole of the north. The four letters typify the four corners of all beginning. Apt is the name of the goddess of the Great Bear, and of the four corners. Here the secret of the mystery is that Jhvh was represented by the beast that went on all fours, whose name was written with four letters, and who was a figure of four. Apt the genetrix is the abode of the four corners. The four corners at first represented by the four legs of Apt, the beast, were afterwards depicted by a goddess bending over the earth and resting upon her hands and feet, or on all-fours." Ellipses added.

X WHEEL OF FORTUNE

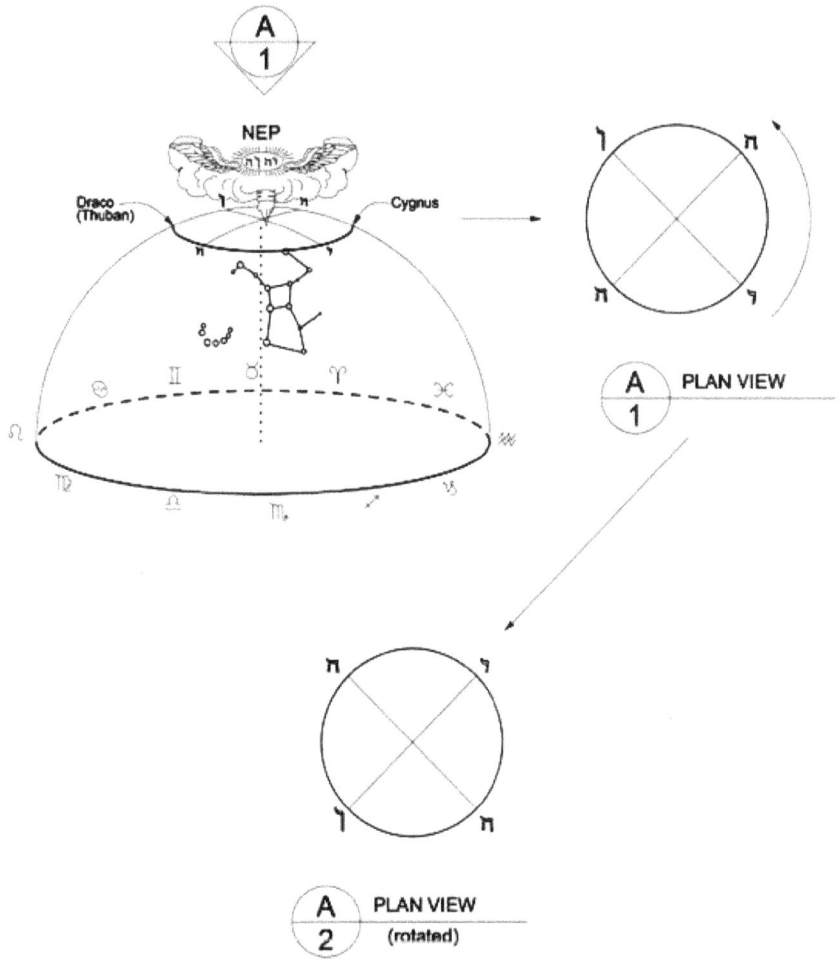

[Figure 5: The four letters of the Divine Name as they relate to the four circumpolar quarters anciently ascribed to the Genetrix. Orthogonal view hypothesized to correspond to Schweighardt's illustration of the Invisible College.]

Different time cycles are suggested by the directionality attributed to the wheel's "motion." The order of letters of the Tetragrammaton (JHVH), and the reading of the English letters as "T-O-R-A" indicate a counterclockwise direction, and thus the diurnal axial rotation of Earth, as well as the annual motion of the Sun along the ecliptic [Figure 6].

Reading the English letters clockwise as R-O-T-A, or T-A-R-O is suggestive of the retrograde motion of the NCP about the NEP throughout the 24-26,000 year "cycle of eternity," or Great Year [Figure 7].

CELESTIAL ARCANA

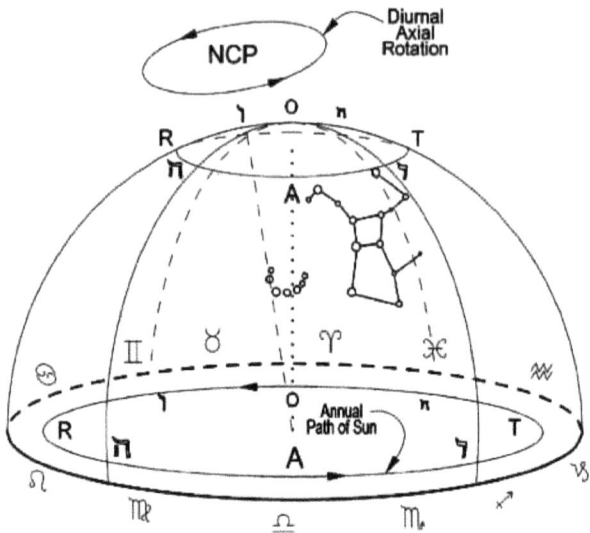

[Figure 6: "T-O-R-A / J-H-V-H" as representative of the diurnal axial rotation of Earth and annual path of Sun.]

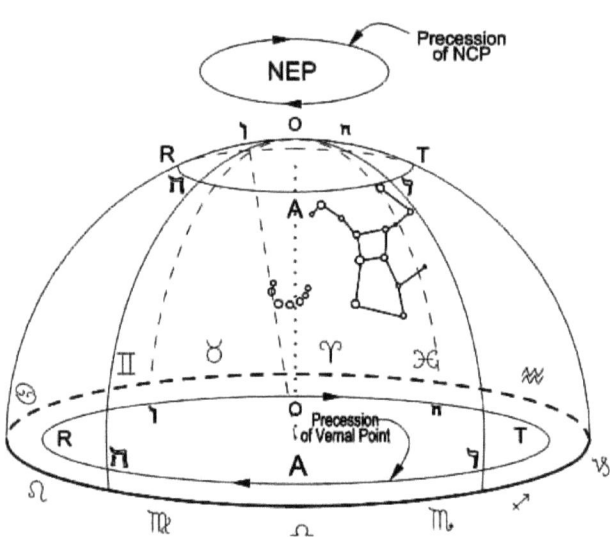

[Figure 7: "T-A-R-O / R-O-T-A" as indicative of the retrograde motion of the NCP about the NEP throughout the Great Year.]

The "O," "T," and "A" spokes of the wheel are each labeled with the three metals of alchemy: salt, mercury, and sulfur, respectively. The glyph on the "R" spoke could be interpreted to signify the element water, but it is unclear exactly what the wavelike symbol represents, as typically there is no fourth element to the alchemical triad of sulfur, salt, and mercury [Figure 8].

X WHEEL OF FORTUNE

Furthermore, the alchemical symbol for water is an inverted triangle, the double-wave form (♒) generally being recognized as the symbol for Aquarius. In the present case, the two readings are not mutually exclusive, as Aquarius was considered the nocturnal source of inundation ("abyss of source")[297] during the Age of Taurus.

According to Case, this symbol is identical to the astrological sign for Aquarius, and represents the alchemical process of dissolution, which is considered to be fundamental to the Great Work.[298] As previously mentioned, the Great Work involves the spiritual evolution of Humanity as a whole. Within the cycle of precession, the Age of Aquarius can be seen at one level to mark the descent of Humanity's consciousness into a mode concerned primarily with the material aspects of existence. Dissolution at this level can be seen as the dissolving of the purely spiritual aspects of consciousness into the materialism of the present day. This sets the stage for a purging and purification, such that the bonds of wealth and power are gradually realized and relinquished by degrees, in the ascent from the depths of the Abyss.

Interpretation of this locus on the wheel as symbolizing the abyss of source makes additional sense when we consider the fact that Cepheus and Cygnus (the sphinx of the Rider-Waite trump) are paranatellons of Leo, and Ursa Major (the hermanubis of the Rider-Waite trump) is a paranatellon of Aquarius [Figures 1 & 9].

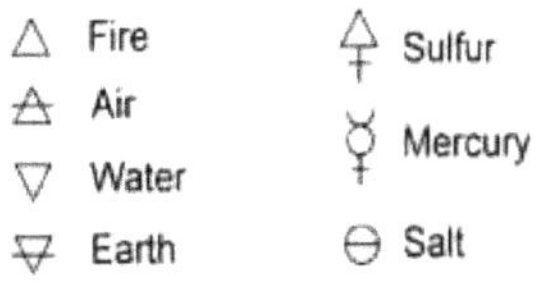

△ Fire	⚨ Sulfur
△ Air	
▽ Water	☿ Mercury
▽ Earth	⊖ Salt

[Figure 8: The Alchemical elements]

The fourfold division of the wheel according to an "elemental" classification calls to mind the Mesoamerican tradition of dividing the Great Year into four World Ages of Fire, Air, Earth, and Water. In the case of the alchemical elements, sulfur represents the active force present in any given phenomenon, salt the passive force, and mercury the mediating force which brings the two opposed forces together, providing a medium for manifestation. Note that the position of the active sulfur corresponds to the locus of the NCP during the Golden Age of Leo/Cancer, and the passive salt to the station of the pole that marks the nadir of the Great Year during the Ages of Aquarius and Capricorn. The mediating mercurial station appears as

---

[297] Massey. *Ancient Egypt, the Light of the World.* Book 5. Page 280, 300.
[298] Case. *The Tarot: A Key to the Wisdom of the Ages.* Chapter 14.

the sphinx, which demarcates the position of the NCP during the Age of Sagittarius/Scorpio – the Spring of the Great Year. The other point of equipoise is found in the sign of Taurus, when the station of the pole was localized in the region of Ursa Major, corresponding to the "R" spoke of the wheel. Here the double-wave glyph is interpreted to represent the abyss of source as Aquarius, during the Age of Taurus [Figure 10].

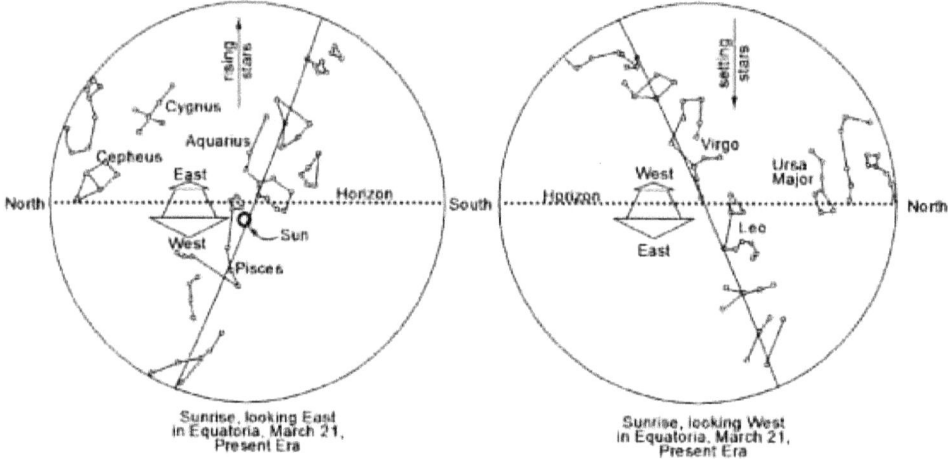

[Figure 9: Sunrise looking east and west in Equatoria on the vernal equinox, present era.]

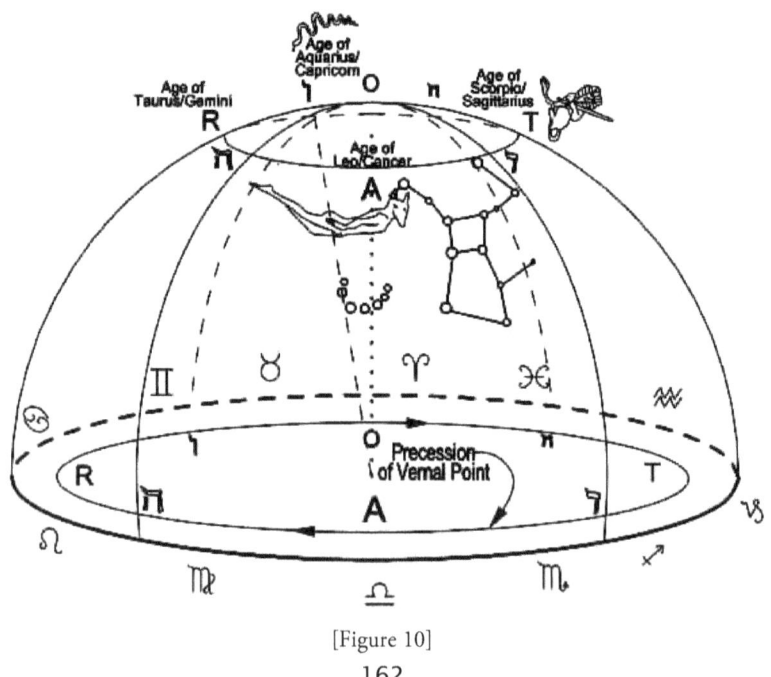

[Figure 10]

X WHEEL OF FORTUNE

Crowley appends a footnote in his discussion of this card[299] wherein he adds the numeric values of the letters *Kuf* (20 or 500) and *Pei* (80), arriving at 100=*Kuf*=Pisces. According to this formula, *Kuf* and *Pei* are the initials of the *kteis* (yoni) and *fallos* (lingam), with *Kuf* being ascribed to the Wheel of Fortune, and *Pei* to the Tower. *Kuf* is thus indicative of the feminine *kteis*, yoni, stone, and grail. The *Pei* is the phallus, lingam, sword, and elixir of immortality.

The formula is an apparent attempt to portray the Tower as the force of precessional change, and the Wheel of Fortune as the receptacle into which this force is applied; the interaction between the two characterizes the resultant Age.

Another way of looking at this equation would be to consider the Tower as representative of the old Age that is destroyed, and when added to the Wheel of Fortune – which represents (in part) the NCP locus of the current Age, and has the corresponding planetary value – one arrives at the value of the current Age. According to this model, there is one alternative to Crowley's assignation of values that can satisfy the same equation: instead of Mars being associated with the Tower (which would imply the destruction of the Age of Aries), we apply the value of Jupiter for the Piscean Age. The value of the Wheel of Fortune is thus seen to change depending on the particular Age it represents: in the case of the Piscean Age, that of Jupiter; but in our present Aquarian Age, the value would be that of Saturn. Thus, the Tower is assigned to Jupiter (*Kuf*=20, or 500), and Fortune is assigned to Saturn (*Tav*=400). If we follow the same formula, and add our values for the Tower and Fortune trumps, we arrive at 900, which equates to *Tzadik*, and Aquarius. It should be noted that this system does not work when attempting to apply it to other Ages besides Pisces and Aquarius.

Although Crowley apparently viewed himself as living in the beginnings of the new Aeon of Horus, which he specifically equates with Aquarius, one would not be incorrect in considering the late 19th and early 20th centuries to be the end of the Piscean Age. Technically speaking, the vernal Sun is still within the physical confines of the constellation Pisces, although an equal spacing of zodiacal decans does not correlate perfectly by any means with the actual constellation. Perhaps Crowley's equation – which is solvable utilizing values for both the Piscean and Aquarian Ages alludes to the fact that there is no hard demarcating point wherein one Age changes to the next.

---

[299] Crowley. *The Book of Thoth*. Page 89.

XI Justice

THE PROMINENT DEPICTION of the scales, as well as Waite's acknowledged transposition of this card with that of Fortitude, suggests the implied attribution to be that of Libra. If we assume a vantage point from outside the celestial sphere with a north-south orientation, and relegate the scales themselves to Libra, the following correspondences will be noted: the figure's crown to Corona Borealis; the semicircular contour of the backdrop to the circumpolar heptanomal region, and the sword to the celestial pole. In this configuration, the central meridian of the figure corresponds to the ecliptic pole [Figure 1].

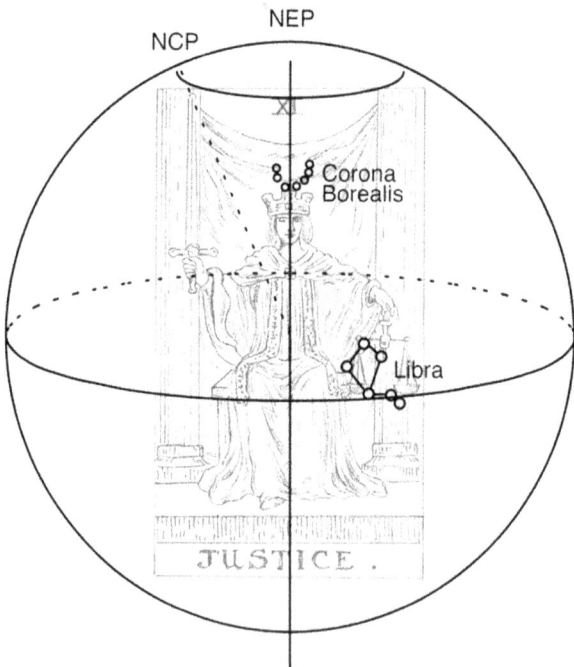

[Figure 1: A proposed reading of the Rider-Waite Justice trump, as it relates to key elements of the celestial sphere. Author's rendering.]

Alternatively, taking a view from within the celestial sphere, and justifying the figure's meridian with the celestial axis rather than the ecliptic axis, the same correlations hold true, except that instead of the celestial pole, the sword now equates to the ecliptic pole [Figure 2].

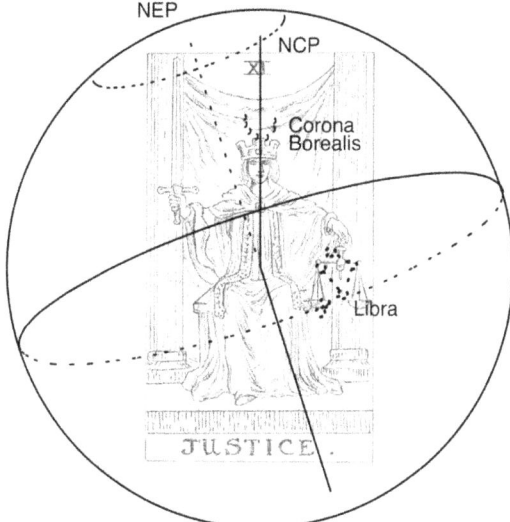

[Figure 2: An alternative reading of the Rider-Waite Justice trump, as it relates to key elements of the celestial sphere. Author's rendering.]

The allusion to Corona Borealis, and the implied zodiacal reference to Libra relate to Justice as framed within the ancient Egyptian concept of the *maat*, or point of equipoise, as seen from alternately stellar and solar chronometries. In both instances, *maat* is associated with justice and balance as manifest in the dynamic equipoise of the celestial mechanics of time cycles, whether they be at the diurnal, monthly, yearly, or precessional level. In the stellar mythos of Egypt, the position of the *maat*, or judgment hall was to be found in the circumpolar heavens as the NCP. The heptanomis was conceived of as a 7-armed balance, with the ecliptic axis serving as the central heavenly support [Figure 3]. Corona Borealis was considered the primary locus of equipoise within the precessional circle, and its 7 stars – with ring-like shape and central Clavis Corona – were seen as a kind of fractal representation of the larger circle of the heptanomis [Figure 4].[300]

---

[300] Massey. *Ancient Egypt, the Light of the World*. Book 9. Page 602: "The seven polestars themselves did not form one constellation, but the crown would be figured typically as a group of stars that told the story in the customary way, even as we find it in Corona Borealis. Moreover, to the naked eye the constellation of the Crown, consisting of seven large stars, would present a picture of the other seven—the crown of stars upon the summit of the mount, which is so prominent in the eschatology. [...] This is where the balance was then erected at the place of judgment in the circumpolar maat, and also at the point where the crown of life

XI JUSTICE

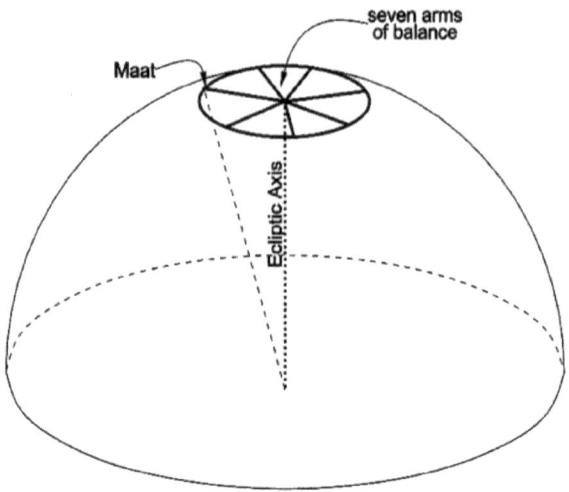

[Figure 3: *Maat* as the circumpolar point of equipoise.]

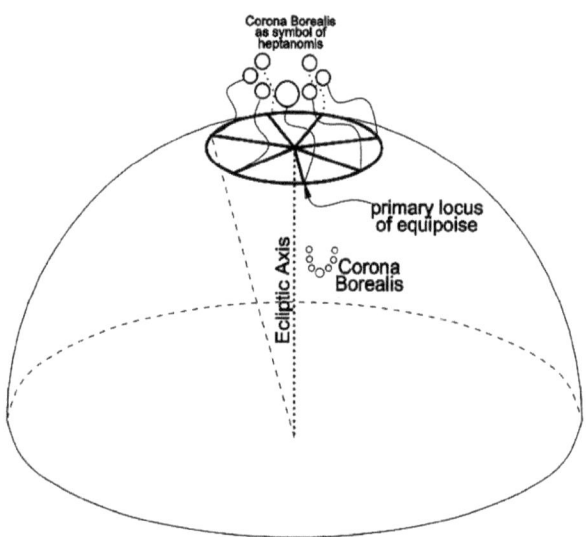

[Figure 4: Corona Borealis as a "fractal" representation of the heptanomis.]

---

was conferred upon the spirits perfected at the summit of the mount. It is also said of the glorified elect, 'He followeth Shu and calleth for the crown. He arriveth at the Aged One on the confines of the mount of glory where the crown awaiteth him.' This is the eternal crown in the eschatology, which had its origin in the seven never-setting stars of the mythology. In the Kabbalah it is the Crown of Crowns pertaining to the Aged in which he had incised the forms and figures of the primordial kings who reigned aforetime in the land of Edom, but who could not preserve themselves and consequently passed away, 'one after the other.'"

The celestial *maat* was also observed at the level of the year, and was fundamentally connected to the annual flood cycle of the Nile. During the Age of Taurus, the season of inundation commenced when the Sun was passing through Cancer and Leo, and the point of equipoise between flood and drought was subsequently reached when the Sun progressed into Libra:

> The four months of the water-season, the first of the three tetramenes, began with the lion, and ended with the scorpion. The inundation reached its point of equipoise coincidently with the entrance of the Sun into the sign that was figured as the Balance or the Scales.[301]

Massey suggests that the two levels of representation of *maat* – the one circumpolar, and the other essentially ecliptic and equinoctial – relate to the sequential stellar and solar phases of the Egyptian celestial eschatology:

> In the stellar mythos Anup had been the judge, with the seat of judgment at the place of equipoise, which was then at the celestial pole. In the solar mythos this was shifted to the vernal equinox, and the mount of glory to the east. An ideal of justice, truth, and righteousness, imaged by the balance or scales, was postulated as established and eternal in the heavens as the reign of law, and there was an annual attempt to make that justice visible and veritable on earth.[302]

> The *maat* was a double law court, first erected for Anup at the pole; but in the solar myth the place of equipoise was changed, and the *maat* was represented where the annual or periodical assize was held. This was at the point of equinox, which was at one time imaged in the sign of the Scales.[303]

This association of *maat* and justice with Libra has persisted into our current Age, even though the equinoctial points of equipoise have nominally shifted to Aquarius and Leo.

Waite associates Astraea – Greek goddess of purity and justice – with this trump.[304] During the transition from the Age of Aries to that of Pisces (circa 0 CE), Ovid writes in his *Metamorphoses* of Astraea having forsaken Earth – as during the Iron Age, moral decay had become so rampant:

> Hard steel [Iron] succeeded [Bronze] then:
> And stubborn as the metal, were the men.
> Truth, modesty, and shame, the world forsook:
> Fraud, avarice, and force, their places took.
> Then sails were spread, to every wind that blew.
> Raw were the sailors, and the depths were new:

---

[301] Massey. *Ancient Egypt, the Light of the World*. Book 5. Page 297
[302] Ibid. Book 6. Page 344.
[303] Ibid. Book 10. Page 678.
[304] Waite. *The Pictorial Key to the Tarot*. Part I. Section 2; Justice.

XI JUSTICE

> Trees, rudely hollow'd, did the waves sustain;
> E're ships in triumph plough'd the watry plain.
> Then land-marks limited to each his right:
> For all before was common as the light.
> Nor was the ground alone requir'd to bear
> Her annual income to the crooked share,
> But greedy mortals, rummaging her store,
> Digg'd from her entrails first the precious oar;
> Which next to Hell, the prudent Gods had laid;
> And that alluring ill, to sight display'd.
> Thus cursed steel, and more accursed gold,
> Gave mischief birth, and made that mischief bold:
> And double death did wretched Man invade,
> By steel assaulted, and by gold betray'd,
> Now (brandish'd weapons glittering in their hands)
> Mankind is broken loose from moral bands;
> No rights of hospitality remain:
> The guest, by him who harbour'd him, is slain,
> The son-in-law pursues the father's life;
> The wife her husband murders, he the wife.
> The step-dame poyson for the son prepares;
> The son inquires into his father's years.
> Faith flies, and piety in exile mourns;
> And justice [Astraea], here opprest, to Heav'n returns.[305]

Edmund Spenser, 16th century English poet, refers to Astraea's departure during the Iron Age (which generally corresponds to the Ages of Aries and Pisces, in the present model) in his epic poem *The Faerie Queene*, circa 1596 CE:

> But when she parted hence, she left her groome
> An yron man, which did on her attend
> Alwayes, to execute her stedfast doome,
> And willed him with *Artegall* to wend,
> And doe what euer thing he did intend.
> His name was *Talus*, made of yron mould,
> Immoueable, resistlesse, without end.
> Who in his hand an yron flale did hould,
> With which he thresht out falshood, and did truth vnfould.[306]

Talus, in ancient Greek legend was said to be associated with a bronze bull that was given to king Minos of Crete. Massey equates Minos with Mena, the bull of the mother – that same mythic king whose wife Pasiphae was discussed

---
[305] Ovid. *Metamorphoses*. Book 1. Bracketed comment added.
[306] Spenser, Edmund. *The Faerie Queene*. Book 5: Canto 1.

in the chapter on the Hierophant as representative of the Full Moon in Scorpio during the Taurean equinox. Talus can thus be seen as somehow representative of the Age of Taurus; perhaps as an anthropomorph of the celestial axis and pole during the Age of Bronze, as he was said to possess only a single vein – running from his head to his ankle, into which a single bronze nail was fixed.[307] When Astraea leaves her groom in this scenario, it is the point of equipoise shifting upon the circumpolar mount, from the Bronze to the Iron Age [Figure 5].

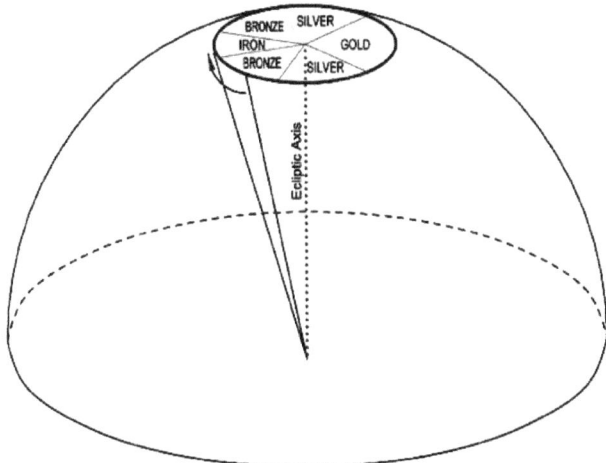

[Figure 5: The precessional shifting of the circumpolar *maat* from the Bronze to the Iron Age.]

The precessional shifting of the Ages from a period of "higher" moral standards to one of corruption and inversion is described in the following passage from *The Faerie Queene*:

> For that which all men then did vertue call,
> Is now cald vice; and that which vice was hight,
> Is now hight vertue, and so vs'd of all:
> Right now is wrong, and wrong that was is right,
> As all things else in time are chaunged quight.
> Ne wonder; for the heauens reuolution
> Is wandred farre from where it first was pight,
> And so doe make contrarie constitution
> Of all this lower world, toward his dissolution.[308]

---

[307] Apollodorus (or Pseudo-Apollodorus). *Biblioteca*. 1.9.26.: "Putting to sea from there, they were hindered from touching at Crete by Talos. Some say that he was a man of the Brazen Race, others that he was given to Minos by Hephaestus; he was a brazen man, but some say that he was a bull. He had a single vein extending from his neck to his ankles, and a bronze nail was rammed home at the end of the vein." Web. <theoi.com/Text/Apollodorus1.html>

[308] Spenser, Edmund. *The Faerie Queene*. Book 5: "The Legend of Artegall or Ivstice".

XI JUSTICE

Here we have shifted some 180° from the Golden Age, to find ourselves in a world turned on its head. The fact that Spenser is speaking of a literal celestial phenomenon connected to the precessional cycle is made clear in the following stanzas:

Ne is that same great glorious lampe of light,
That doth enlumine all these lesser fyres,
In better case, ne keepes his course more right,
But is miscaried with the other Spheres.
For since the terme of fourteene hundred fyeres,
That learned *Ptolomaee* his hight did take,
He is declyned from that marke of theirs,
Nigh thirtie minutes to the Southerne lake;
That makes me feare in time he will vs quite forsake.
And if to those *Ægyptian* wisards old,
Which in Star-read were wont haue best insight,
Faith may be giuen, it is by them told,
That since the time they first tooke the Sunnes hight,
Foure times his place he shifted hath in sight,
And twice hath risen, where he now doth West,
And wested twice, where he ought rise aright.
But most is *Mars* amisse of all the rest,
And next to him old *Saturne*, that was wont be best.[309]

What is being referred to in the first stanza is the fact that the declination of the Sun from its prior solstitial point between Cancer and Gemini circa 160 CE[310] had shifted south[311] through the precessional course such that the Sun at that same ecliptic point some 1,436 years[312] later corresponded to midsummer, and therefore to a lower angle of incidence of solar radiation [Figure 6].

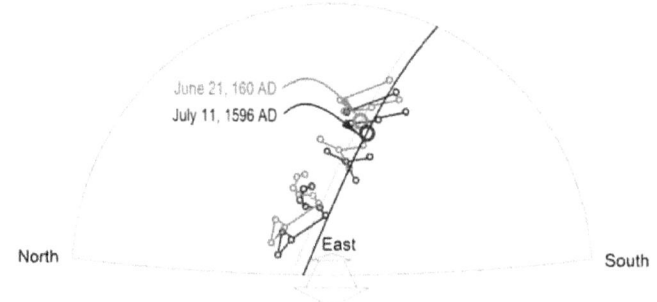

[Figure 6]

---

[309] Ibid.
[310] Ptolemy's era, and likely around 10 years after his *Tetrabiblos* had been published.
[311] According to Spencer, by approximately 30 minutes, or a half degree.
[312] Or roughly 1,400, answering to Spenser's "fourteene hundred fyeres."

The second stanza refers to the concept of the Sun "rising" in the course of the year at the spring equinox, and then "setting" at the autumnal equinox. From this perspective, what is being referenced is the precessional shifting of the points of equinox such that what was once the vernal point (where the Sun "rises") becomes the autumnal point (where the Sun "Wests"), and vice versa. This amounts to 1.5 precessional cycles prior to some unspecified Age, as for the Sun to have twice "risen" where it currently "sets," and vice-versa, the most recent occurrence would be associated with 180° of precessional rotation into the past. The incidence prior to that would entail an entire 360° precessional rotation into the past from that point . If we consider that unspecified Age as having been Spenser's own Piscean era, 1.5 precessional cycles into the past would roughly correspond to the Age of Virgo/Leo, some 36,000 years prior, as when they first "tooke the Sunnes hight" [Figure 7].

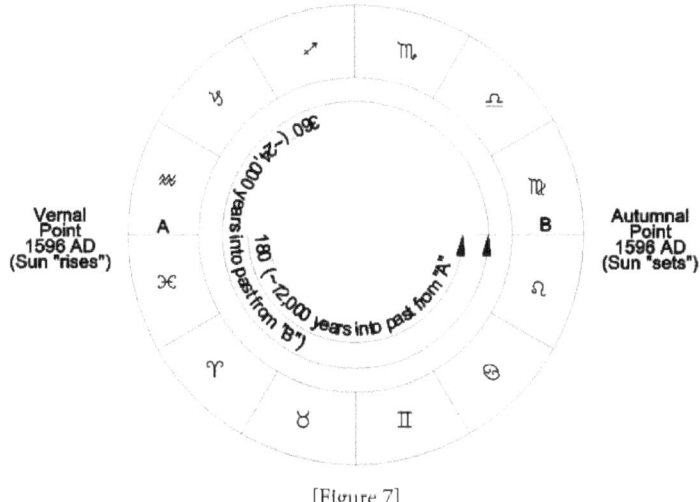

[Figure 7]

Spenser's allusion to Astraea as the "Virgin sixth in her degree" in the following stanza refers to Virgo as sixth house of the tropical zodiac:

Now when the world with sinne gan to abound,
Astraea loathing lenger here to space
Mongst wicked men, in whom no truth she found,
Return'd to heauen, whence she deriu'd her race;
Where she hath now an euerlasting place,
Mongst those twelue signes, which nightly we doe see
The heauens bright-shining baudricke to enchace;
And is the *Virgin*, sixt in her degree,
And next her selfe her righteous ballance hanging bee.[313]

---

[313] Spenser, Edmund. *The Faerie Queene*. Book 5: Canto 1.

XI JUSTICE

This association of Astraea with the Celestial Virgin may be intended to reflect the shifting of the point of equipoise as originally associated with the annual flooding of the Nile from Libra to Virgo, during the transition from the Age of Taurus to that of Aries, during which period ancient Hellenic culture began to thrive.

In this connection, it is interesting to consider the possibility that Astraea (as Libra/Virgo) leaving Talus (as Taurus) behind signifies the shifting of the point of equipoise or *maat* from Libra to Virgo, in the transition from the Age of Taurus to that of Aries. Astraea leaves the earth celestially as the constellation Libra gradually rises throughout the course of precession from its original locus of *maat* on the horizon, to be supplanted by the next sign in the retrograde motion of the equinox in the Great Year [Figure 8]. Her fabled return to Earth is thus to be expected some 7,000 years hence, when Libra will join Scorpio at the vernal point – and as we shall see further on, this cyclical phenomenon of the return of the vernal point to the place of *maat* is symbolically alluded to in the Judgment trump.

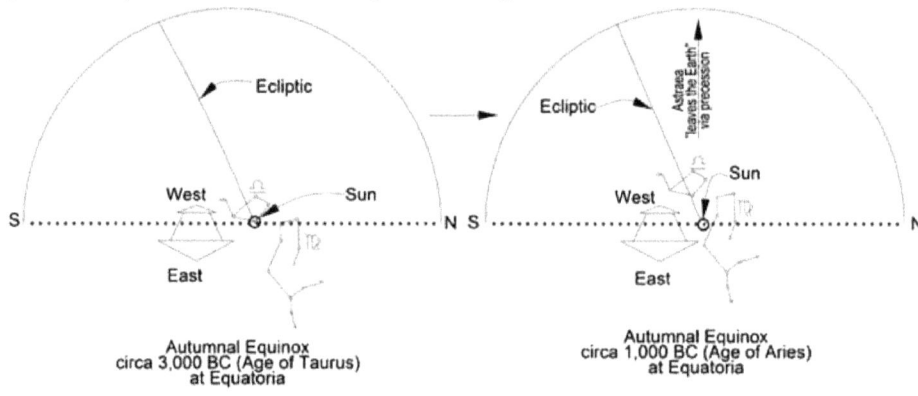

[Figure 8]

One may postulate some implied connection between this trump and the High Priestess, based on the supposed reference to Corona Borealis in the figure's crown. In fact, Case arrives at a similar conclusion through a different path, noting that since 11 (Justice) reduces to 2 (The High Priestess) via "theosophical addition" (11 = 1+1 = 2), Justice represents an aspect of Wisdom (The High Priestess) that

> analyzes, separates things into their component parts, weighs and measures, and so discovers the workings of nature. Her purpose is the adjustment of man to his environment, and the modification and improvement of all the conditions of that environment, through the intelligent direction of natural forces according to their laws.[314]

---

[314] Case. *An Introduction to the Study of the Tarot*. Chapter 8.

Justice is a zodiacal trump, and it should therefore be expected that it not be subject to the vagaries of the "blind," in contrast to the 7 planetary trumps. All versions of the *Sefer Yetzirah* referenced in the "Origins and Attributions" chapter agree that Libra is ascribed to the letter *Lamed*. We recall, however, that the positions of Justice and Strength were juxtaposed in earlier decks, such as the *Tarot de Marseille*. According to Case:

> In the exoteric Tarot, Key 8 is Justice and Key 11 is Strength. This blind does not mislead those who know the attributions of the signs of the zodiac to the letters of the alphabet. *Just why it was employed at all may be difficult to understand now.* Yet it does serve to emphasize the fact that Keys 8 and 11 represent two aspects of the operation of a single power, which power is the creative imagination symbolized by the Empress.[315]

As to why this purported obfuscation may have been employed, there could be any number of reasons. One might also question whether this was meant intentionally as a "blind" at all, or if there could be an alternative explanation.

Massey suggests *Lamed* to have been derived from a pictograph of the hinder-part of a lion, rather than an ox-goad, as Case and others commonly assert.[316] In the Book of Judges, we read that Shamgar – an Israelite judge – repelled the Philistines' advances with an ox-goad:

"Shamgar [..] slew of the Philistines six hundred men with an ox goad: and he also delivered Israel."[317,318]

In the same book Samson – also an Israelite judge, during a later period – is said to have similarly repelled the Philistines with the jawbone of an ass: "And he found a new jawbone of an ass, and put forth his hand and took it, and slew a thousand men therewith."[319]

We have previously established the connection of Samson's slaying of the lion with the heliacal rising of Sirius, and it is likely that his defense of the Israelites against the Philistines with the ass' jaw is a similar mythic permutation of this celestial event that was so sacred to the Egyptian culture the Israelites were attempting to distance themselves from. If we follow Massey's lead, and posit *Lamed* to have been derived from the pictograph of the hinder-part of a lion, the connection between Shamgar and Samson as anthropomorphs symbolizing the heliacal rising of Sirius becomes evident: it

---

[315] Case. *The Tarot: A Key to the Wisdom of the Ages.* Chapter 15.
[316] Massey. *A Book of the Beginnings.* Section 13. Page 121.
[317] Judges 3:31. KJV.
[318] Shamgar is connected Kabbalistically with *Lamed*. Additionally, the numerical value of *Lamed* is 30, which is said Kabilistically to correspond to the gematrial value for the Israeli tribe of Judah (the lion), to which Leo is ascribed See Ginsburgh. *The Hebrew Letters.* Pages 186-189.
[319] Judges 15:15. KJV.

XI JUSTICE

is suggested that Shamgar represents the harbinger of the inundation during an earlier period, when the summer solstice occurred with the Sun in Leo, and Samson represents a later figure of the same function, when the solstice had shifted precessionally from Leo to Cancer. Thus, when the solstice and therefore life-saving floods occurred in the sign of Leo, the hinder-part of the lion was instrumental in fighting back the power of drought (the Philistines). When the solstice had shifted to Cancer, the jaw of an ass would be the logical weapon of choice, as the ass was a zootype for that constellation.

All of this is mentioned to suggest a possible astromythological connection as to how or why the 8th and 11th trumps were inverted. During the Age of Taurus, the hinder-part of the lion was associated with the West, and the setting Sun, since it was the tail of Leo that was visible as night fell during the summer solstice.[320] In those days, Libra was considered the point of equipoise of the inundation. The floods began when the Sun entered Leo, and ended after it had passed into Scorpio. The midway point occurred more or less with the Sun in Libra (the sidereal constellation, not the tropical sign). *Thus, the sign of the balance was originally used to denote the midpoint of the flood season, not the midpoint of the year as a whole.*[321]

Today, the west is commonly associated with the autumnal equinox, just as the east is associated with the spring equinox. As the tropical astrological system has become erroneously fixed with Libra denoting the autumnal equinox, one can see how a certain confusion between the 8th and 11th trumps may have arisen: the sign (Libra) that was once a symbol of equipoise during the summer season – when the summer was anciently truly ruled by Leo – has now become associated with the midpoint of the year as a whole. Compound this with the fact that the autumnal equinox now occurs with the Sun in Leo, not Libra, and it becomes evident that there are multiple layers of possible reasons for these two trumps to have been so closely associated as to become juxtaposed one way or the other throughout the various tarot attribution systems.

---

[320] Massey. *Ancient Egypt, the Light of the World.* Book 10. Pages 671-672.
[321] Ibid. Book 5. Page 297.

XII The Hanged Man

THIS TRUMP RECALLS the Nordic god Odin,[322] who was said to have hung himself from the world tree (Yggdrasil) in order to fathom the runic alphabet. The depiction of the Hanged Man in Nicolas Conver's 1760 "Ancient *Tarot de Marseille*" deck with a single eye [Figure 1] may be a reference to this association, as Odin was known as the one-eyed god, who had sacrificed the other for wisdom obtained from Mimir's well, located in the nether regions towards the base of the Tree.[323]

[Figure 1: Nicolas Conver's 1760 version of the Hanged Man trump. Author's rendering.]

---

[322] Gertrude Moakley implied such a connection: "It is possible that the 'one-eyed' theme, which occurs again in The Cocktail Party, is also connected with the Hanged Man. Odin, who paid the price of an eye to obtain all wisdom, is one of the Hanged Gods of Frazer. His remaining eye is the Sun, another of the Trumps, and perhaps it is suggested by the nimbus of the Hanged Man." From an article entitled "The Waite-Smith 'Tarot,' A Footnote to the Waste Land." *Bulletin of the New York Public Library*, 1954. Volume 58. Pages 471-475.

[323] This may well be an "accident" of the lithographic process, since other cards in this deck also have missing patches. Whether intentional, or accidental, the association between Odin and the Hanged Man still holds.

Another image that suggests a connection between the Nordic god and Hanged Man trump is Lorenz Frolich's "Sacrifice of Odin," published in 1895 [Figure 2]. Note the fylfot arrangement of the legs, and the overall pose and composition, which are essentially congruent with the Hanged Man (inverted).

[Figure 2: The Sacrifice of Odin, by Lorenz Frolich; 1895. Public Domain. Image source: Wikipedia]

Odin as Godfather is analogous to the Hebraic Ancient of Days. Whereas the Semitic form of this god (El) was represented by the bull,[324] and by inference the Age of Taurus, the Nordic version is associated with Mercury – suggesting the Age of Gemini, or prior still, the Age of Virgo. The possibility that these Mercurial associations of the Hangagod may indicate the survival of a mythos from as early as 12,000 BCE during the Golden Age of Virgo is entirely within the acceptable boundaries of the alternative Egyptological school of thought.

Examples of Ice Age cave paintings (which can be considered as precursors to "proto-writing") go back at least as far as the Magdalenian Period (15-10,000 BCE),[325] which does leave the possibility open for the interpretation of the Hangagod's runic epiphany as referencing some type of primordial development of writing during the Age of Virgo, when the polestar was localized in the constellation Lyra.

---

[324] Caquot, André; Sznycer, Maurice. *Ugaritic Religion*. Page 12.
[325] *Art Encyclopedia*. Web. <visual-arts-cork.com/prehistoric/cave-painting.htm>

THE HANGED MAN

According to Manly P. Hall:

> In the light of the secret philosophy of the Egyptian initiates, W. W. Harmon, by a series of extremely complicated yet exact mathematical calculations, determines that the first ceremonial of the Pyramid was performed 68,890 years ago[326] on the occasion *when the star Vega for the first time sent its ray down the descending passage into the pit*. ... While such figures doubtless will evoke the ridicule of modern Egyptologists, they are based upon an exhaustive study of the principles of sidereal mechanics as incorporated into the structure of the Pyramid by its initiated builders…By the Egyptians *the Great Pyramid was associated with Hermes, the god of wisdom and letters and the Divine Illuminator worshiped through the planet Mercury*. ... The Great Pyramid was supreme among the temples of the Mysteries. In order to be true to its astronomical symbolism, it must have been constructed about 70,000 years ago. It was the tomb of Osiris [Orion], and was believed to have been built by the gods themselves, and the architect may have been the immortal Hermes. *It is the monument of Mercury, the messenger of the gods, and the universal symbol of wisdom and letters.*[327]

This is quite suggestive with respect to the aforementioned putative correlation between the Hangagod myth and the development of writing during a period in which the polestar was in the region of Lyra. Although Hall places the construction of the Great Pyramid some 2.5 precessional cycles in the past, one need only go one half-cycle back in order to meet the conditions of Vega as polestar forming a 27° angle with respect to the horizon of Earth, and thus penetrating into the monument's subterranean chamber [Figure 3]. The rationale for going back some 65,000-70,000 years as opposed to 14,000-16,000 is unclear, nonetheless the association of the "first" alphabet with Mercury and the Age of Virgo/Lyra still obtains.

In Greek mythology, Lyra represents the lyre given by Hermes (Mercury) to Orpheus, who was subsequently transubstantiated into the swan of Cygnus. Orpheus was enlisted with the Argonauts, along with Cepheus, as well as Hercules, Autolycus and Eurytus (sons of Hermes – cf. Lyra), Euphemus and Eriginus (sons of Poseidon, whose son is Cycnus/Cygnus), sons of Dionysus (cf. Bootes), as well as Augeas, son of the Sun. Amongst ancient sources,[328] there are some 84-85 Argonauts enumerated.[329]

---

[326] The geographical location of Egypt and the angle of the shaft to the subterranean chamber (approximately 27°) indicate that the shaft serves as a conduit for radiations of the polestar. As Vega was polestar circa 12,000 BCE, if using the standard 25,920 years as the value for length of the Great Year, the correct date would be around 63,840 BCE, which would equate to 65,768 years ago as of 1928 CE – the year Hall's book was published.

[327] Hall. *The Secret Teachings of All Ages*. Pages 110, 117, 190. Italics and ellipses added.

[328] See Apollonius Rhodius: *Argonautica*; Pseudo-Apollodurus: *Biblioteca*; Hyginus: *Fabulae*.

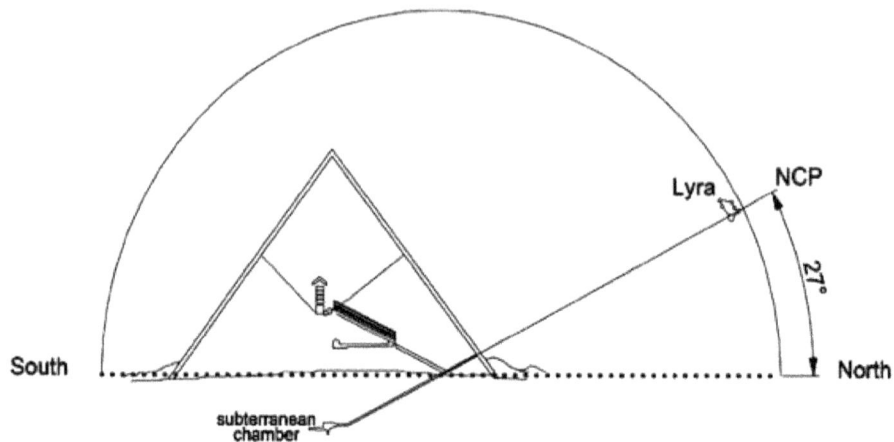

Cross-Section of Great Pyramid
looking west, circa 12,000 BC.
Giza, Egypt

[Figure 3]

The stars of the ancient Argo Navis are equated with the Argonauts' ship.[330] This now defunct constellation is currently broken into three discrete units: Vela (the sail), Carina (the keel), and Puppis (the deck), but at one time it was the largest constellation depicted on celestial maps. In the 17th century, "the Ark" was a common title for this constellation.[331]

The main southern circumpolar equivalent to Vega is Canopus. This star is located in what is now known as Carina, the keel of the ship. Mathers[332] and Crowley[333] both make reference to a somewhat obscure reading of the

---

[329] The notion that this could be tied to a 72 + 12/13 arrangement – thereby indicating a relationship to the Shemhamphorasch and the zodiac – merits further investigation.

[330] Allen. *Star Names: Their Lore and Meaning.*

[331] Ibid.

[332] From Mather's late 19th century 6=5 Adeptus Major Ritual of the Golden Dawn as reprinted in *The Book of the Concourse of the Watchtowers* (Cicero): "Therefore, also, thou wilt understand that the Twelfth Key of the Tarot, commonly called 'The Hanged Man' would be more properly called 'The Drowned Giant' and its position [should be] horizontal rather than perpendicular. In this position, the lower side of the Key represents the Bed of the Waters and the upper side the Keel of the Ark of Noah, floating above the Drowned Figure. Or, in Egyptian Symbolism, the Baris or Sacred Barque of Isis; whilst the Figure is One of the Bound and Drowned followers of the Evil Forces; though yet again, in another sense, it may represent the Body of the Slain Osiris, in the Pastos, sent down the Nile to the Sea: Whilst above is the Keel of the Baris or Barque of Isis in which she travelled to seek him." Bracketed comment added.

[333] Crowley. *The Book of Thoth.* Page 97: "Note on the Precession of Aeons. 'The Hanged Man' is an invention of the Adepts of the I.N.R.I. – J.A.O. formula; in the Aeon previous to the Osirian, that of Isis (Water), he is 'The Drowned Man.' The two uprights of the gallows

THE HANGED MAN

Hanged Man trump which apparently arose as early as the middle Ages, wherein the card is rotated 90° from its upright position, such that the two vertical posts of the gallows are now horizontal, with the upper post representing the keel of the Ark, and the lower that of the bottom of the ocean (Crowley), or bed of the waters (Mathers).

The ancient Greeks associated the constellation Eridanus with the river Okeanos ("Ocean").[334] If we interpret Crowley's "bottom of the ocean" and Mather's "bed of the waters" to refer to Eridanus, an interesting correlation emerges when the celestial sphere is oriented horizontally, as shown in [Figure 4].

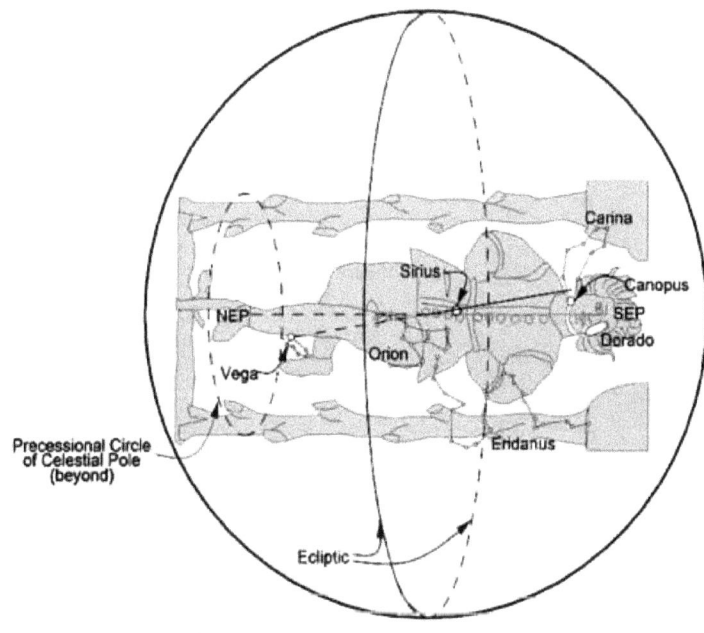

[Figure 4]

Here the lower (left) post of the gallows corresponds to Eridanus, and the upper (right) to the region of Carina – the keel of the Argo – the main star of which is Canopus. What lies between – i.e., the "drowned giant" – corresponds to the ecliptic axis, with the head of the "giant" or hanged man corresponding to the SEP, near the constellation Dorado:[335] the solar "Overman" has become inverted; the giant Odin has hung himself by the foot. This is not to be read as a literal global polar inversion, where the

---

shewn in the Medieval packs were, in the parthenogenetic system of explaining and ruling Nature, the bottom of the Sea and the keel of the Ark."

[334] Allen. *Star Names: Their Lore and Meaning.*

[335] Note also that Orion is centrally justified along the axis in this configuration.

geographic North becomes the South, and vice versa – a concept deriving in part from an apparent misunderstanding of precession – rather is to be taken as a reference to a diametrically opposed locus within the circumpolar precessional circle [Figures 5 & 6].

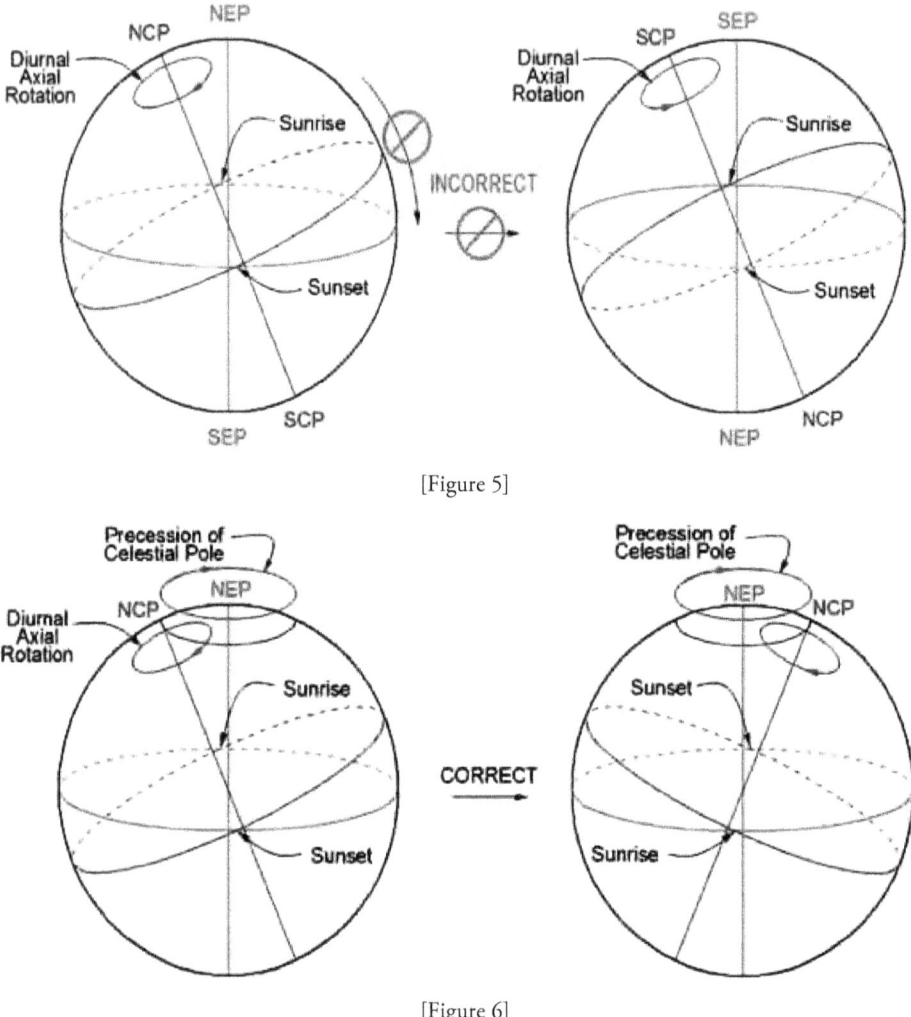

[Figure 5]

[Figure 6]

The first written records which mention Allfather Odin as such are the Icelandic Eddas, of which there are the poetic and prose forms. The former date from approximately the 10th century CE, whereas the latter are attributed to the historian Snorri Sturluson, and dated to approximately 1220 CE.[336] This timeframe, from 900-1200 CE, corresponds to the period during which

---

[336] Faraday, Winifred. *Divine Mythology of the North.*

THE HANGED MAN

the polestar was located between Thuban of the constellation Draco, and our current Alpha Ursae Minoris of the "Little Dipper," thus comprising a portion of the Age of Pisces[337] – during which it can still be said that we nominally exist.

Now of the numerous appellations ascribed to Odin, two of them are Geirtýr ("Spear God"), and Geirvaldr ("Spear Master"). If this is taken to reference Odin as master of the shifting of the celestial pole throughout the long Ages of precession, this connects the Allfather with both the NEP as the eternal and unchanging solar source of all that we know – in which we live, move, and have our being – as well as the ever-changing NCP. This combination of eternal and mutable aspects within a single deity could go a long way towards explaining the many names ascribed to this multifaceted Nordic god.

Interestingly, the two flanking stars of this interstitial circumpolar region of 900-1200 CE – Thuban and Polaris – are both associated with dogs or wolves; in the case of Thuban, we have the Arabic Al Dhib ("the wolf"), and Polaris was deified as Anubis the jackal by the ancient Egyptians. These two "canine" stars buttressing the NCP during the Age under consideration suggest an association with Odin's wolves Gere and Freke, which are often depicted to either side of the Allfather, such as in the following illustration by Carl Emil Doepler (1882) [Figure 7].

Perhaps this station of the pole is the source for the symbol of the winged caduceus associated with Hermes, as it is flanked on both sides by portions of the serpentine constellation Draco, and Ursa Minor was known at least as early as 600 BCE as the "wings of the dragon," i.e. the wings of Draco. Thus, the winged serpentine motif of the caduceus can be seen as representative of a stellar analog as witnessed in the nocturnal circumpolar heavens by the ancients some 3000 years ago. It is yet visible to us today as an astral co-type of the original, as the pole still resides between the two protrusions of Draco.

Here, then, we have our diametrically opposed circumpolar locus to Vega, and all of these considerations taken together suggest a conception of the Hanged Man trump as a kind of precessional marker where the head corresponds to the Golden Age or Summer of the Great Year, and the feet to the Iron Age – or Great Winter [Figure 8]. Although the hanged man is represented as inverted, he is only to be considered as such from our current perspective, which is as we have seen associated with the precessional nadir. From his own vantage point, there is no suffering, only the sublime

---

[337] Thierens ascribed Pisces to this trump, and in this connection it seems notable that Waite considered his interpretation to be one of the more "valid" ones. See Waite's Introduction to Thierens' *General Book of the Tarot*.

consciousness associated with the Golden Age, represented by his glowing nimbus.

[Figure 7: "Spear Master" Odin, enthroned with Gere, Freke, Huginn, and Muninn, as illustrated by Carl Emil Doepler, 1882. Image source: Wikipedia]

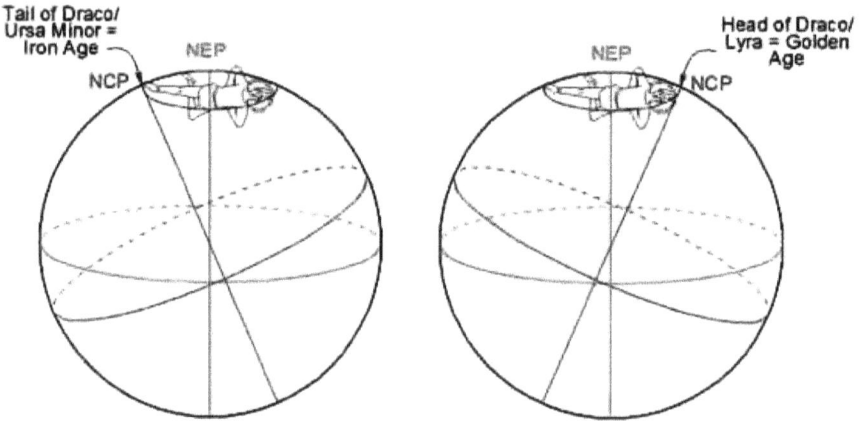

[Figure 8: The Hanged Man as circumpolar precessional symbol.]

XIII Death

EARLIER EXAMPLES OF this trump – such as the *Tarot de Marseille* version – portray Death as a reaping skeleton with scythe [Figure 1], but Waite suggests that this imagery is outdated.[338] Accordingly, we are presented with an image of Death derived from the Apocalypse of St. John, where the rider of the "pale horse" bears "a black banner emblazoned with the Mystic Rose, which signifies life."[339] The Mystic Rose referred to here is undoubtedly the one of Rosicrucian import, which not only is a symbol of Life, but also of the Sun.[340,341] It is the Rose of Sharon, where Sharon translates to "plain" or "level place" – thus giving Rose of the Horizon. The Mystic Rose and the Cross together form the hieroglyph of the formula of the Universal Medicine.[342]

[Figure 1: The Jean Dodal *Tarot de Marseille* Death trump. Circa 1701-1715. Public Domain. Image source: Wikipedia]

---

[338] Waite. *The Pictorial Key to the Tarot*. Part 2, Section 2 (XIII Death).
[339] Ibid.
[340] Albert Pike. *Morals and Dogma*. Pages 290-291.
[341] J.C. Cooper. *Encyclopedia of Traditional symbols* (1978). Pages 141-142.
[342] Manly P. Hall. *The Secret Teachings of All Ages*. Page 466.

[Figure 2: The Rider-Waite Death trump. Author's rendering.]

There is some degree of conflation between the Mystic Rose and the Lily. Sometimes the two flowers form a complementary pair, and other times are blended into the same symbol as the "rose of the lily," or more anciently, the "lily-lotus." It is evident that both flowers correspond to different aspects of the Sun. According to Massey:

> One frequently meets with proofs that the ancient symbolism survived more or less in the secret societies. For example, Jacob Bohme, who was one of the *illuminati*, observes, "We must be silent concerning the times of the ancients, whose number shall stand open in the Rose of the Lily." And he further remarks, "Those who are ours will know what I mean."

> Here is an allusion to the two times of the Two Truths, whose perfect flower-symbol was the lily-lotus of Egypt. The lily-lotus, the *sushenin*, or *sushen*, was the flower of the Two Truths and two colours, the breather in and out of the waters. Isis was said to have conceived by smelling this flower. So Gabriel, the announcer, offers the lily to Mary at the time of her conception. The Greek muses were said to speak with the lilied voice of the gods. The lily-lotus, or rose of the lily, is the only flower really identifiable in the Hebrew *Bible*.[343]

---

[343] Massey. *The Natural Genesis*. Section 3. Page 179.

XIII DEATH

The "Two Truths" here refers to the two times of the year marked by the solstices in the Age of Taurus, when the waters of life were "breathed in" or "inspired" at the summer solstice with the Sun in Leo, and "breathed out" or "expired" when the Sun had reached Aquarius 6 months later. The "two colors" are white and red (or black),[344] where white corresponds to Horus of the inundation (summer season), and red/black to Sut – the "evil" brother, who is the personification of the Sun in the abysmal regions (autumnal/winter seasons). The impregnation of Isis upon the smelling of the lily-lotus can thus be seen as a reference to the passage of the Sun (the Lily) through the regions of the heavens occupied by Sirius (Isis) during the summer months of inundation in the Age of Taurus.

But the concept of the two truths also relates to the dialectic nature of the equinoxes, as is evident in the Egyptian figure of the Hall of two truths as the *Maati*, or Judgement Hall, which was a representation of the two equinoctial points of equipoise in the year.[345] The western or autumnal equinoctial mount corresponded to the entrance into Amenta, its appellation being "the Gate of Fair Entrance," and the eastern equinoctial mount was called "the Gate of Fair Exit" (from the underworld).

At the entrance to the valley of Amenta, or the valley of Death, there was said to be a walled-up doorway that led to the underworld. It was the first in a series of twelve doors which were required to be passed through to demarcate the Sun's passage through the lower 6 signs of the zodiac, as described in the Egyptian *Book of Hades*. The mystery of how to enter the walled-up doorway was given by Horus (i.e. the Sun) who served as the "door in the stone," which revolved to open at the Gate of Fair Entrance when the "open sesame"[346] was spoken.[347]

---

[344] Black and red were both ascribed to Sut (Massey. *Ancient Egypt, the Light of the World*. Book 1. Page 366; Massey. *The Natural Genesis*. Book 3. Page 178), presumably at different phases of the eschatology, although no attempt is made here to trace which ascription came first.

[345] Massey. *Ancient Egypt, the Light of the World*. Book 12. Pages 731-732.

[346] In this connection, it is interesting to note that *sushenin*, the Egyptian lily-lotus is an anagram for "sunshine," which raises the question whether the original "open sesame" may in fact have been rendered as some form of "open *sushenin*."

[347] Massey. *Ancient Egypt, the Light of the World*. Book 4. Page 227: "At the entrance to the mysterious valley of the Tuat there is a walled-up doorway, the first door of twelve in the passage of Amenta. These twelve are described in the *Book of Hades* as twelve divisions corresponding to the twelve hours of darkness during the nocturnal journey of the Sun…Here is the mystery: how to enter where there is no door and the way is all unknown? It is explained to the manes how divine assistance is to be obtained…Thus Horus [i.e., the Sun] was the door in the darkness, the way where no entrance was seen, the life portrayed for the manes in death. The secret entrance was one of the mysteries of Amenta. It was known as 'the door of the stone,' which name was given to their Necropolis by the people at Sut, the stone that revolved when the magical word or 'open sesame' was spoken." Bracketed comment and ellipses added.

This mysterious entrance, or first door of Amenta is otherwise known as the "Ante-Chamber of the Tuat," which corresponds to the setting Sun on the western horizon. According to E. A. Wallis Budge:

> In the first division of the "Book of Gates of the Tuat," according to the sarcophagus of Seti I, we see the horizon of the west, ⸺, or the mountain of the west, divided into two parts, ⸺, and the boat of the sun is supposed to sail between them, and to enter by this passage into the Tuat.[348]

The lily-lotus – and by inference the Mystic Rose – can also be connected to the ecliptic axis and circumpolar astronomes via the Hindu conception of these elements as comprising Mount Meru, the celestial mountain, of which Massey says:

> The earliest geocentric mount would be a figure of station, in the midst of the stellar revolution, which became a type of the pole; and this natural genesis would lead up to the symbolical mountain of Meru, Alborz, or Eden...The pole, or polar region, is Meru...Mount Meru is said to be 84,000 *yojanas* in height, having the shape of an inverted cone, and being 32,000 *yojanas* in diameter at the top, and only 16,000 at the base. It is considered to form the central point of *Jambu-Dvipa*, the island of the Rose-Apple Tree, and to be "like the Seed-Cup of the Lotus of Earth," the leaves of which are formed by the various *dvipas*. For the mount is also described as a lotus rising up out of the waters, the lotus being an early type of emergence from the liquid element.[349]

Here the "Seed-Cup of the Lotus" is a reference to the shape formed by the precessional circle of the NCP as centered on the NEP, and the leaves that comprise the dvipas or "islands" correspond to the various circumpolar stations of the pole [Figure 3].

We may gain some insight regarding the use of the inverted pentagonal form of the Mystic Rose as seen in the Rider-Waite trump by examining how the pentagram is used in the Serer culture of Senegal. As we have seen, this was a symbol of Sirius in ancient Egypt, and so it is in the Serer religion,[350] in addition to serving as a diagram of the Universe.[351] The two crossed lower diagonals of the pentagram are said to represent the intersection of the axes of the four cardinal directions of the Universe,[352] and the line formed by connecting this intersection with the apex of the pentagram is therefore to be seen as the Universal Axis.

---

[348] Budge. *The Book of Gates*. Chapter 2.
[349] Massey. *The Natural Genesis*. Section 9. Page 29.
[350] Berg, Elizabeth L., & Wan, Ruth, "*Cultures of the World: Senegal*". Page 144.
[351] Faïk-Nzuji. *Tracing memory: A Glossary of Graphic Signs and Symbols in African Art and Culture*. Pages 27, 114-115.
[352] Ibid.

XIII DEATH

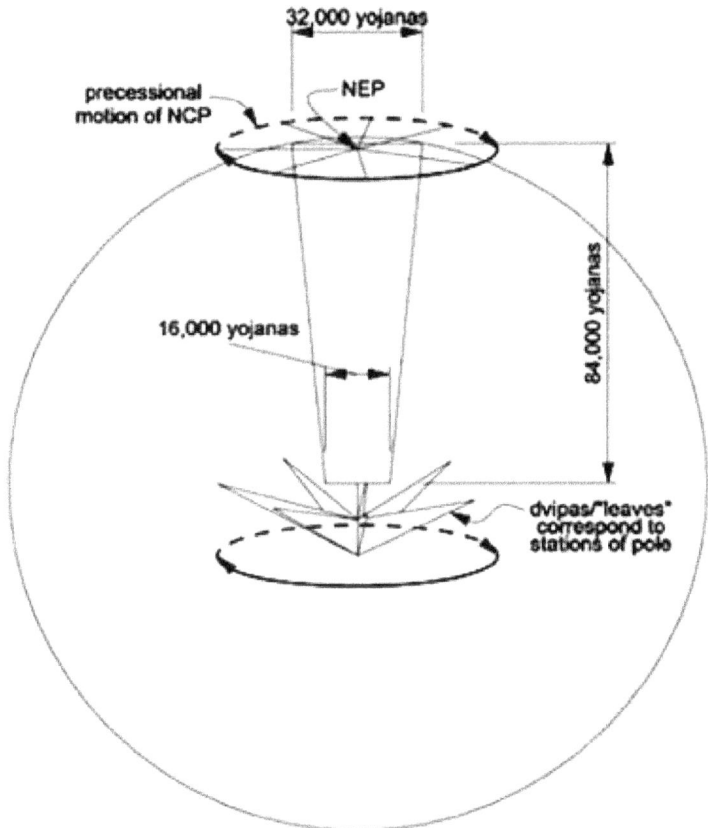

[Figure 3: Author's Rendering of One Form of Mount Meru.]

Henry Gravrand presents a 3-dimensional diagram of this concept [Figure 4] in his *La Civilisation Sereer*.[353] Here we see the crossing point of the cardinal Universal Axes labeled as "*L'homme*" ("Man"), and it is evident that the whole diagram is a representation of the celestial sphere. Here "Man" occupies the central point of the sphere of Heaven, as seen from a geocentric perspective. We see the progression of the seasons depicted as the commencement of the growth of vegetation in the quadrant of spring to its maximum at the height of the summer quarter, when the rains bring the waters of life. This is followed by the autumnal winds, and Death is figured at the equinox as depicted by the gravestones. The Dead reside in the underworld, which corresponds to the winter regions of the abyss, where "nourished by sacrifices, they emit the energies." These energies are shown returning to the living at the equinoxes via the ever-living Ancestral Spirits

---

[353] Gravrand, Henry. "*La civilisation sereer*", Volume 2 : *Pangool*. Page 216.

who reside at the lower pole. It is unclear what the nature of energy return is for the autumnal equinox, perhaps in the form of guidance of the newly deceased from the Ancestral spirits, but the way of return via the spring equinox is through the reincarnation of the Cyid, or "disembodied souls." Contrapuntal to the Ancestral Spirits, at the upper pole resides the supreme deity Roog, which is the Universal Source of all the energies [Figure 5].

This model may have been derived from a prior period when the originators of it resided in sub-equatorial Africa, as Gravrand's diagram indicates – by the implied directionality of seasonal progression – the upper pole to correspond with the SEP/SCP, and the lower pole to the NEP/NCP. The connection between Gravrand's 3-dimensional construct, and the 2-dimensional pentagram is illustrated in [Figure 6].

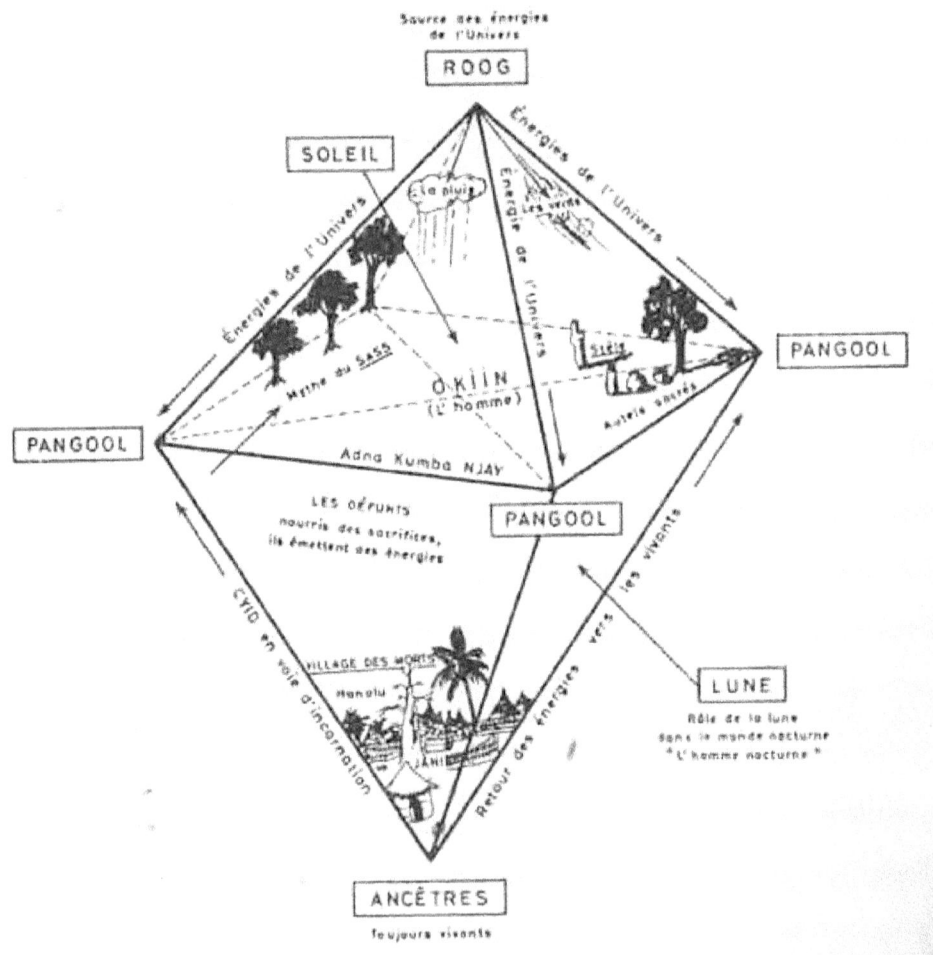

[Figure 4: The Serer Model of the Universe, by Henry Gravrand.]

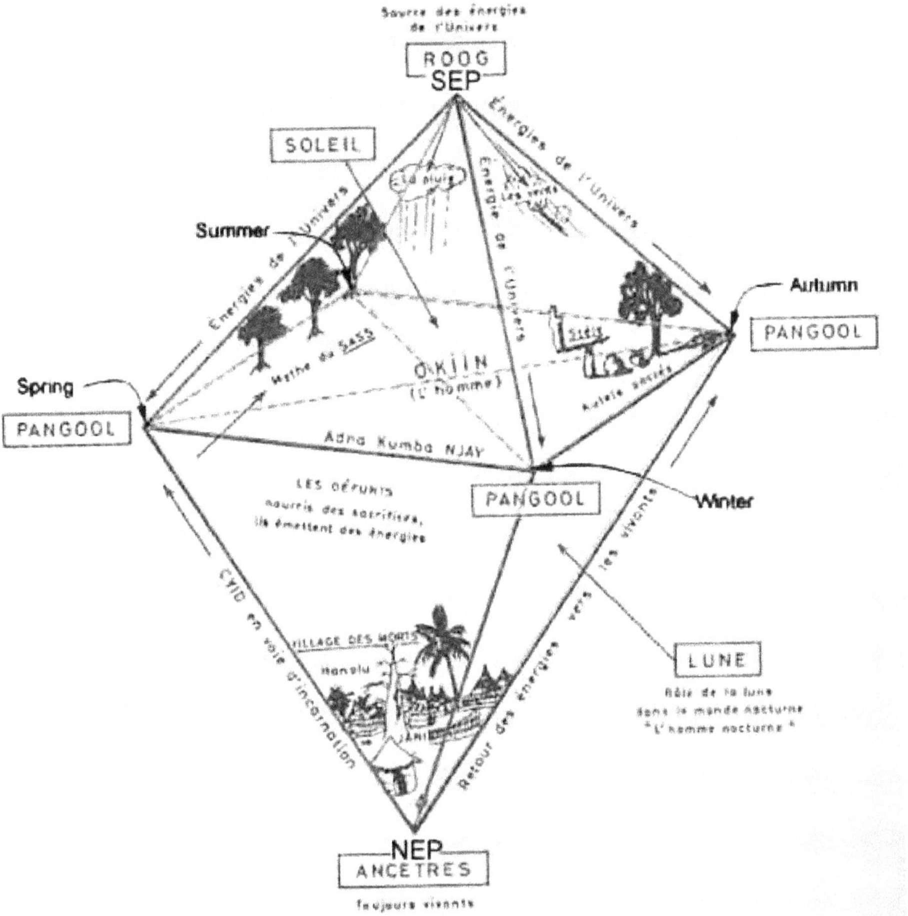

XIII DEATH

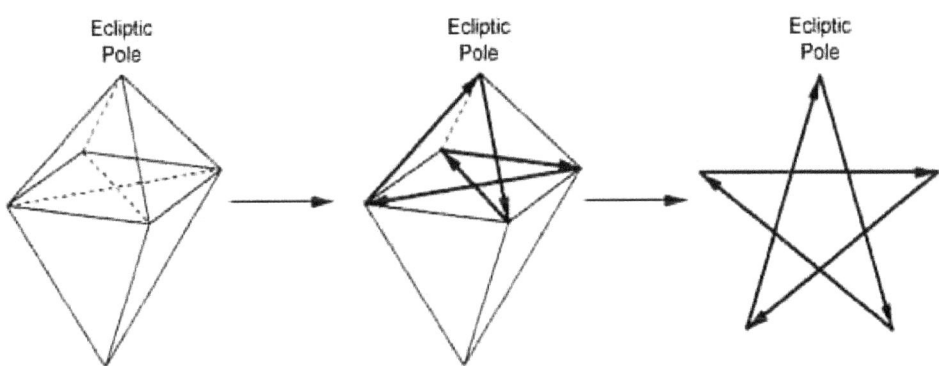

[Figure 5: The Serer Model of the Universe, with Seasons shown. Author's rendering.]

[Figure 6: The Pentagram as Symbol of the Celestial Sphere. Author's rendering.]

A similar use of the pentagram as symbol of the celestial sphere is seen in the "Rosicrucian Lamen" worn by the adepts of the *Roseae Rubeae et Aureae Crucis* (Red Rose and Gold Cross), the symbolism and rituals of which were developed by Samuel Liddell Macgregor Mathers [Figure 7]. Here the four fixed zodiacal signs represent the cardinal directions of the Universe, as well as the four primary elements, and owing to the placement of the signs, the apex of the star is interpreted to represent the NEP, corresponding to the fifth element, or "spirit."

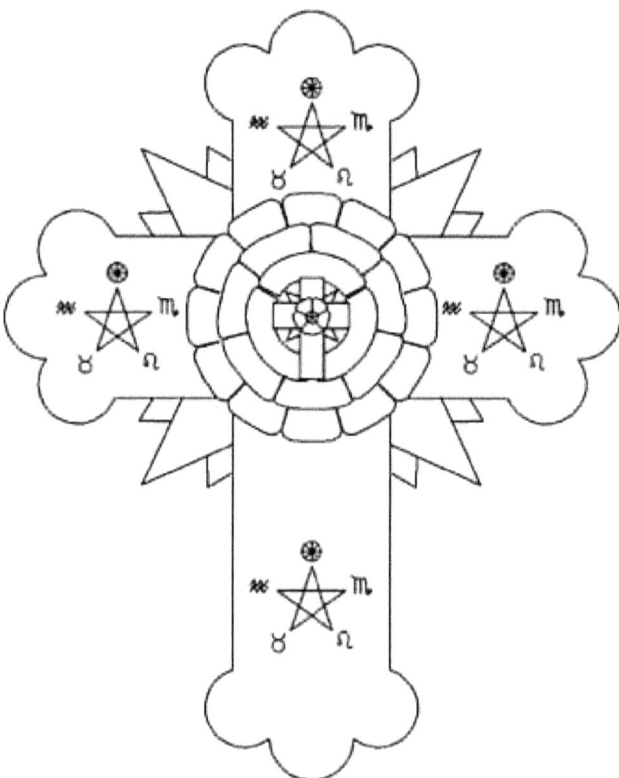

[Figure 7: A Rosy Cross Lamen. Author's rendering.]

The inverted pentagram of the Mystic Rose which the skeletal rider bears on his standard in the Rider-Waite trump may therefore be a clue indicating the necessity for inversion of the celestial sphere in order to decipher the astral tableau presented therein. From such a vantage point – looking from "within" the celestial sphere – we can tentatively identify the following constellations and their symbolic counterparts within this trump: Ara as the flag; Corona Borealis as the crown on the ground, and the horse's head as the tail of Scorpio. Pisces lies to the right in this configuration, indicated by the pope in

XIII DEATH

front of the towers that flank the Sun (see [Figure 8]).[354] The message: a changing of the Ages; the associated death of the Piscean Aeon indicated by the Pale Rider's inexorable approach towards the pontiff.

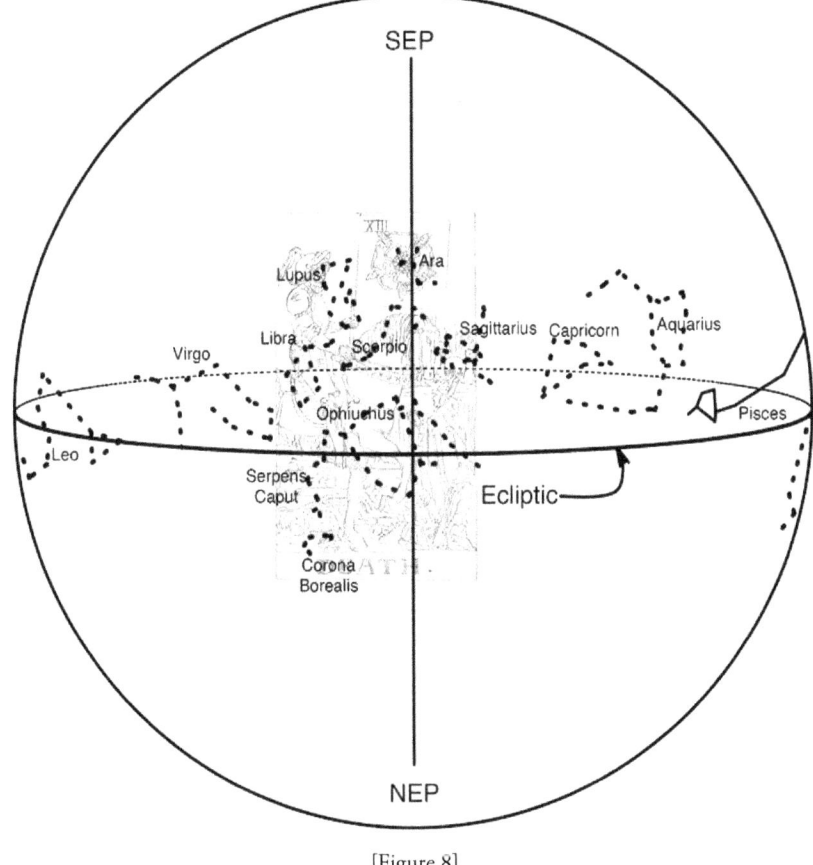

[Figure 8]

The appearance of the papal figure in this trump not only signifies the aforementioned Piscean connotations, but is also a direct reference to the Hierophant or Pope of Trump V. As we have seen, one of the fundamental associations of that card is to the Silver Gate of Humanity, located near the constellation Taurus, where the ecliptic intersects the Galactic Equator. The other gate to which the Hierophant holds the key lies within the diametrically opposed region of Scorpio.

The connection of the apocalyptic horseman of Death to Scorpio is indirectly referred to in the Revelation of St. John, where it is written that

---

[354] The pope's connection to Pisces is evident in the symbolism of the papal mitre, which represents a fish with open mouth (see Massey. *The Natural* Genesis. Book 7. Page 450).

four beasts were seen "in the midst" of the throne, as well as "round about the throne" which was "set in heaven." The beasts in the midst of the throne are possibly a reference to the 4 quarters of the circumpolar Mount of Heaven. The beasts surrounding the throne are interpreted to signify the four cardinal directions of the Universe, as represented by Leo (lion), Taurus (calf), Aquarius (man), and Scorpio (eagle).

At the summit of the Mount – or on the throne – are the 7 lamps of fire, which are the 7 spirits of the Lord, corresponding to the 7 stations of the pole arrayed about the central ecliptic axis. These are also congruent with the 7 seals to which the immortal Lamb who was slain holds the keys. The slain Lamb is of course a symbol of the end of the Age of Aries, and is one form of what was at that time the newly emergent Piscean Christos as a figure of the ever-renewing solar force of spring which had moved from Aries (the lamb) to Pisces (the Ichthus) in the course of precession. Here the immutable ecliptic axis is portrayed as the central throne upon which the unified godhead rests and manifests itself in various mutable forms throughout the precessional cycle, where each Age is represented by one of the 7 seals [Figure 9].

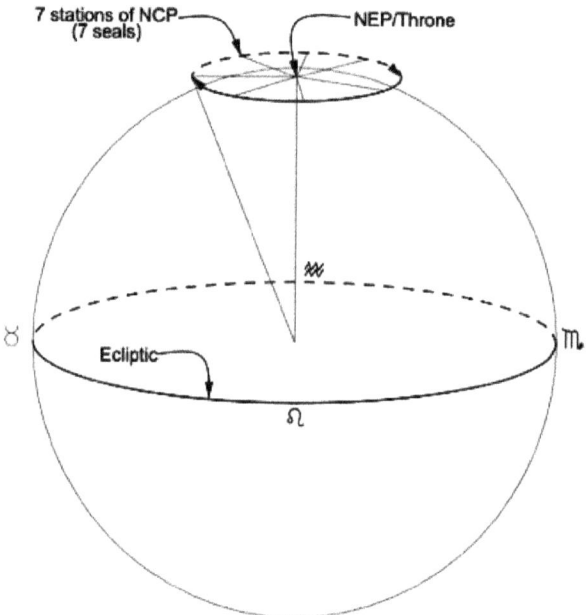

[Figure 9: The ecliptic axis as "throne of God," with the heptanomis as the "7 seals."]

The fourth beast (the eagle/Scorpio) reveals to John a pale horse (sometimes rendered as yellow or greenish-yellow) upon which Death rides, "and hell followed with him." As the horse is symbolically used to represent the vehicle of the Sun's motion throughout the course of the year, this is

XIII DEATH

interpreted to refer to the entrance of the Sun into Scorpio during the autumnal months, and the hell that follows Death can be seen as the train of abysmal winter signs trailing behind in the wake of the Scorpion. This refers to the condition of autumnal equinox during the Age of Taurus. Massey suggests that much of the imagery used in these passages of Revelation are reworked portions of the Egyptian *Book of the Dead*,[355] and this would explain the Taurean overtones to much of the symbolism.[356]

Crowley ascribes Scorpio to this trump, and indicates that it is the completion of the Strength/Lust trump.[357,358] The original notion of Scorpio completing Leo derives from the ancient Egyptian divisioning of the year into 3 months of flooding (inhalation) and 9 months of drying (exhalation), where the flood season began with the Sun in Leo and ended with the Sun entering Scorpio. This sign was associated with Death through its heliacal occurrence during the autumnal period of subsidence (Sun in Scorpio), as well as its acronychal rising (Sun in Taurus) during the typhoon season in the Age of Taurus. The celestial aspect of the autumnal subsidence was rendered mythologically as Osiris entering the valley of Amenta (or the valley of Death). Here Scorpio was the gate to Hell, and it was here that Typhon tore Osiris into 14 pieces.[359] This refers to a parsing of the 6 months of autumn and winter into the 14 lunar mansions (half of the annual total of 28) through which Osiris must pass before becoming born again upon the vernal equinoctial Mount.

In ancient Egyptian astrology, the 2nd house – or house of Death – was associated with this sign. According to Fagan:

> The 2nd house, which the Greeks designated *Haidou pyle*, the door of Hades or the gates of hell, was related with the constellation Scorpio and its ruler Mars. This house was identified with Tartarus, a place in the bowels of the earth, even lower than hell, where Typhon and the Titans were imprisoned by Zeus.

---

[355] Book 11 of Massey's *Ancient Egypt, the Light of the World* ("Egyptian Wisdom in the Revelation of John the Divine") is devoted to a thorough examination of this notion.

[356] The symbolism of the Book of the Dead appears to have been developed, or at least codified in its "final form" when the vernal point was in Taurus, or at the very latest when it was transiting from Taurus to Aries. This rationale is based on the preeminence of Taurean elements in the symbolism, as well as the fact that the earliest known fragments of the book appear carved on sarcophagi of the 11th dynasty (2134-1991 BCE [see Bunsen. *Egypt's Place in Universal History*. Vol 5. Page 128.]), which corresponds to the transition from Taurus to Aries.

[357] Crowley. *The Book of Thoth*. Page 100.

[358] His reasoning for this involves the sequence *XI Strength, XII The Hanged Man, XIII Death*; the explanation being that the Hanged Man links the two flanking complementary trumps. This line of reasoning will not be pursued further, as the unsatisfactory nature of this order has already been noted in that it necessitates the use of the zodiacal "double loop" – a device that has no observational basis in Nature.

[359] Massey. *A Book of the Beginnings*. Section 16. Page 249.

The Egyptians symbolized the winds as wriggling serpents, and Typhon (Egyptian, Set) was the typhoon or violent hurricane. If the reader consults a star atlas he will observe that the constellation Serpens the serpent and Ophiuchus the serpent-bearer are immediately above that of Scorpio, which in demotic script has a representation of a serpent for its ideogram. [...] In Egyptian astrology the 2nd house indicated the agony of death, wars, woundings, violence, imprisonment, upheaval, revolt and rebellion.[360]

But Scorpio was not always associated with negative qualities such as these, and in fact, used to connote the very opposite:

> The scorpion was not a type of evil in the zodiac. It represented Isis-Serkh who fought for Horus *when the birthplace was in Scorpio*. A fragment of the myth survives in the Ritual. It is the merest allusion, but suffices to show that in the wars of the solar god (Horus or Ra) with the enemy Apap, Isis-Serkh joined in the battle and was wounded. The passage is confused but, as rendered by Renouf, it runs: "Apap falleth; Apap goeth down. And more grave for thee is the taste (tepit) than that sweet proof through the scorpion-goddess (Isis-Serkh) which she practiced for thee, in the pain that she suffered." When the summer solstice was in the sign of Leo the autumn equinox occurred in Scorpio, and it would be then and there the scorpion-goddess gave proof of her sympathy and suffering on behalf of Horus or of Ra in the latter mythos.[361]

The "birthplace" mentioned above is an apparent reference to the birthplace of Horus in the inundation; in other words, the time referred to is the Age during which the summer solstice, and thus the flooding season, was localized in Scorpio. This corresponds to some 13,000 years ago, when the vernal point was to be found in the Golden Age of Leo. The "wounding" of Isis-Serkh is therefore an allusion to Scorpio having fallen from its primordial solstitial position as bringer forth of the waters from the deep during the Age of Leo, to its autumnal locus during the Age of Taurus, in which the symbolism shifted to take on the Typhonian characteristics associated with the deadly Khamsin winds (Sun in Taurus; full Moon in Scorpio), and autumnal harbinger of the drought season (Sun in Scorpio). The dualistic nature of this sign, which is alternately associated with death and betrayal, as well as renewal and regeneration, can at one level be explained as the result of the accumulation of various disparate qualities through time from the primordial Age of Leo to our current Aquarian Age, some 13,000 years, or half of a precessional cycle, later.

As our solstitial and vernal points have shifted some 90° via precession since the Age of Taurus, the question arises as to whether there is any

---
[360] Fagan. *The Solunars Handbook*. 1970. Page 12.
[361] Massey. *Ancient Egypt, the Light of the World*. Book 5. Page 297. Italics added.

XIII DEATH

contemporary symbolic resonance between the pairing of Scorpio with the concept of Death. Since this constellation currently marks the winter solstitial point, it may be said that within the cycle of precession, the Osiris has passed from the entrance of the valley of Amenta down to its lowermost point. Here we have reached an equilibrium point between putrescence and regeneration as the Sun pauses for 3 days at its solstitial nadir of winter before commencing its climb from this lowest point of the valley of Death; hence we see a mixing of both elements of Life and Death in the same card.

Crowley's emphasis on the fish symbolism in his version of this Atu presumably derives from his attribution of it to the Hebraic letter *Nun*: "This card is attributed to the letter Nun, which means a fish; the symbol of life beneath the waters; life traveling through the waters."[362] It is possible to trace an Egyptian origin to the Hebraic association of *Nun* with the fish. In ancient Egypt, Nun was the god of the waters of the Abyss, and the *nun* was the Abyss, which comprised the signs of Pisces, Aquarius, and Capricorn during the Age of Taurus. The primary fishes of the *nun* during that Age were constellated in Piscis Austrinis and Pisces.

The inclusion of both Piscean and Scorpioan elements in the Rider-Waite and Harris-Crowley trumps may ultimately be read as an allusion to the 90° of precessional rotation from the Age of Taurus to our current Aquarian Age, where Pisces/Piscis Austrinis represented the primordial *nun* during the Age of Taurus, and Scorpio represents the same abysmal source – i.e., the winter solstice – in the Age of Aquarius. We thus see the accumulated symbolism of some 6,000 years contained within a single trump, and this may have been one of the reasons Crowley considered it "as of greater importance and catholicity that would be expected from the plain Zodiacal attribution [of Scorpio],"[363] as well as "a compendium of universal energy in its most secret form."[364]

---

[362] Crowley. *Book of Thoth*. Page 99.
[363] Ibid. Page 101. Bracketed comment added.
[364] Ibid.

XIV Temperance

PAPUS ATTRIBUTES THE Hebraic letter *Nun* to this trump, as well as the constellation Scorpio. In *Le Tarot Divinatoire* ("The Divinatory Tarot"),[365] he depicts the "avatar" of this card as wearing an Egyptian style headdress, suggestive of the lily-lotus [Figure 1].

[Figure 1: The figure in Papus' Temperance trump, showing the lily-lotus headdress. Author's rendering.]

According to Waite:

> In his last work on the Tarot, Dr. Papus abandons the traditional form and depicts a woman wearing an Egyptian headdress. The first thing which seems clear on the surface is that the entire symbol has no especial connexion with Temperance, and the fact that this designation has always obtained for the card offers a very obvious instance of a meaning behind meaning, which is the title in chief to consideration in respect of the Tarot as a whole.[366]

---

[365] Papus. *Le Tarot Divinatoire*.
[366] Waite. *The Pictorial Key to the Tarot*. Part I. Section 2 – Class I: The Trumps Major, otherwise Lesser Arcana.

The lily-lotus was connected with fertility and the season of inundation, when the child Horus was said to emerge from his bed of reeds upon a papyrus stalk – the lotus-headed messiah come to bring forth the waters of salvation; or Jack climbing the stalk, personifying the Sun/Sirius increasing in declination during the summer months to slay the giant of drought.[367] The same flower graces the headdress of Hapi, god of the Nile and fertility [Figure 2]. A notable difference between the Egyptian god's lily-lotus headdress and that of the figure in the Temperance trump is in the number and angularity of the flowers depicted. Hapi wears numerous lilies upon his head; some drooping, some erect, as if to signify the totality of phases of inundation from inception, when the lilies begin to rise with the increase in moisture; to the height of the flood season when the aquatic forces of virility are at their maximum; to the gradual diminishing of the river's levels and hydrostatic pressure, when – once again – the lotus begins to drop. We can thus read the depiction of a single flaccid lotus in the Papus trump as a reference to either the beginning or ending phase of the inundation. The fact that the figure pours from a higher urn to a lower one suggests a decrease in potential energy, and therefore the autumnal or winter seasons, during which the floodwaters subside.

[Figure 2: Hapi (Aquarius), God of the Nile. From *The Gods of the Egyptians*. E. A. Wallis Budge. Chapter XIX: *Miscellaneous Gods*. Author's rendering.]

The Hebraic "N," or *Nun* (נ), may have its precursor in the Egyptian hieroglyph of the same name:

Nun ↱ ○ , a god.

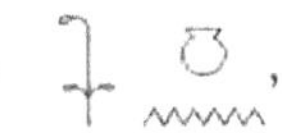

[Figure 3: The Egyptian Hieroglyph *Nun*. From *An Egyptian Hieroglyphic Dictionary*. E. A. Wallis Budge. Page 354. Author's rendering.]

Here we have symbols representing the lily-lotus (or possibly the papyrus shoot), the urn, and the water/wave glyph, all of which can be read as

---

[367] Massey. *Ancient Egypt, the Light of the World*. Book 5. Page 252.

XIV TEMPERANCE

references to the inundation, and the three elements together form the hieroglyph for the Egyptian god of the inundation, Nun, who prefigured the later Hapi. Massey obliquely touches upon the connection between the letter "N" as the Hebraic *Nun* and the flooding of the Nile:

> In the hieroglyphics, the "n" and "m" permute with the running water for the sign of both. In Hebrew, the letter *mem* means a wave. The Egyptian symbolism as an ideograph of water is a visible wave /\/\/\ /\/\/\; the plural of this "n" forms the "m," or *mem*. The Hebrew *mem*, in pronunciation, has the twofold character of the dull labial and the strong nasal sound corresponding to the Egyptian duality of the water sign, and the unity of origin for both letters.[368]

This unity of origin is the *Nun* as source of the waters of the abyss – the subterranean wellspring of which was constellated as Piscis Austrinis in the Taurean Age. During the summer months of inundation, the waters of the abyss corresponded to this very region of the zodiac presided over by Aquarius – acronychal counterpart to the signs of Leo and Crater (Krater) [Figure 4]. The urn as accouterment of Nun as indicated by the Egyptian hieroglyph is thus found to be constellated in the Aquarian vessels and celestial Krater, revealing the fluvial root of the symbolism, in general. According to Massey:

> The urn was a figure of the inundation. Aquarius was called the constellation of the Urn by the Arab astronomers. We shall understand the sign of "Krater" better if we take it as an extra-zodiacal image of the urn, which not only represented the inundation and its bounty, but also the abyss of source from which the welling waters came. The two urns are followed by the two vases at a later stage. Howsoever poured out, water was the primary means of fertilization. When the goddess pours out a libation from her vase – or two divine personages from two vases – on the water plant or shoot of palm, the signification is the same as when the wet-nurse Hathor suckles Horus as a child or Neith the crocodile as a calf. According to the most primitive imagery in Egypt, the waters of the inundation issued from the Mother-earth as the water-cow, the wateress in the primordial abyss or water source. But when the sky was looked to as a source of water, heaven was represented as the milch-cow, and the river flowing from the highest source was imaged as the Milky Way. Thenceforth there were two cows. The cow of earth was the water-cow, and the milch-cow was the cow of heaven. The water-cow of earth was constellated in the stars of the Great Bear, the milch-cow of heaven in the group now known as Cassiopeia, or the Lady in the Chair, which was the earlier constellation of the Haunch or Meskhen as a figure of the birthplace when the birth was typical of life in water.[369]

---

[368] Massey. *A Book of the Beginnings*. Section 13. Pages 121-122.
[369] Massey. *Ancient Egypt, the Light of the World*. Book 5. Page 286.

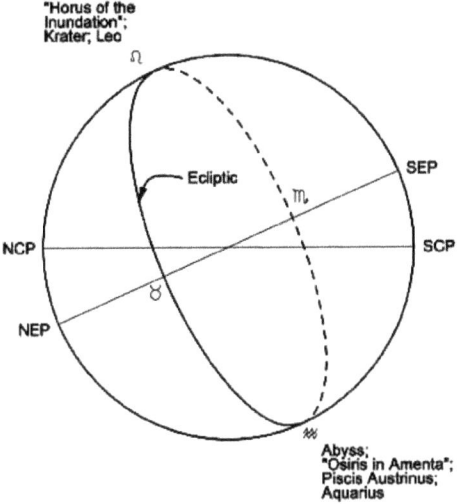

[Figure 4: Leo/Krater and Aquarius/Piscis Austrinus as Diurnal and Nocturnal Sources of Inundation during the Age of Taurus.]

As Cassiopeia was at one time a figure of the heavenly source of inundation, another reading of the urn symbolism of the Rider-Waite and Papus trumps suggests the upper urn to correspond to that constellation, and the lower urn to the constellation Krater. The stream flowing between the two would thus represent the Galactic Equator or Milky Way, the ultimate cosmic source for the celestial waters of inundation [Figure 5].

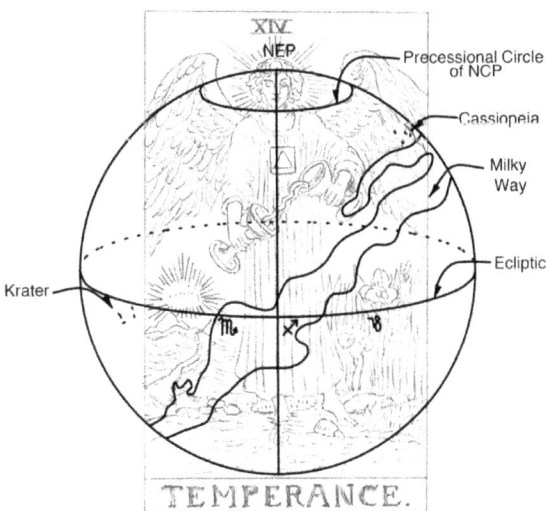

[Figure 5: The two urns of inundation (Cassiopeia and Krater) and the *Nun* (abyss/Milky Way) correlated with the Rider-Waite Temperance trump. Author's rendering.]

XIV TEMPERANCE

The centrality of Sagittarius in this configuration is notable, since the *Nun* as abyss of source in our current Age corresponds to the three nocturnal signs of inundation: Scorpio, Sagittarius, and Capricorn, with Sagittarius / Scorpio denoting the lowest point of the abyss.

The *meskhen* as constellated in Cassiopeia may refer to the acronychal rising of this antiscion of Pisces during the Age of Gemini/Taurus, when the inundation of summer occurred when the Sun was in Virgo/Leo. At this point in time, the Egyptian mode of chronometry was presumably stellar, and so the "place of birth" of the waters of the inundation still primarily circumpolar. During this ancient phase, the ancestors of the Egyptians dwelt closer to the equator, and the earth and heaven were one, in the sense that the circumpolar regions (heaven) were always to be seen at the horizon (earth).

As the NCP was essentially at the same longitude as the NGP in that Age, the Milky Way was always to be seen encircling Earth very close to its equatorial apex no matter what the season, the great rainbow bridge leaning to the north during the fall and the south during the spring, reaching its apex in the solstitial months of inundation and drought [Figure 6].

At Equatoria, this effect was even more pronounced, and the deity of the celestial pole – instead of occupying the lofty circumpolar regions as typified after the Egyptian northern migration may have originally been elevated above the horizontal celestial pole in the figure of the arched galactic goddess Neith, who was amongst other things known as the goddess of the bow.[370]

Neith was one of the early forms of the goddess of the north[371] (i.e. NCP), also known as Uat, goddess of the papyrus reed, who was replaced later by Sefekh,[372] or Seshat – mistress of writings[373] – whom Massey identifies with the 7 stars of Ursa Major,[374] as well as the 7 colors of the rainbow.[375] Seshat is connected with the Pedjeshes, or "stretching of the cord" ceremony in which the Egyptian priests would orient the axes of the temples with respect to certain stars in Virgo (Spica) and Ursa Major (Eta Ursa Majoris). The *uat*, or papyrus scepter can thus be seen as a figure of the celestial pole as constellated in the stars of the Great Bear, perhaps also doubling as representative of the NGP, as the two poles shared the same longitude during the Age of Taurus. Massey equates this scepter with the flag-flower, or iris:

> [The flag-flower] answered to the papyrus sceptre, the *uat* held in the hand of the Egyptian goddesses. The woman composed of flowers is called the

---

[370] Massey. *A Book of the Beginnings.* Section 1. Page 13.
[371] Ibid. Section 8. Pages 320-321.
[372] Ibid. Section 8. Page 362.
[373] Ibid. Section 17. Page 286.
[374] Ibid. Section 17. Page 344.
[375] Ibid. Section 8. Page 362.

rainbow, that is Iris, and the flag-flower is the Iris. Thus, the woman enchanted from blossoms identified by the Iris leads us to see a personification of the reed as the instrument of the written letters invented by Gwydion in Britain and by Taht in Egypt.

The uat or papyrus sceptre is identical by name with the goddess of the north, Uat, the earlier Kheft, our Kêd. And this goddess of the papyrus reed is replaced by Sefekh [Seshat] as mistress of the writings and consort of Taht. Sefekh reads number seven, which identifies her primarily with the seven stars and with the seven colours of the rainbow, or Iris. It was for this goddess of the rainbow, the seven colours, that Gwydion formed the horse (or horses) on which she was to ride forth as mistress of the writings or as the feminine word.[376]

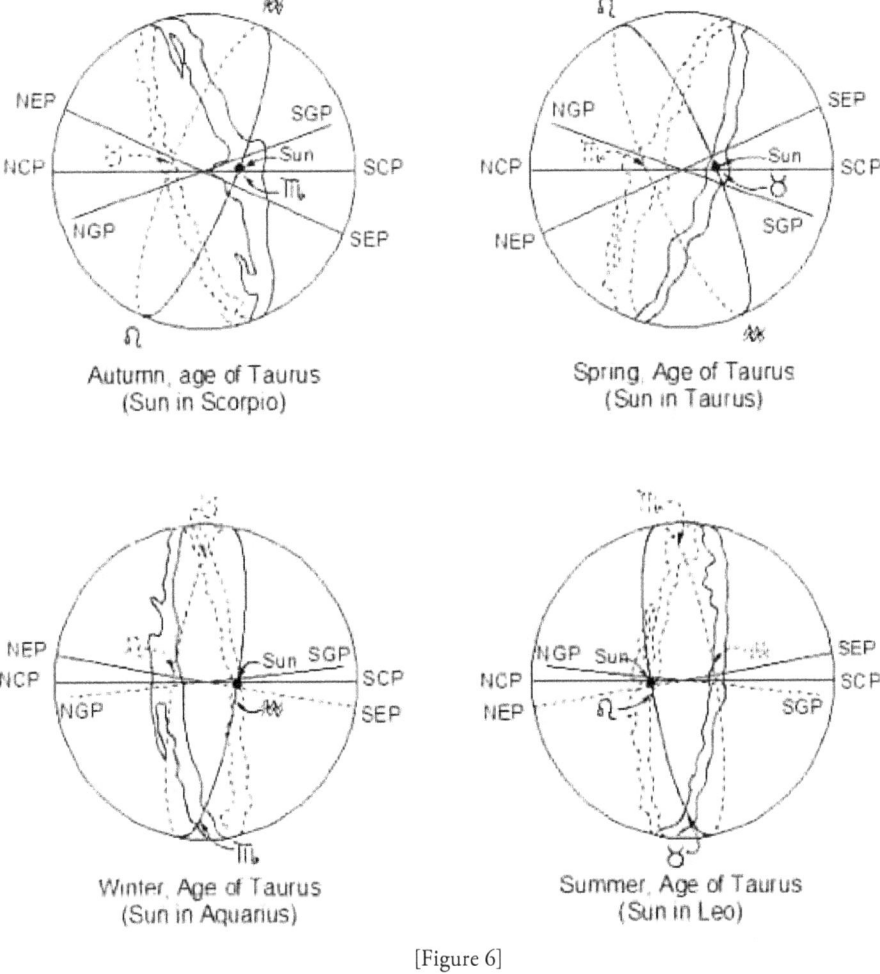

[Figure 6]

---

[376] Ibid. Bracketed comment added.

XIV TEMPERANCE

Iris – the Greek goddess – is often depicted wielding an arrow-tipped caduceus [Figure 7]; a figure of the pole similar to the Hermean caduceus, which has previously been suggested (in the Hanged Man chapter) to be connected with the astronome localized between the tail of Draco and the Lesser Bear, locus of the NCP during the Age of Aries. In the case of Iris, the circumpolar locus is suggested to be the antecedent astronome demarcated by Eta Ursa Majoris and Coma Berenices, corresponding to the Age of Taurus. The connection between Iris and the Galactic Equator as primeval "bow in the sky" is also suggested by the mythological parentage of this goddess, whose mother is said to be Electra, and father, Thaumas. As Electra is associated with the Pleiades, and Thaumas was a centaur, the celestial correlation points to the mother as figure of the crossing of the ecliptic and Galactic Equator in the region of the Pleiades/Orion, and the father as the complementary point in Sagittarius, the centaur. The offspring, as resultant force of the parental complementarity is figured as the galactic arch, which connects the Gate of the Gods to the Gate of Humanity. This aspect of bridging, or joining together may be reflected in one of the suggested derivations of the goddess' name from the Greek *eirô*, "I join":

> Her name is commonly derived from erô eirô; so that Iris would mean "the speaker or messenger" but it is not impossible that it may be connected with eirô, "I join," whence eirênê; so that Iris, the goddess of the rainbow, would be the joiner or conciliator, or the messenger of heaven, who restores peace in nature.[377]

[Figure 7: Iris. Detail from an Attic red-figure pelike, middle of 5th century BCE. From Agrigento, Sicilia. Public Domain. Image source: Wikipedia]

If we were to observe our Galaxy from outside its virtually illimitable extents, we would see that there is indeed a literal bridge at the cosmic scale which connects these two gates of gods and mortals. This occurs between two

---

[377] *Dictionary of Greek and Roman Biography and Mythology.* Page 621.

parallel arms of the galactic spiral; namely, the Orion and Sagittarius arms. The point of contact between the Sagittarius Arm and the galactic bridge represents the Gate of the Gods, located further "in," towards the Galactic Center and the star Mu Sagittarii in the constellation Sagittarius. The other point of crossing occurs at the Gate of Humanity, located further "out," in the direction of Orion [Figures 8 & 9].

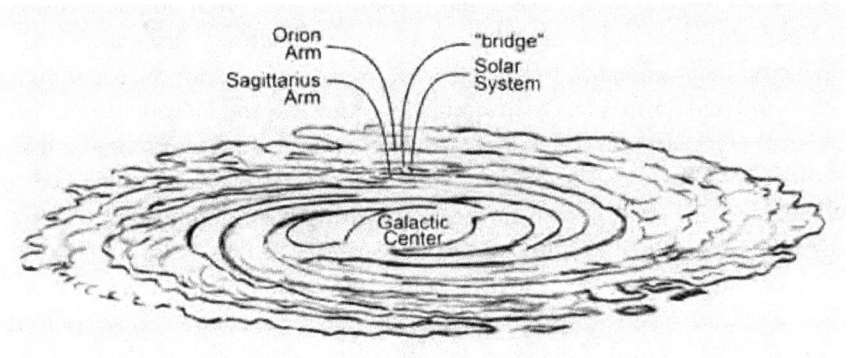

[Figure 8: The Milky Way Galaxy. Author's rendering.]

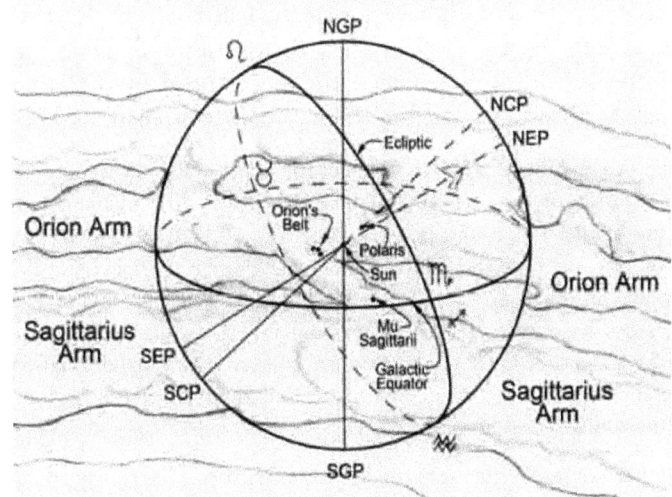

[Figure 9: The Bridge Between the Orion and Sagittarius Arms of the Milky Way As Seen in Relation to the Celestial Sphere. Author's rendering.]

What could have caused the formation of this cosmic bridge? It is known that the Orion Nebula is a star nursery that gave birth to both Sirius and our Sun, and in fact, contains the closest black hole to Earth. Perhaps these two ends of the bridge demarcate two black holes, whose mutual gravitational forces are pulling complementary stellar streams from the parallel arms of

XIV TEMPERANCE

Perseus and Orion. In this context, the two urns of Temperance may be interpreted to represent these "stargates," and the stream which pours from one to the other is the cosmic bridge, Iris – the one who joins Earth to Heaven; humanity to the gods.

Crowley differs from Papus in his Hebraic alphabetic ascription for this trump, assigning it *Samech* rather than *Nun* (see [Table 3] in the Chapter on Origins and Attributions; this book). The issue of correspondence between Hebraic alphabet and Major Arcana trumps has already been discussed as having no well-defined factual framework for its support, but the equation has been proffered so frequently in occult literature via the likes of Papus, Levi, Thierens, Crowley, etc., that it remains a subject which must be discussed, albeit one potentially leading to the mental "lacerations" described by Waite,[378] as one finds oneself in the discomfiting position of trying to solve a puzzle using pieces from different but similar sets.

According to the Gra, Short, Long, and Saadia versions of the *Sefer Yetzirah*, *Samech* corresponds to Sagittarius, and if we confine ourselves to the constraints of ascription dictated by using only the zodiacal, planetary, and elemental categories of the Hebraic alphabet, *Samech* and Sagittarius appear to be the best fit for the present model. If we entertain the notion put forward by some that the ancient alphabets were in fact partially derived from the figures and symbols as seen in the zodiac, the confusion of *Samech*/Sagittarius, and *Nun*/Scorpio with Temperance may be explained by the fact that these two signs, along with Libra, used to form a single constellation:

> According to Achilles Tatius, the sign of the scales was likewise known as the claw of the scorpion; Greek chelai, the claws; also El Zubanan, the claws, consisting of the stars α and β Libra, form the 16th Arab Manzil. Following this hint, we turn to the various ancient Egyptian zodiacs and find that Sagittarius, or the Centaur, is portrayed with a scorpion's tail. Thus, the scorpion has its claws in Libra and its tail in Sagittarius, which shows that there was once a scorpion of the western quarter extending through three of the present signs, in accordance with the four quarters of the beginning, on which the zodiac was founded.[379]

Crowley associates his version of this trump ("Art") with the multi-breasted Diana of Ephesus, who is the Greek representation of the ancient Egyptian Dea Multimammae [Figures 10 & 11], primordial genetrix of the stellar eschatology associated with the circumpolar heavens as well as the

---

[378] Waite. *The Pictorial Key to the Tarot*. Part I. Section 1: "But the aspects of history, as seen through the lens of occultism, are not as a rule decorative, and have few gifts of refreshment to heal the lacerations which they inflict on the logical understanding."

[379] Massey. *The Natural Genesis*. Book 11. Pages 196-197.

inundation. Identified at various phases throughout the course of precession with the sow Rerit (possibly corresponding to the constellation Bootes), the great water monster Ta-urt (Draco), or the Cow of Ursa Major, this multi-breasted suckler is just one of many faces of Isis as Great Mother – who no matter what phase; whether stellar, lunar, or solar – is always fundamentally tied to the feminine source of the waters of life.

[Figure 10: Dea Multimammae from Drummond's *Oedipus Judaicus*. 1866.]

[Figure 11: Dea Multimammae as Aquarius from Drummond's *Oedipus Judaicus*. 1866. Page xxxiii ("Observations on the Plates Annexed to the Oedipus Judaicus"). Annotation added.]

Thus far, we have considered the "bow in the sky" as representative of the arch of the Milky Way in general, but Massey gives an illuminating treatment of this subject which relates the bow to Sagittarius, the inundation, and the

XIV TEMPERANCE

Hebraic legend in Genesis, in which the Elohim place the bow in the sky as a covenant with humanity:

> And God [Elohim] said, This is the token of the covenant which I make between me and you and every living creature that is with you, for perpetual generations: I do set my bow in the cloud, and it shall be for the token of a covenant between me and the earth.[380]

According to Massey:

> The three months' inundation of the Nile is the fact of facts enshrined once and for ever in the zodiac. There the three water-signs are figured twice over, in relation to the Sun and to the full Moon, the bringer forth of the waters as the lunar genetrix. Thus in the fixed year the month Mesore (June 15) is named from the rebirth of the waters, corresponding to the Sun's entrance into the Crab, the first of three water-signs on that side of the zodiac. The Scales show the point at which the three months' flow was suspended. The Scorpion is the sign of exhalation, disappearance, and drying up; and in the next sign appears the bow of the Archer. This sign is called Nephte which in Egyptian means breathed; nef being breath, wind, or the sailor. The archer in the Hermean or lunar zodiac is Shu, who is the god of breath and air; and his bow is the sign of the ended inundation. Also the bow is set in the ninth sign from the sign of Pisces, the last of three water-signs, and the bow and number nine are synonymous. Further, we can tell exactly how the bow got into the cloud of the Hebrew version. The Akkadian name of the ninth month, Can Ganna, is the cloud. In the Hermean zodiac it is the month of Nephte the cloudy; and this was the month of fog, mist, and vapour in Egypt. The bow therefore is in the sign of the cloud; the month is the "cloud" in Akkadian by name; and so the bow in the cloud (month) is the sign of the Archer. The month Tybi or Tebi, modern Toubeh (Nov. 17, sacred year), the month of the Archer, is named from tebu, to draw water. In the ancient calendar instructions are given for filling the cisterns in this month, when the sinking Nile was in its most clarified condition.[381]

All of these correspondences refer to the Age of Taurus, during which the summer months of flooding corresponded to Cancer, Leo, and Virgo, and the point of equipoise to Libra, Scorpio, and Sagittarius. Here the richness and complexity of the symbolism as an evolving eschatology throughout precessional time becomes evident. On the one hand, the trump can be read as a relictual symbol for the point of equilibrium of the flood season during the Age of Gemini/Taurus, when the Sun had reached its autumnal locus of Sagittarius. On the other hand, we can read the symbols as pertaining to our current Age, in which an interpretation of the *nun* or abyss as nocturnal

---
[380] Genesis 9:12-13. King James Version.
[381] Massey. *The Natural Genesis*. Book 11. Pages 210-211.

source of the waters of inundation, constellated in Sagittarius, antipode of the summer Sun in Gemini is suggested. With these considerations in mind, here are some thoughts regarding the symbolism and structure of the Rider-Waite trump as it can be seen to apply to the celestial sphere:

The solar emblem crowning the angel's head symbolizes the North Ecliptic Pole. Below this, the circumpolar heptanomal region is signified within the contour of the brachial portion of the wings, with the "square and triangle of the septeniary" figuring as central motif. The waters of life, corresponding to the milky way, and more specifically to the region of the dark rift in Sagitta, are signified by the "…pouring [of] the essences of life from chalice to chalice."[382] Inclusion of the iris – if read as an allusion to the Greek goddess of the same name – would further support this association, as she is said to signify the rainbow (i.e. Bifrost, the rainbow bridge), which has previously been discussed as a symbol of the Milky Way [Figure 5].

According to Waite, the natures of the essences being poured from chalice to chalice are indicated by the placement of the angel's feet: one being on the earth; the other "upon the waters."[383] As we shall see, an analogous symbolical device is used in trump XVII (The Star) to indicate the zodiacal placement of its avatar. In the present instance, the water in which the figure's right foot is placed represents the nature of the elemental qualities of Scorpio, while the land on which the other foot rests corresponds to the earthy nature of Capricorn; between the two resides the fiery Sagittarius.

To one side of the figure, a Sun can be seen rising behind mountains. This "great light, through which a crown is seen vaguely"[384] may refer to the Galactic Center, and in a sense can be read as the same "sun" which appears between the pillars in trump XIII (Death). Additionally, if we consider the dualistic representation of the mount of the equinox/heptanomis as discussed previously, this refulgent "crown" can be taken as the circumpolar keystone Clavis Corona, which is an antiscion of Sagittarius, rising above the horizon.

---

[382] Waite. *The Pictorial Key to the Tarot*. Part II. Section 2; Temperance.
[383] Ibid.
[384] Ibid.

XV The Devil

TRUMP 17 ("IPEO") from the Sola-Busca deck appears to be an early form of the Devil card [Figure 1]. Here we see a bat-winged figure facing a pole or tree, atop which there is a cherubic or angelic head. The figure clasps his hands towards this rather grim solstice tree, as though in thanks. The vertical extents of the monk can be seen to represent the celestial axis, his hands that of the ecliptic axis, and his dangling belt of spheres that of the ecliptic itself [Figure 2]. The references here all clearly point to the winter solstice, and thus (from a tropical perspective) Capricorn.

Waite refers to the devil in his version of this trump [Figure 3] as a Baphometic figure.[385] According to Massey, Baphomet (Mete) was sacred to the Templars, and was a representation of the Great Mother, as Ta-urt, the female hippopotamus.[386] He suggests this zootype to be associated with Ursa Major, but a slightly different reading may shed a little more light on this equation. The figure of Ta-urt as depicted in the circular zodiac of Denderah appears to associate her with Draco, Cepheus, and the NEP [Figure 4].[387] The tip of her "mooring-post" points to the circumpolar region of Bootes and Corona Borealis. This was the astronome assigned to the male hippopotamus Sut, who represented the primary station of the polar heptanomis, and was the son/consort of the female hippopotamus Ta-urt.[388,389] This was the polar

---

[385] Waite. *The Pictorial Key to the Tarot.* Part II. Section 2; The Devil.

[386] Massey. *The Natural Genesis.* Book 9. Page 14.

[387] This interpretation may also shed some light on Crowley's reference to Baphomet (i.e., Ta-urt) as "Father Mithras" [Crowley. *The Spirit of Solitude*], for Mithras is cognate with the Sun [Massey. *A Book of the Beginnings.* Section 20. Page 474], whose key celestial "functionaries" are the ecliptic, and the ecliptic axis/poles.

[388] Massey. *Ancient Egypt, the Light of the World.* Book 9. Page 623.

[389] Although this pair later became figured in the constellations of the greater and lesser bears, the "original" progenitors (i.e., Adam and Eve) can be traced at least as far back as the heptanomal astronome of the hippopotamus, about which both male and female cotypes revolved in the Age of Cancer. It should also be noted that Ta-urt's mooring-post is almost certainly intended to highlight the star Spica in Virgo, below Bootes and Corona Borealis, as this star was the original fiducial point of the Hypsomatic zodiac. However, this does not preclude the additional interpretation of the mooring-post as precessional marker for the

locus during the Age of Cancer, circa 8,000 BCE, when the Great Flood was said to have occurred by Berosus.[390,391] The ideogram of Ta-urt with mooring-post may thus represent a fixed precessional marker, with preeminence being given to the Cancerian Age.

[Figure 1: The Sola-Busca "Ipeo" Trump. Author's rendering.]

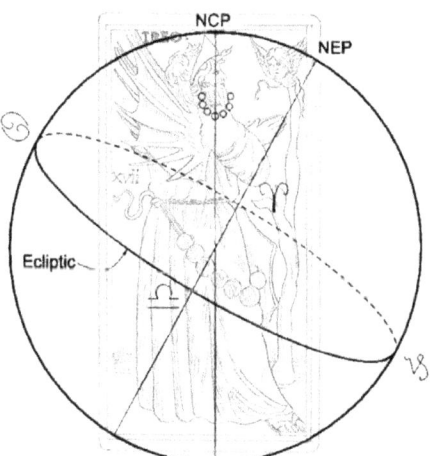

[Figure 2: A proposed correlation between key elements of the Celestial Sphere, and the Sola-Busca Ipeo trump. Author's rendering.]

Cancerian Age as mentioned above, especially since its tip points more specifically to the circumpolar region of Bootes/Corona Borealis – as opposed to Spica – which is considerably lower, on the ecliptic.

[390] Massey. *The Natural Genesis*. Book 11. Page 193. See my explanation as to how this refers to the Cancerian Age further on in this chapter.

[391] Ibid. Book 12. Page 340. See my explanation as to how this refers to the Cancerian Age further on in this chapter.

XV THE DEVIL

[Figure 3: The Rider-Waite Devil Trump. Author's rendering.]

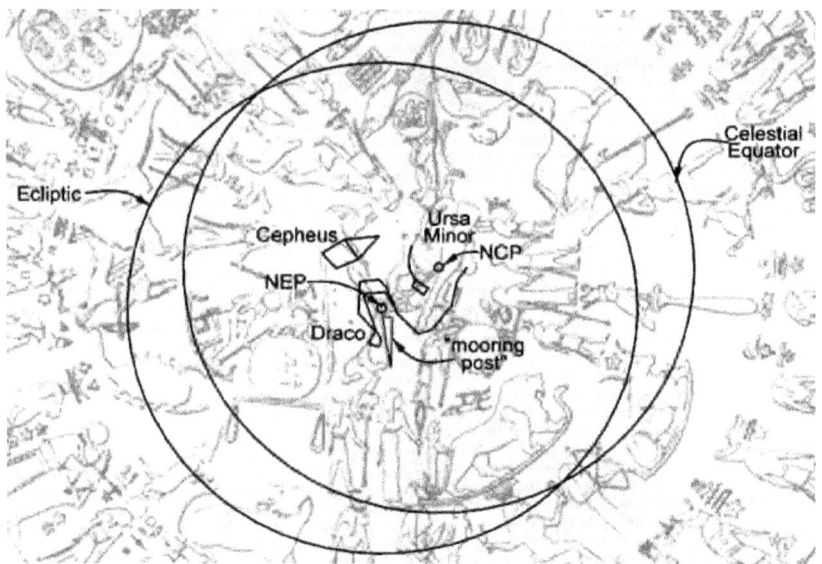

[Figure 4: The "mooring-post" of Ta-urt from the circular zodiac of Denderah, here interpreted to represent the NEP.]

In Volume I of *The Secret Doctrine*, Blavatsky makes the following comment regarding Baphomet during her discussion of the various transformations and differentiations of the all-pervading, "etheric" substance (known in its various forms as the great Archaeus, Akasa, or the Astral Light, within Theosophy) throughout the vast geological cycles of time, which she termed as "Rounds":

> In short, none of the so-called elements were, in the three preceding Rounds, as they are now. For all we know, FIRE may have been pure AKASA, the first Matter of the Magnum Opus of the Creators and "Builders," that Astral Light

which the paradoxical Eliphas Levi calls in one breath "the body of the Holy Ghost," and in the next "Baphomet," the "Androgyne Goat of Mendes."[392]

And in a footnote to this, she elucidates further:

> Eliphas Levi shows it very truly "a force in Nature," by means of which "a single man who can master it . . . might throw the world into confusion and transform its face"; for it is the "great Arcanum of transcendent Magic."... The Author says of the great Magic Agent – "This ambient and all-penetrating fluid, this ray detached from the [Central or "Spiritual"] Sun's splendour ... fixed by the weight of the atmosphere [?!] and the power of central attraction ... the Astral Light, this electromagnetic ether, this vital and luminous caloric, is represented on ancient monuments by the girdle of Isis which twines round two poles and in ancient theogonies by the serpent devouring its own tail, emblem of prudence and of Saturn" – [emblem of infinity, immortality, and Kronos] – "Time" – [not the god Saturn or the planet]. "It is the winged dragon of Medea, the double serpent of the caduceus, and the tempter of Genesis; but it is also the brazen snake of Moses encircling the Tau. Lastly, it is the devil of exoteric dogmatism, and is really the blind force [it is not blind, and Levi knew it], which souls must conquer in order to detach themselves from the chains of Earth; for if they should not, they will be absorbed by the same power which first produced them and will return to the central and eternal fire."[393]

Let us work through some of the issues alluded to in these statements.

Astral Light

The first point is that Levi associates Baphomet (who according to him is a form of the Holy Spirit) with the "Astral Light," which is detached from the Sun's splendor, and fixed by the weight of the atmosphere and the power of central attraction. This seems to be a reference to the stellar influences of the circumpolar regions. They are detached from the Sun's splendor in the sense that during the day they cannot be seen, only being visible nocturnally, although they are electromagnetically active upon Earth day and night, without ceasing. Their influences are not deflected by the atmosphere as they would be if the angle of incidence of the radiations were shallow with respect to the pole, but are trapped within after relatively easy penetration by virtue of their steep angle (in this sense, "fixed by the weight of the atmosphere"), as well as the constant and fixed area of bombardment of their radiations resulting from their central locus within the turning sphere of stars; i.e., the "power of central attraction" [Figure 5].

---

[392] Blavatsky. *The Secret Doctrine*. Volume I. Page 253.
[393] Ibid. Bracketed comments Blavatsky's.

XV THE DEVIL

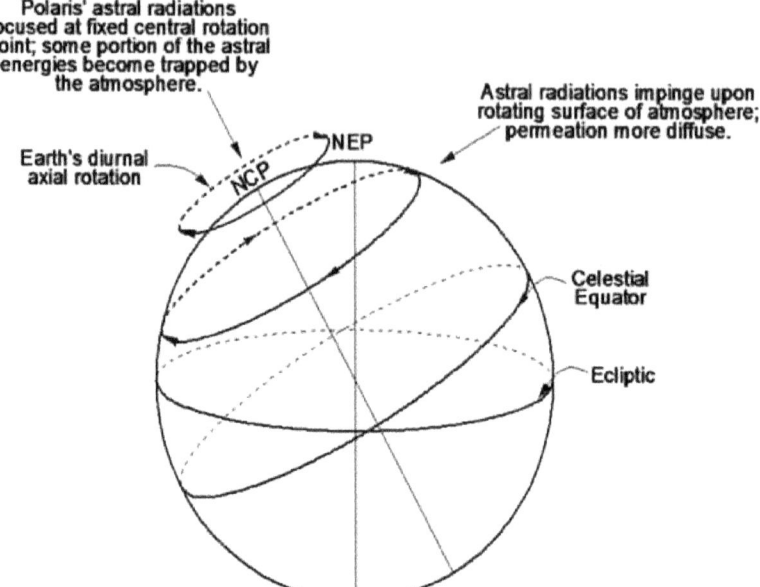

[Figure 5: Polaris as the "Astral Light [...] fixed by the weight of the atmosphere and power of central attraction."]

Girdle of Isis

He further goes on to say that the Astral Light is represented by the girdle of Isis, which he describes as twining around two poles. This so-called "girdle" can be interpreted as the mysterious influence that binds the celestial pole to the ecliptic pole, as the former circles the latter throughout the eternal precessional Ages.

The girdle,[394] or knot of Isis is very similar in shape to the so-called *sa* of Ta-urt [Figure 6], and the two may in fact be different representations of the same concept. The girdle of Isis was a symbol of the goddess' blood, strength, and power, and was often worn as an amulet of protection.[395] The *sa* was also associated with protection,[396] and as Ta-urt preceded Isis as the Great Mother of the waters,[397] it is reasonable to posit a correlation between the *sa* and the girdle of Isis, wherein the girdle represents a later type of the earlier *sa*.

Ta-urt is often depicted holding the *sa* in the same manner as she holds the Egyptian ideogram for "mooring post" in the circular zodiac of

---

[394] The girdle of Isis is sometimes referred to as the buckle or knot of the girdle of Isis. See *The Book of Talismans*. Page 66.
[395] Thomas, Pavitt. *The Book of Talismans*. 1922. Page 66.
[396] Ibid.
[397] Massey. *A Book of the Beginnings*. Book 2. Page 366.

Denderah.³⁹⁸ The "mooring-post" is a representation of the heavenly pole;³⁹⁹ whether the ecliptic or celestial pole is signified is not altogether clear, as the spatial organization of the Denderah Zodiac does not rigorously conform to the actual planisphere constellation arrangement. Nonetheless, the imagery suggests the "mooring post" to be indicative of the NEP, to which the jackal (Ursa Minor) is tethered [Figure 4].

[Figure 6: The *sa* of Ta-urt (70), and the Girdle of Isis (72). Source: *The Book of Talismans, Amulets, and Zodiacal Gems*. William Thomas, Kate Pavitt. 1922. From plate 5: Egyptian Talismans.]

The connection between the *sa*, the "mooring-post," and the "girdle" is further evidenced when we consider the fact that *sa* represented the astral fluid, or blood of the gods. This substance was able to be transmitted to mortals, and when the gods were in need of a fresh supply, it was obtained exclusively from the circumpolar fount of healing – the so-called Lake of Sa at the celestial pole.⁴⁰⁰

Ouroboros

As discussed in the section on the Magician, the ouroboros is among other things a symbol for the ecliptic, inclusive by its very nature of the ecliptic pole itself, as that is the center point around which the snake can be said to enclose itself. Other connotations may be relevant as well, such as the self-contained generation of transmutational energies from within, catalyzed perhaps by solar forces.

Dragon of Medea

The allusion to the winged dragon of Medea can be seen as a reference to the interstitial polar region between Thuban and Polaris, since that astronome corresponds to the location of the NCP during the Age of Aries, and the

---

³⁹⁸ The device Ta-urt is depicted as holding in the circular zodiac of Denderah is the Egyptian ideogram for "mooring post." See *An Egyptian Hieroglyphic Dictionary*. Page cxxxviii. E. A. Wallis Budge.

³⁹⁹ Massey. *Ancient Egypt, the Light of the World*. Book 6. Page 396: "The mooring post was an image of the pole, to which the stellar ark or solar bark was fastened by the cable, as it made the voyage round the starry mount."

⁴⁰⁰ Massey. *Ancient Egypt, the Light of the World*. Book 3. Page 175.

XV THE DEVIL

whole tale of the Golden Fleece can be read as an astromythological dissertation on the movement of the equinoctial Sun into the vernal point of Aries. In this case, Medea – who is in love with the solar hero Jason – represents the feminine (i.e. "nocturnal") counterpart of the diurnal (i.e. "masculine") solar vernal point, as the corresponding circumpolar locus of the NCP during the Age of Aries [Figure 7]. As noted previously, one of the ancient appellations of Ursa Minor is the "wings of the dragon," thus the characterization of the circumpolar locus between Polaris and Thuban as a winged dragon is fitting.

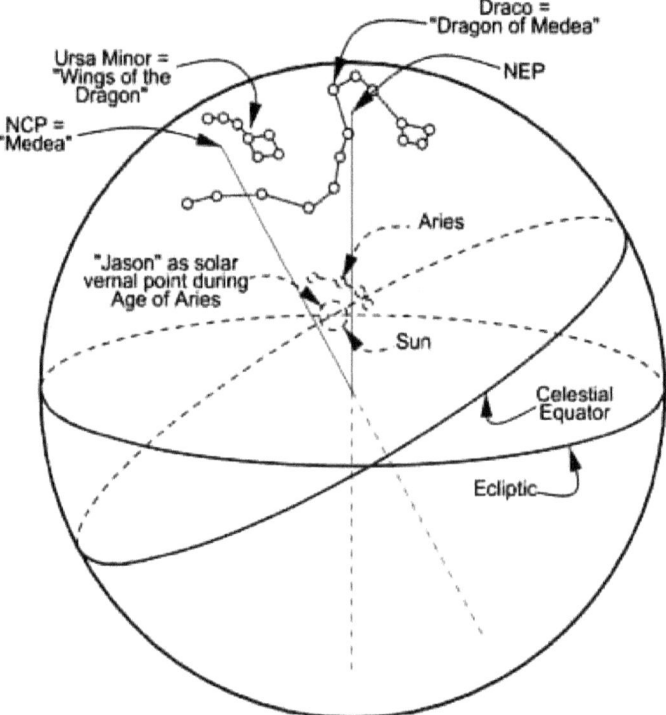

[Figure 7: Medea and Jason as anthropomorphs of the NCP and vernal point, during the Age of Aries.]

Caduceus

We have discussed the "double serpent of the Caduceus" as a symbol for the rotation of the two dragon-associated loci of Thuban and Polaris around the NCP during the Age of Aries [Figure 8]. Similarly, the "tempter of Genesis" and "brazen snake of Moses" can both be seen to represent a similar construct referencing the circumpolar astronome more closely associated with Thuban, and the Ages of Taurus and even Gemini. This being said, the

Caduceus can also be seen on a higher timescale to represent the double spiral of the two antithetical circumpolar loci represented by Polaris,[401] and Beta/Gamma Draconis,[402] eternally spiraling around each other, centered on the ecliptic axis, as the whole higher cosmic sphere moves along the direction of this axis throughout precessional time [Figure 9].

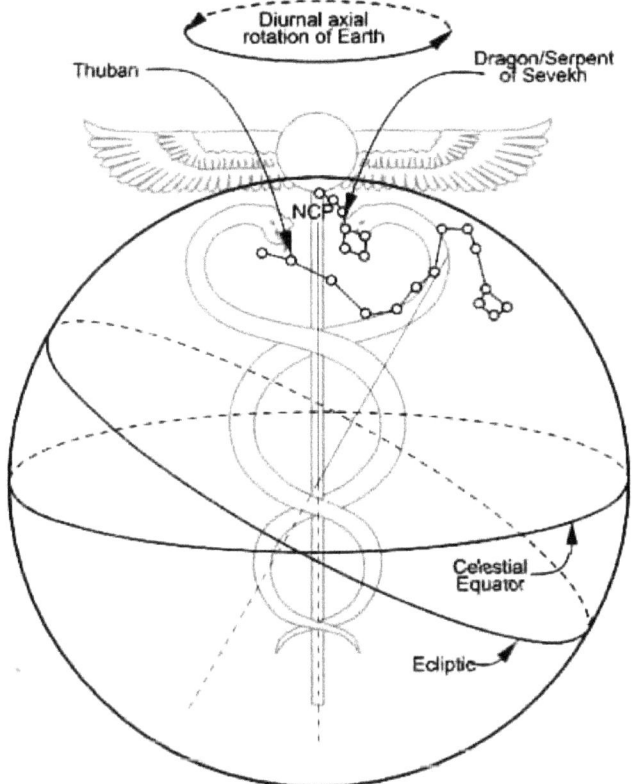

[Figure 8: The Caduceus as double spiral of Draco and Ursa Minor about the celestial pole.]

Pliny the Elder makes an interesting reference to Typhon in his *Natural History*. In Chapter 25 ("Of Comets or blazing stars, and cœlestiall prodigies, their nature, situation, and diverse sorts"), while discussing the various characteristics and types of comets, he says the following:

> A terrible one likewise was seene of the people in Æthyopia and Ægypt, which the king who raigned in that Age, named *Typhon*. It resembled fire, and was

---

[401] The central head of the 7-headed dragon, Sevekh-Kronus, associated with the middle portion of the precessional nadir, and the Age of Aquarius.

[402] The head of the dragon (Draco), corresponding to the midpoint of the precessional apex, and Golden Age of Leo.

XV THE DEVIL

plaited or twisted in maner of a wreath, grim and hideous to be looked on; and no more truly to be counted a starre, than some knot of fire. Sometimes it falleth out, that the Planets and other stars are bespread all over with hairs.[403]

Pliny does not offer a date for this occurrence, but it was presumably considerably earlier than his era (23-79 CE) at the beginning of the Piscean Age.

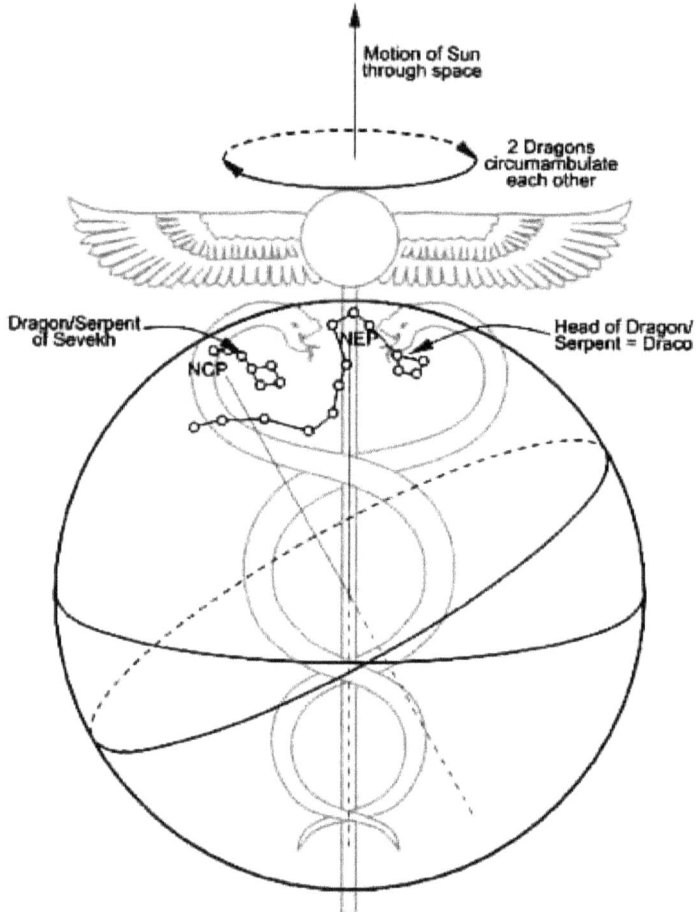

[Figure 9: The Caduceus as double spiral of Draco and Ursa Minor about the NEP throughout the course of precession.]

Here we see Typhon connected to some cometary event, and the question necessarily arises as to whether some periodic cometary incursion may be implied in Blavatsky's reference to the "approach of Typhon." Plato speaks of

---

[403] Pliny the Elder. Natural History, Book 2. Philemon Holland translation, 1601.

such periodic events in his Timeaus. In the following passage, Critias recounts the visit of Solon (638-558 BCE) to the Egyptian city of Sais:

> To this city came Solon, and was received there with great honour; he asked the priests who were most skillful in such matters, about antiquity, and made the discovery that neither he nor any other Hellene knew anything worth mentioning about the times of old. On one occasion, wishing to draw them on to speak of antiquity, he began to tell about the most ancient things in our part of the world-about Phoroneus, who is called "the first man," and about Niobe; and after the Deluge, of the survival of Deucalion and Pyrrha; and he traced the genealogy of their descendants, and reckoning up the dates, tried to compute how many years ago the events of which he was speaking happened. Thereupon one of the priests, who was of a very great Age, said: "O Solon, Solon, you Hellenes are never anything but children, and there is not an old man among you." Solon in return asked him what he meant. "I mean to say, he replied, that in mind you are all young; there is no old opinion handed down among you by ancient tradition, nor any science which is hoary with Age. And I will tell you why. There have been, and will be again, many destructions of mankind arising out of many causes; the greatest have been brought about by the agencies of fire and water, and other lesser ones by innumerable other causes. There is a story, which even you have preserved, that once upon a time Phaeton, the son of Helios, having yoked the steeds in his father's chariot, because he was not able to drive them in the path of his father, burnt up all that was upon the earth, and was himself destroyed by a thunderbolt. *Now this has the form of a myth, but really signifies a declination of the bodies moving in the heavens around the earth, and a great conflagration of things upon the earth, which recurs after long intervals; at such times those who live upon the mountains and in dry and lofty places are more liable to destruction than those who dwell by rivers or on the seashore. And from this calamity the Nile, who is our never-failing saviour, delivers and preserves us.* When, on the other hand, the gods purge the earth with a deluge of water, the survivors in your country are herdsmen and shepherds who dwell on the mountains, but those who, like you, live in cities are carried by the rivers into the sea. Whereas in this land, neither then nor at any other time, does the water come down from above on the fields, having always a tendency to come up from below; for which reason the traditions preserved here are the most ancient.[404]

Thus, according to Plato, the ancient Egyptians were aware of the periodicity of the destruction of civilizations via the agency of atmospheric and gravitational perturbations caused by the cyclic "invasion" of extra-solar cometary bodies. Moreover, they may have possessed a deeper understanding of such phenomenon, by virtue of their unique geographic location along the Nile, which enabled them to survive such catastrophes, and therefore keep more extensive records of these events.

---

[404] Plato. *Timaeus*. Italics added.

XV THE DEVIL

Typhon as Eclipse

The following passage from Plutarch's *On Isis and Osiris* illustrates the tendency of the ancients towards allegory regarding the various observable celestial manifestations and cycles:

> Some make an allegory out of the rule of the eclipses, for the Moon is eclipsed at her full, when the Sun holds the station opposite to her when she falls into the shadow of the earth, in the same way as they tell Osiris did into the coffer; and she herself, upon the thirtieth conceals and puts out of sight, yet does not altogether destroy, the Sun, as neither did Isis Typhon. And when Nephthys conceives Anubis, Isis adopts him, for Nephthys signifies what is under the earth and invisible; Isis, what is above ground and visible; and the circle touching these, called the Horizon, and common to both, has been named Anubis, and is figured as a dog; for the dog has the use of his sight both by night and by day; and Anubis appears to have the same office with the Egyptians that Hermes has with the Greeks, being both infernal and celestial. Some, however, think that Anubis signifies Time, wherefore as he brings forth all things out of himself, and conceives all things within himself, he gets the title of Dog. Besides, the votaries of Anubis celebrate a certain mystery, [A passage is lost here, containing a description of this rite, in which it is evident a dog played the principal part.] and in old times the dog enjoyed the highest honors in Egypt. But when Cambyses had slain the apis and cast him out, nothing approached, or tasted of the carcass, except the dog, so he lost his place of the first, and the most honored of all the other animals. *And there are some that think he is the shadow of the earth into which the Moon passes when she is eclipsed, and they call him Typhon.*[405]

Isis was anciently associated with the star Sirius, but there are numerous other instances where she symbolizes the Moon, and here Plutarch maintains that Isis merely signifies that which is "above ground and visible."

Each of these three faces of the Sothic genetrix can be explained via the "two truths" of the ancient Egyptian model of time and cosmos. From this perspective, each face relates to a particular aspect of the Egyptian Sothic year and its associations with inundation and drought at the two solstitial extremes, as well as the equinoctial dyad of planting and harvesting.

Thus, Isis as Sirius references the heliacal rising of that star which heralded the annual inundation during the Taurean Age.

Isis as the Moon – or more particularly, the Full Moon – refers to the same solstitial locus of the year (summer), although this time as witnessed through the nocturnal lens of the dark mother's underworld turned downside up; the Full Moon being testament to the safe passage of the Osirian Sun

---

[405] Plutarch. *On Isis and Osiris*. Paragraph XLIV. Italics added.

through the underworld, and the arrival of the inundation through some mysterious force somehow connected to the mesmeric increase of her radiance [Figure 10].

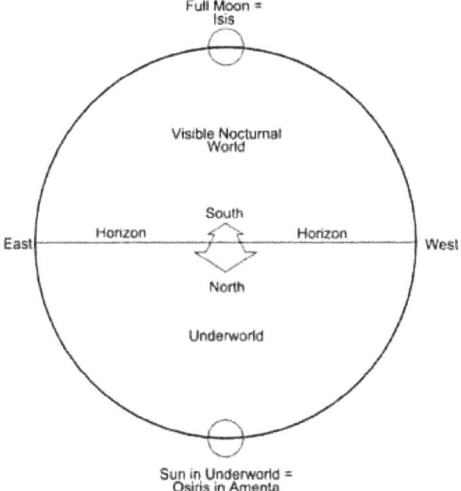

[Figure 10: The Full Moon (Isis) as testament to the Sun's (Osiris') safe passage through the Underworld.]

Finally, Isis as that which is "above ground and visible" refers to the very act itself of observing the heliacal rising of Sirius during the season of inundation [Figure 11], as by definition that is exactly what one is observing – namely, that which is visible immediately preceding the entrance of the solar king, whose refulgence outshines the other luminaries during his diurnal reign.

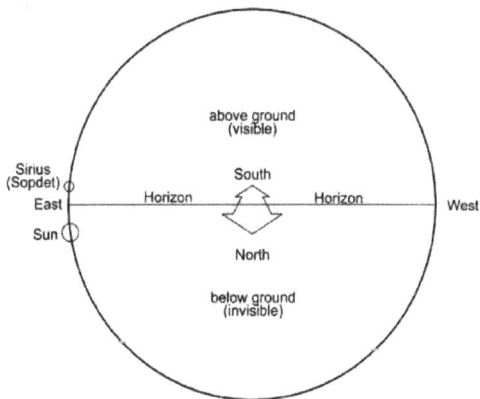

[Figure 11: Isis (Sirius) as "that which is above ground and visible."]

XV THE DEVIL

Also implicit within this passage from *On Isis and Osiris* are the multiple celestial and deific connotations associated with Anubis, for as we shall see, this god appears in several different guises in addition to the one commonly associated with the jackal-headed figure of the "weighing of the heart" ceremony. In this solemn post-mortem procedure – as illustrated in the following reproduction from the 18th Egyptian Dynasty, circa 1500 BCE – the heart of the deceased is said to be weighed against the "feather of *maat*," which represents impartial or objective justice [Figure 12].

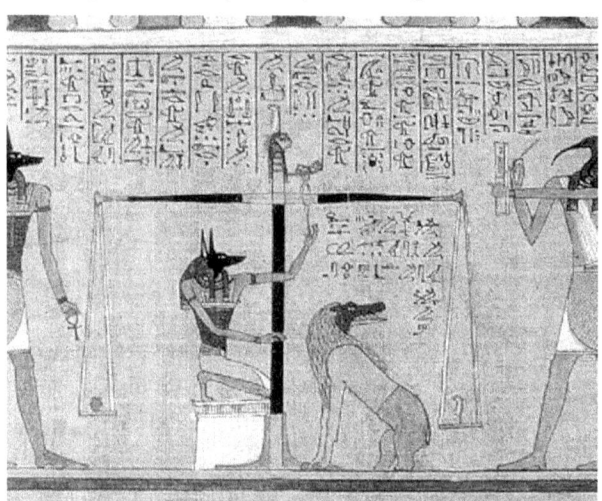

[Figure 12: Weighing of the heart Ceremony from the Hunefer Manuscript. Image source: Wikipedia.]

In this depiction, the deceased is led to the place of judgment by Anubis (on the left), with Taht (Thoth) – the ibis-headed scribe associated with the Moon as nocturnal representative of the Sun – flanking the scene on the right. In the words of Massey:

> The balance was set in the midst before them, and they were weighing the evil deeds against the good deeds, the great god Taht (Aan) recording, with Anup giving the word to his colleague. These are the prototypal "two witnesses" stellar and lunar for the Father and Son in the solar mythos. Taht-Aan was the witness for Horus, the only-begotten son of the father. In the mythos, which preceded the eschatology, Taht-Aan was the light of the world as the god whose luminary was the Moon. Read doctrinally, he was not the true light, but he came that he might bear witness to the true light.[406]

Whereas Taht is in this context associated with the Moon, Anubis (or Anup) is to be seen as representative of the polestar:

---

[406] Massey. *Ancient Egypt, the Light of the World.* Book 11. Pages 705-706.

In the stellar mythos Anup had been the judge, with the seat of judgment at the place of equipoise, which was then at the celestial pole. In the solar mythos this was shifted to the vernal equinox, and the mount of glory to the east.[407]

There is no overt solar element in this scene; it all takes place in the underworld. Here the hidden places of the heart will be revealed. The lunar witness to the right side of the balance – with the lurking Ammut "demon" ready to devour the soul if it should be found wanting – may answer to the esoteric doctrine of the Moon as vampiric feeder upon the undeveloped soul substance of humanity; a concept which has been developed along parallel lines both by Ouspensky and Blavatsky.

But how could it be that Anubis, as polestar, could come to be associated with the horizon, as Plutarch indicates? The answer is to be found in the effect of the spheroidal geometry of Earth as it relates to the subjective appearance of the stars relative to the latitude from which observation is made. In the northernmost inhabited latitudes, the polestar's unchanging rotational locus of astral permeation is seen to occupy the uppermost portion of the celestial dome, where it truly can be said to be the "crown of heaven." As one traverses down towards the equator, this once vertical reference point begins to lean towards the horizontal [Figures 13-15]. This phenomenon may in fact go a long way towards explaining the various references to the so-called "bent mountain" of heaven, such as the Culhuacan of the Aztecs of Mesoamerica [Figure 16]. From the perspective of regions in Africa which Massey refers to as Equatoria (i.e. the lands along the equator proper), the circumpolar mount lies recumbent, and looking to the east or west from this vantage point, one can witness the turning of both extremities of the celestial poles, as two great wheels of some heavenly chariot of the gods.

Egypt being somewhat north of the equator by 30 or so degrees of latitude, the NCP – while not completely horizontal, as at the equator – is nonetheless considerably "bent" towards the horizon, such that through the course of the night – rather than the lofty wheeling in the sky of the mount of heaven associated with the more northerly latitudes – the turning of the pole describes a shape more akin to a hillock off to the left, as viewed from an eastward facing vantage point. It is in this sense that Anubis, as polestar, can be said to be associated with the horizon, for at this latitude, he is the only representative of the starry host who keeps his place faithfully as the steadfast dog of heaven, chasing his tail on the low-angled hill of the pole. In the Taurean Age, when the Sun was in the sign of Cancer during the beginning of the season of inundation, Sirius was rising heliacally, and as the planet Mercury never strays far from the Sun in its cycles as seen from Earth, it

---

[407] Massey. *Ancient Egypt, the Light of the World.* Book 6. Page 344.

XV THE DEVIL

became associated in addition to Sirius with the annual flood. Mercury was thus known as the faithful companion of the "Dog-Star," and possibly in part due to the association of Anubis (Anup) with the horizon as the low-angled polestar, Mercury became associated with Hermes-Anubis (Hermanubis), announcer of the inundation and guide of the Sun through the Underworld.[408]

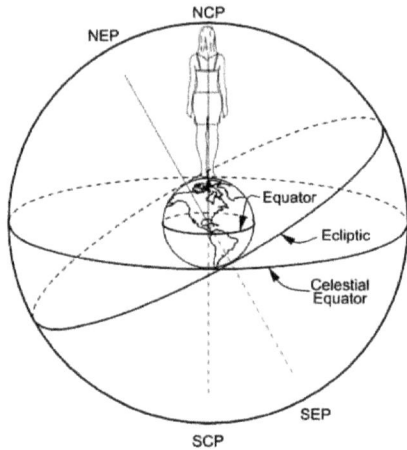

Orientation of Firmament as seen from
North Pole, facing East

[Figure 13]

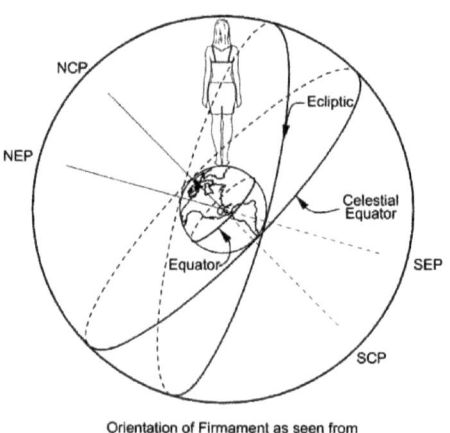

Orientation of Firmament as seen from
45 degrees North of Equator, facing East

[Figure 14]

---

[408] Massey. *The Natural Genesis*. Book 10. Pages 103-104.

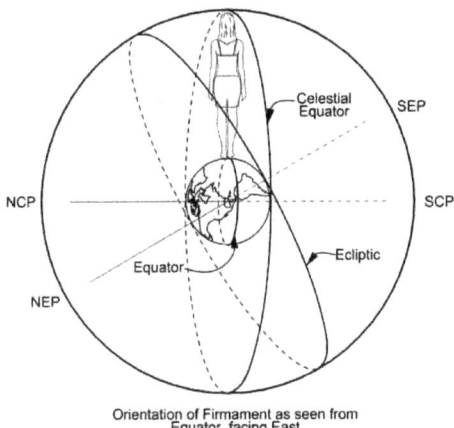

[Figure 15]

[Figure 16: Culhuacan. Author's rendering]

Taking the aforementioned considerations into account, we are thus able to harmonize the seemingly disparate roles of Anubis as divine intercessor of the deceased, personification of the polestar, guardian of the horizon, as well as his conflation with the planet Mercury. And yet there is the additional association mentioned by Plutarch: "And there are some that think he is the shadow of the earth into which the Moon passes when she is eclipsed, and they call him Typhon."[409]

This fear of eclipse, as evidenced by the association of the phenomenon with the Typhonian monster is spoken of in the following passage from Massey's *Luniolatry: Ancient and Modern*, which also contains an oblique reference to the connection of eclipse with the polestar:

> Now the most thrilling and fearsome act of the lunar drama was during the period of eclipse. There is something very weird, uncanny, and unked, in the projection of the earth's shadow across the luminous face of the Moon. To the primitive mind it was the crocodile above, or the dragon, swallowing the orb of light, or Sut swallowing his father Osiris. An eclipse was the meal-time of the monster. An eclipse was the scene of the great battle between Horus and

---

[409] Plutarch. *On Isis and Osiris*. Paragraph XLIV.

XV THE DEVIL

Sut, or Horus and the Dragon, and the great battle was identical with that of our George and the Dragon. The same struggle between the powers of light and darkness is portrayed in the Book of Revelation when the woman clothed with the Sun, and the Moon under her feet, is about to bring forth her man child, and the great dragon of eclipse stands before her ready to devour the child as soon as it is born! In the oldest astronomy the years were reckoned by the eclipses, as it was in Egypt, China, and India. And the most ancient type of time or Kronus, as Egyptian, is Sevekh, the crocodile-headed god, that is, the dragon of eclipse who annually swallowed the Moon containing the Lord of Light or his infant Image.[410]

The dragon Sevekh-Kronus was associated with the circumpolar region during the Age of Taurus/Aries. The seven heads of the dragon corresponded to the seven stars of Ursa Minor.[411] We thus have a fairly solid reference here to the connection between the phenomenon of eclipse as conceived of by the ancients, and the northern circumpolar regions associated with the polestar. The question still remains as to what mechanism of eclipse the ancients may have had in mind that would account for this conflation between the polestar, and "dragon of the eclipse."

One possibility is that a method was being employed similar to the "siting of the temple," associated with the goddess Seshat, and the so-called "stretching of the cord" ceremony. As mentioned previously, the periodicity of full eclipse occurs only once every 18 years, which corresponds to the amount of time it takes for the lunar nodes to retrograde a full 360° along the ecliptic. In order to track this cycle, it is necessary to establish some fixed point along the ecliptic from which to make reference. In the stretching of the cord ceremony, this fixed ecliptic locus was determined to be the star Spica, in the constellation Virgo. True celestial north was then found and mapped onto the ground by tracing the arc that connected the stars Spica and Eta Ursa Majoris. This celestial arcsegment can be continued on up into the tail of

---

[410] Massey. *Luniolatry: Ancient and Modern.*

[411] Massey. *The Natural Genesis.* Book 6. Page 349: "Kefa, the Beast in the Abyss, became the Goddess of the Great Bear and Mother of the revolutions or cycles of Time. Sevekh did duty as her Dog (Lesser Bear, or Dragon)... In the 'Chapter of stopping the Crocodiles' which come to take the mind of a Spirit from him (presence of mind) in Hades, the Swallowers are Eight in number; and in the 17th Chapter the Seven Spirits, or Genii, who are stationed behind the constellation of Ursa Major in the Northern Heaven, are called the Crocodiles. According to my interpretation we have to look on the Seven Stars of the Lesser Bear as representative of these elementary gods, who were Seven as the heads of the Dragon, but who were also one as a constellation represented by Sevekh. For example, at the centre or the zodiac of Denderah we see the Hippopotamus and the Dog, Jackal, or Fox. These two were a form of Sut-Typhon. "The Little Bear," says Dupuis, "was also known as the Fox." The Egyptian Fox was the Fenekh-type of Sut, the Fox-dog. Thus the Two Bears represent the Mother and Son at the centre of all."

Draco (wherein resides Thuban, the polestar during the Age under consideration), and even further to the star alpha Ursa Minoris, our current polestar. It would make sense, therefore, for the ancients to define as their prime "eclipse meridian" this very locus on the ecliptic, as it corresponded exactly with the station of their polestar at the time. Perhaps it is though this, or some similar mechanism, that the dragon of the eclipse became associated with the dragon of the pole in the ancient system of stellar time reckoning [Figure 17].

Amphitrite

Amphitrite as consort to Poseidon and mother of Triton is indirectly associated with the region of Capricorn, as Poseidon is said to be connected to Delphinus (paranatellon of Capricorn); and Triton – as half fish, half man – can be seen as a form of Oannes, or Capricorn.[412]

But this is only a vague intimation of what may be a deeper connection to an original Egyptian source of the legend of Amphitrite, and how she came finally to be wed to Poseidon (or Roman Neptune) via the intercession of his faithful dolphin, Delphinus.

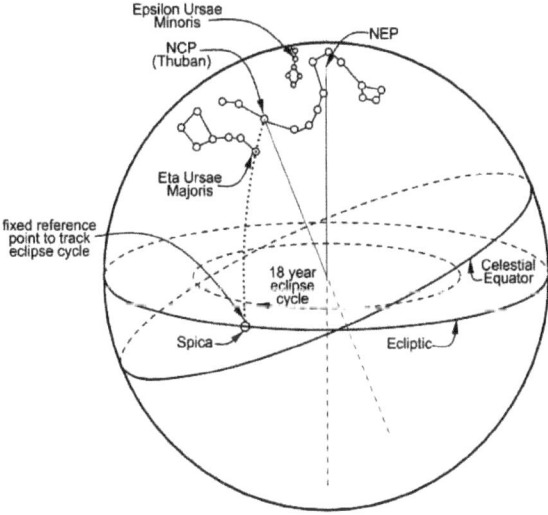

[Figure 17: A proposed mechanism explaining the ancient association of the "dragon of eclipse" with the "dragon of the pole."]

Amphitrite – Greek goddess of the sea – is often depicted with crab pincers adorning her temples [Figures 18 & 19], which is to be read as a reference to the zodiacal sign of Cancer, and the associated heliacal rising of

---

[412] *Encyclopedic Theosophical Glossary*. "Poseidon" and "Triton" entries.

XV THE DEVIL

Sirius during the month of July – the first of the three months of inundation in ancient Egypt.

According to Blavatsky, Amphitrite is an "early form of Venus," and here the underlying meaning of the myth begins to take shape with respect to the natural truths of universal celestial phenomena and cycles. What is undoubtedly being referenced via the poetic language of myth is in this case fundamentally concerned with the Venus/Sirius heliacal cycle, as witnessed from the perspective of the season of inundation; an idea first conceived by the ancient Egyptians, and later carried on in the mythology of the classical Greeks and Romans. A consideration of the male counterpart of this "moaning" goddess of the sea should help to flesh out this hypothesis.

Poseidon – or Neptune, as the Romans would have it, is mentioned by Massey towards the end of the following passage from *The Natural Genesis*, in which the division of the celestial sphere into upper, middle, and lower portions – designated as "sky," "earth," and "waters" (of the "abyss") – is discussed:

> But, in Vedic cosmology there are three skies – the upper, the middle, and the lower. The three divisions are elsewhere stated to be the sky, earth, and waters, and these are the same three regions as those of the Egyptian and Chaldean solar mythologies. Vishnu passes through these three regions in three strides; his three footprints being figured in the twenty-third lunar mansion, Sravana. The three footprints equate with the trident symbol; and in some astronomical works, the Sakalya, for example, the trident is depicted for Sravana instead of the three footsteps. Plutarch observes that the trident is the symbol of the third region of the world which the (mystic) sea possesses, situated below the heaven and the air (or earth). The trident is a type of the male triad, and is assigned to the supreme one of the three. This may be Siva in one cult, or Vishnu in another. The god of the third region, the abyss of the waters, was the Af-Ra, in Egypt; Yav or Hea in Assyria; Javeh or Jah in Israel; Vishnu in India. Khnef (or Num) was likewise a form of this solar god; nef being the sailor, the Neptune of the Romans, and the British nevvy. This was the Sun of the waters, the darkness, the abyss; the god who completed the circuit round, the protector by night, the seer unseen.[413]

Here Neptune is unveiled as the Full Moon, which – being in the sign antipodal to the Sun – serves as the nocturnal representative of its diurnal twin, and is thereby known as the "Sun" of the abyss, or the "nocturnal Sun."

As the solar orb entered the "water signs" of Cancer/Leo/Virgo, bringing on the annual flood of the Nile, so the Full Moon as nocturnal twin was seen in the opposite signs of Capricorn/Aquarius/Pisces. Neptune, as the Full Moon which is the nocturnal Sun of the abyss is able finally to wed

---
[413] Massey. *The Natural Genesis*. Book 8. Page 526.

Amphitrite / Venus / Bennu / Sirius via the "intercession" of Delphinus / Capricorn, all of which is – at one level – a metaphor for the heliacal rising of Sirius conjunct Venus in the sign of Cancer, during the months of inundation, while the Full Moon (Neptune) resides in the opposite sign of Capricorn. This essentially dates the mythic temporal reference to sometime in the Age of Aries; say, around 2,000 BCE, although a variability of +/- 1,000 years would appear to be acceptable [Figure 20].

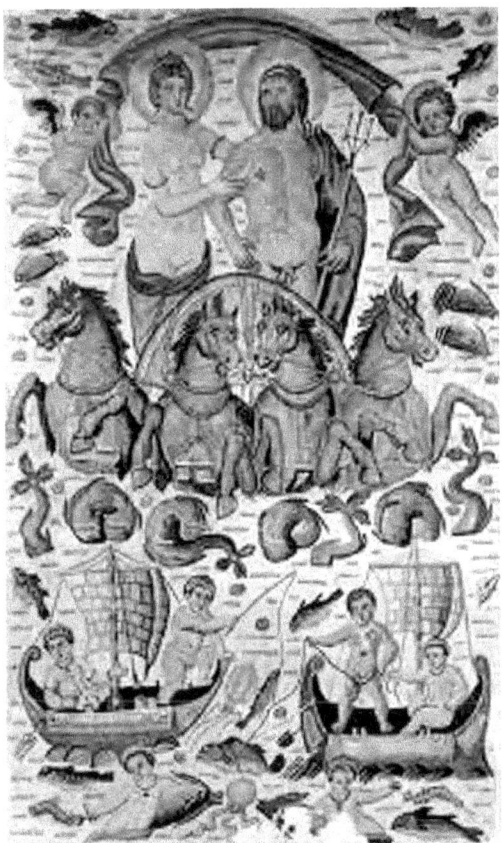

[Figure 18: Triumph of Poseidon and Amphitrite. Detail of a mosaic from Cirta, Roman Africa (ca. 315–325 CE, now at the Louvre). Image source: Wikipedia.]

[Figure 19: Amphitrite on an ancient Greek Coin. Obverse: Head of Amphitrite wearing crab headdress. Reverse: Crab. Author's rendering.]

XV THE DEVIL

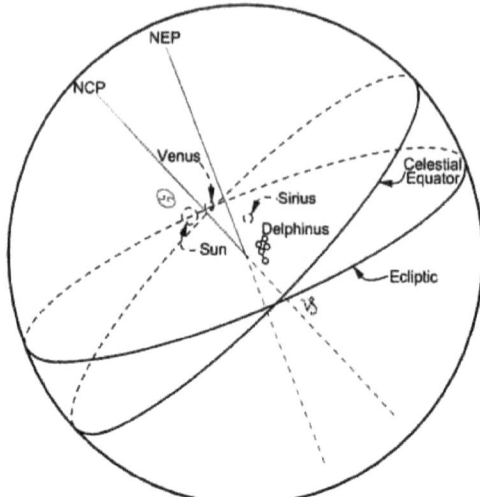

[Figure 20: The heliacal rising of Sirius and Venus during the summer, circa 2,000 BCE]

Durga Kali

The mention of Durga Kali in connection to this sign implicitly references Durga as consort of Shiva,[414] as well as Kali, which is one of the seven "tongues" of Agni, who in turn is equated with Shiva. This multifarious god, both terrible and beneficent, known as both creator[415] and destroyer[416] – accomplishing the one through the other as a function of the "economy of nothingness" – is associated with the Hindu Mount Meru, which corresponds to the northern circumpolar region of the heavens.[417]

Considering the profusion of appellations and qualities ascribed to this deity, as well as his association with the circumpolar mount, it seems reasonable to posit a correlation with this god to the solar axis of the ecliptic. This is the central "tree of eternity," around which the "tree of time" and dualism (represented by the celestial axis) revolves throughout the precessional cycles. The "tree of eternity" contains within itself all of the separate and seemingly irreconcilable natures of each phase of the "tree of dualism," much as the white light of the Sun contains and harmonizes within it each of the unique wavelengths of the visible spectrum.

---

[414] Shiva is equated with Typhon, among other forces, by Blavatsky.

[415] i.e., the creator of higher states and vehicles of consciousness.

[416] i.e., destroyer of the "lower" desires and perceptions, which are predicated upon a dualistic conception of phenomena.

[417] Massey. *The Natural Genesis*. Book 9. Page 21: "In various mythologies and forms of the mythos the birth-place of creation is in the north. It was so in India as in Egypt. There stood the Mount of Meru as the typical centre of the starry revolution."

Proceeding from this hypothesis, Kali – as one of the seven tongues of Agni/Shiva – would thus correspond to one of the heptanomal astronomes, i.e. one of the seven Ages into which the Platonic Year can be partitioned. Moreover, Kali as the "black tongue" of Agni suggests a reference to the station of the pole during the Age of Darkness known as the Kali Yuga. According to the present model, this corresponds to the current locus, effectively equating Durga with Ursa Minor.

Makara and the Night of Brahma

The amphibious creature referred to as Dagon by Blavatsky, is attributed by her in *The Secret Doctrine* to the constellation Capricorn, which she names as Makara, and is said to be related in some "occult" fashion to the so-called Kumaras of the Hindu Puranas. The Kumaras are said to be entities – varying in number according to didactic modality from seven to four – which Brahma created with the intent that they should in turn generate their own progeny, but they refused, and are said to be exempt from all forms of attachment. One of their functions is to aid humanity in the realization of the path to enlightenment. The relationship between the Kumara and Capricorn (Makara) is touched upon in the following passage from *The Secret Doctrine*:

> Everyone knows that [Capricorn] is the tenth sign of the Zodiac into which the Sun enters at the winter solstice, about December 21st. But very few are those who know – even in India, unless they are initiated – the real mystic connection which seems to exist, as we are told, between the names Makara and Kumara. The first means some amphibious animal called flippantly "crocodile," as some Orientalists think, and the second is the title of the great patrons of Yogins (See "Saiva Puranas,") the Sons of, and even one with, Rudra (Siva); a Kumara himself. It is through their connection with Man that the Kumaras are likewise connected with the Zodiac.[418]

A little further on, Blavatsky states the following:

> The Kumara (in this case an anagram for occult purposes) are five in esotericism, as Yogis – because the last two names have ever been kept secret…[419]

An anagram is a word formed by transposing certain letters of a different word, comprised of the same or similar letters. In the case of Makara/Kumara, it is evident that the derivation of the latter from the former is the result of the transposition of Ma-Ka to produce Ka-Ma, with the subsequent substitution of a "u" in place of the "a" to produce Ku-Ma. This "flipping" of

---

[418] Blavatsky. *The Secret Doctrine*. Volume II. Page 576.
[419] Ibid. Page 578.

XV THE DEVIL

two adjacent components in fact appears to point to the true "occult purposes" which Blavatsky hints at; namely, a secret revealing through overt obfuscation that Makara – instead of referring to the sign of Capricorn – actually corresponds to the adjacent sign of Aquarius.

The anagrammatic transposition, itself – and Blavatsky's comment about it – are in themselves enough to strongly support this hypothesis, but the following passage from Massey's *Natural Genesis* bolsters it even further:

> It is said by the Vamadeva-Modely that "when the Night of Brahma is approaching, dusk rises at the horizon and the Sun passes away behind the thirtieth degree of Makara (Capricorn) and will reach no more the sign of Mina (Pisces)." These are the three water-signs which represent the negation of time, the place of non-creation, the abyss.[420]

The fact that Massey fails to mention Aquarius, yet specifically calls out Capricorn and Pisces – two zodiacal signs, only – then says "these are the *three* water signs which represent the negation of time" cannot be ignored or dismissed as some accidental omission. Furthermore, it seems clear that Makara is actually Aquarius, since this is the only sign of which it would make sense to say that when the Sun passes behind its 30th degree, it no longer reaches the sign of Pisces, for this is in fact a reference to the retrograde motion of the vernal point of the Sun from Pisces to Aquarius. The degrees of the zodiacal signs are arranged such that their increase is in the direction of the progression of the Sun along the ecliptic throughout the course of the year. Thus, the 30th degree of Aquarius is found adjacent to the 1st degree of Pisces, and as the Sun, in its precessional journey moves out of this 1st degree, reaching "no more the sign of Mina (Pisces)," it necessarily passes into – or "behind"[421] – the 30th degree of Aquarius [Figure 21].

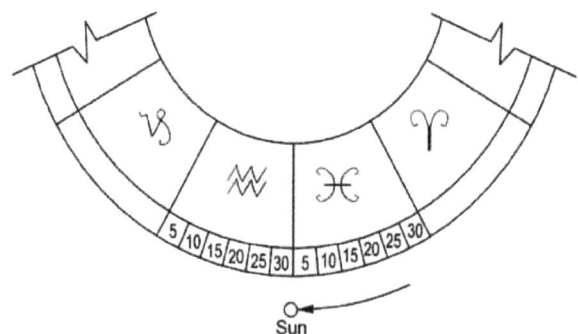

[Figure 21: The Sun "will reach no more the sign of Mina (Pisces)."]

---

[420] Massey. *The Natural Genesis*. Book 11. Page 263.
[421] The Sun is "behind" the 30th degree of Aquarius when it is in the interstitial region between Pisces and Aquarius.

The point at which this transition from Pisces to Aquarius can be said to occur is based entirely upon the system of zodiacal divisioning employed. According to Fagan, the "original zodiac" was the so-called Hypsomatic, or Egypto-Babylonian zodiac, which was sidereal rather than tropical, and in which the initial point was "35° distant from the Pleiades[,] 45° distant from Aldebaran[,] 125° distant from Regulus[, and] 179° distant from Spica[,] which were its fiducial stars."[422] This places the initial point of the Hypsomatic zodiac at what is essentially the juncture between the signs of Pisces and Aries – or more specifically, at 4° 27' Aries, according to Fagan. Using this system as our basis for zodiacal divisioning, in which each of the 12 signs are given an equal section comprising 30°, it will be seen that our current vernal point is located at approximately 5° Pisces. Based on the standard precessional rate of 1° retrograde motion of the vernal point per 72 years, it would thus follow that the "official" date for entrance into the Age of Aquarius is yet some 360 years in the future.

The "Night of Brahma" is described as follows:

> Strange noises are heard, proceeding from every point. . . . These are the precursors of the Night of Brahma; dusk rises at the horizon and the Sun passes away behind the thirtieth degree of Macara (sign of the zodiac), and will reach no more the sign of the Minas (zodiacal pisces, or fish). The gurus of the pagodas appointed to watch the ras-chakr (Zodiac), may now break their circle and instruments, for they are henceforth useless.
>
> Gradually light pales, heat diminishes, uninhabitable spots multiply on the earth, the air becomes more and more rarefied; the springs of waters dry up, the great rivers see their waves exhausted, the ocean shows its sandy bottom, and plants die. Men and animals decrease in size daily. Life and motion lose their force, planets can hardly gravitate in space; they are extinguished one by one, like a lamp which the hand of the chokra (servant) neglects to replenish. Sourya (the Sun) flickers and goes out, matter falls into dissolution (pralaya), and Brahma merges back into Dyaus, the Unrevealed God, and his task being accomplished, he falls asleep. Another day is passed, night sets in and continues until the future dawn.[423]

Whether this apocalyptic description of the Night of Brahma is intended to illustrate a more gradual or a more sudden change in the natural order of life on Earth, or a combination of both, such as what may be experienced as the ongoing reverberations proceeding from some initial cataclysmic (i.e. cometary) event remains unclear; nor was it obviously ever meant to be made clear – at least not during the period in which these things were written.

---

[422] Fagan. *The Solunars Handbook*. Page 7. Bracketed notes added.
[423] Blavatsky. *Isis Unveiled*. Volume 2. Chapter 6. Page 273.

XV THE DEVIL

"Pole Shift"

I am inclined to view Vamadeva-Modeley's description of the Night of Brahma as a metaphorical description of the precession of the equinoxes, as opposed to some sudden and cataclysmic pole shift, the notion of which appears to have been taken up with some degree of credulity. This idea of catastrophic pole shift appears to be the result of the erroneous interpretation and muddling together of various disparate sources of information, such as the aforementioned description of the Night of Brahma, as well as portions of the writings of Herodotus and Plato.

Herodotus, Greek historian circa 500 BCE, writes in his *History* that the Egyptians of his day related to him that

> the Sun had moved four times from his accustomed place of rising, and where he now sets he had thence twice had his rising, and in the place from whence he now rises he had twice had his setting.[424]

This statement by the Father of History has been seized upon by some as clearly referencing a reversal of the poles; that is to say, a shifting of the rotational poles of Earth by 180° through a corresponding displacement of the globe itself in space, such that what is now the north geographic pole becomes the south, and vice-versa – not to be confused with the periodic inversion of magnetic poles which occurs without any significant concomitant motion of Earth's polar geography [Figures 22 & 23].

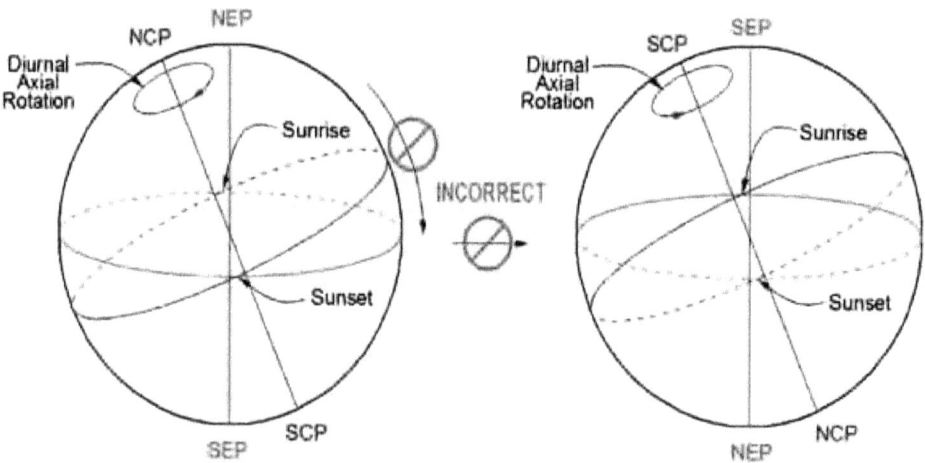

[Figure 22: Incorrect model, involving a geographic pole inversion, which explains how the Sun would appear to rise where it once set.]

---

[424] Herodotus. *History*. Book 2 : Euterpe. Paragraph 142.

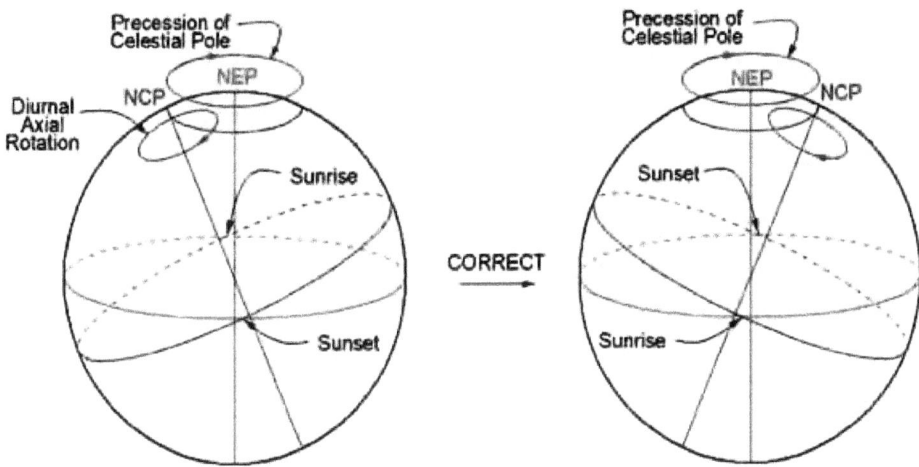

[Figure 23: Correct model, based on verifiable precessional motion, which explains how the Sun would appear to rise where it once set.]

Some of the apocalyptic tinge ascribed to Herodotus' account can be seen to have its roots some four centuries earlier via the writings of such initiates into the Egyptian mystery schools as Plato. To give an example, Plato tells us in his *Politicus* (via the words of the "Eleatic Stranger"):

> There is a time when God himself guides and helps to roll the world in its course; and there is a time, on the completion of a certain cycle, when he lets go, and the world being a living creature, and having originally received intelligence from its author and creator turns about and by an inherent necessity revolves in the opposite direction.[425]

The Stranger then goes on to further describe the effects of this apparent reversal of Earth's rotation:

> Hence there necessarily occurs a great destruction of them, which extends also to the life of man; few survivors of the race are left, and those who remain become the subjects of several novel and remarkable phenomena, and of one in particular, which takes place at the time when the transition is made to the cycle opposite to that in which we are now living…The life of all animals first came to a standstill, and the mortal nature ceased to be or look older, and was then reversed and grew young and delicate; the white locks of the aged darkened again, and the cheeks of the bearded man became smooth, and recovered their former bloom; the bodies of youths in their prime grew softer and smaller, continually by day and night returning and becoming assimilated to the nature of a newly-born child in mind as well as body; in the succeeding

---

[425] Plato. *Politicus*.

XV THE DEVIL

stage they wasted away and wholly disappeared. And the bodies of those who died by violence at that time quickly passed through the like changes, and in a few days were no more seen.[426]

The "resurrection of the dead," spoken of in various places throughout the Bible – most notably by Paul in his Corinthian letters – is very similar to what is described by the Eleatic Stranger in Plato's *Politicus*. In both cases, we see a period of cataclysm and tribulation followed by a period of resurrection and subsequent return to a more or less Golden Age wherein all is made anew and imperishable.

Plato, as an initiate into the Egyptian mysteries was presumably bound at least to some degree by an oath of secrecy, so it would be naïve to think that all of his writings are to be taken at face value, much less to be interpreted literally. Moreover, it is clearly stated by the Eleatic Stranger prior to his recounting of the strange effects associated with the periodic "reversal" of Earth's rotation that it is at some level to be taken as a fairy tale of sorts:

> Listen, then, *to a tale which a child would love to hear;* and you are not too old for childish amusement…There did really happen, and will again happen, like many other events of which ancient tradition has preserved the record…which tells how the Sun and the stars once rose in the west, and set in the east, and that the god reversed their motion, and gave them that which they now have as a testimony to the right of Atreus.[427]

In fact, what appears to be referred to both in the aforementioned passage from Herodotus' *History* as well as the Eleatic Stranger's children's tale is at one level nothing more than the precessional motion of the vernal point throughout the course of the Platonic Year, with particular reference to a point some 14-16,000 years ago (and then, again, some 24-26,000 years prior to that) which was considered a Golden Age. With respect to that time period, it can truly be said that the Sun then rose where it now sets, but only from the perspective of a fixed point of reference. In the words of Massey:

> The priests informed the Greek inquirer [Herodotus] that time had been reckoned by them for so long that the Sun had twice risen where it then set, and twice set where it then arose. This, which was asserted in a previous volume, can only be realized as a fact in nature by means of two cycles of Precession, or a period of 51,736 years.[428]

This point at which "after the completion of a certain cycle," God "lets go" and allows the world to govern itself, without direct intervention of the

---
[426] Ibid.
[427] Ibid.
[428] Massey. *The Natural Genesis*. Book 12. Page 318. Bracketed comment added.

Divine in the affairs of humanity seems to be a reference to the concept of the declining of the spiritual awareness of humanity, the nadir of which is reached during the 3 Winter Ages of the precessional cycle: Pisces, Aquarius, and Capricorn. These "lowest" signs of the Platonic Year, as representatives of the Great Winter can in a sense collectively be referred to as Capricornian, since tropically, this is the sign ascribed to the winter season of the "mundane" 365 day year.[429]

Pan

If we view Nature from the perspective of time-scales and cycles, Pan – as the "all," or Nature's fullness – can be seen to represent the Great Year as a whole, and invoking the Hermetic dictum "as above, so below," Pan also corresponds to the mundane year of 365 days. If from this perspective we reconsider Blavatsky's comment regarding Baphomet (and by extension the Devil trump) that Pan (Nature) had the feet of a goat, and changed himself into a goat at the approach of Typhon, a new insight emerges. Pan – as the entire year – has the feet of a goat, in that his lowermost extremities represent the winter solstitial portion of the year, indicated tropically by Capricorn, the goat.

The changing of Pan into a goat at the approach of Typhon refers to Nature becoming a sacrifice or scapegoat for its own destructive forces – whether seen as the death associated with winter and/or the annual eclipse of the Sun during the course of the mundane year, or the death of humanity's divine awareness during the Great Winter of the Platonic Year, God sacrifices itself unto itself in a ceaseless dance, wherein creation and annihilation are merely 2 sides of a coin that will eventually redeem all.

The Phoenix Cycle

Venus Morning Star – or Lucifer,[430] as she is sometimes called – was known during the 19th and 20th Egyptian dynasties[431] as "star of the ship of the Bennu-Asar."[432] "Asar" is a reference to Osiris, which can be seen to

---

[429] While this may seem to be playing fast and loose with regards to a mixing of tropical and sidereal nomenclature, unfortunately, this is the very thing which has been done in occult literature; a fact perhaps coyly indicated by Blavatsky in her statement that "Every one knows that Capricorn is the…sign of the zodiac into which the Sun enters at the winter solstice…" (*The Secret Doctrine*, Volume 2, Page 576). Clearly, Blavatsky was aware of the phenomenon of precession, and therefore must also have known that the sign the Sun enters during the winter solstice is Sagittarius/Scorpio, the reference to Capricorn being some 2,500 years out of date.

[430] An appellation also ascribed to Mercury, for similar reasons.

[431] Corresponding to a timeframe within the Age of Aries, circa 1292-1077 BCE.

[432] Budge, E. A. Wallace. *The Gods of the Egyptians*. Chapter XIX – Miscellaneous Gods. Section XIII – The Planets and Their Gods.

XV THE DEVIL

represent the constellation Orion, and/or the Sun, depending on the context. Similarly, "Bennu" references Sirius and/or the Moon, so the ship of the Bennu-Asar has a dual connotation wherein the reference can be read as to that of the dyads of Canis Major-Orion, or Moon-Sun.

According to Massey:

> An earlier type of Sirius than the dog was the bennu or nictorax. This was a beautiful water-bird that came to Egypt as a herald of the inundation, and was given the most glorious of extra-zodiacal signs. The bennu was the prototype of the mythical phoenix.[433] In the Egyptian Book of the Dead, the deceased says: "I am the Bennu, the soul of Ra, and the guide of the gods in the Tuat, Let it be so done unto me that I may enter in like a hawk, And that I may come forth like Bennu, the Morning Star.[434]

Here, the term "Morning Star" is likely a reference to Sirius – and possibly inclusive of Venus, as well – as both luminaries shared this appellation.

Both of these examples point to the "Bennu" portion of the celestial ship as referring to Sirius/Canis Major; thus Venus as star of this ship is logically a reference to the heliacal rising of this planet, along with Sirius, during the months of inundation. The following passage from Massey's *Natural Genesis* answers to the other portrayal of the Bennu-Asar as the ship of the Moon and Sun:

> One type of the luni-solar duality of the signs was the "double-seated ship" or boat of the Egyptian gods. This reappears in Babylonia. It is said, "In the month Tebet, Venus is the spark (star) of the double ship." The constellation of the Sea-goat is the zodiacal sign of the month Tebet. Now when the Sun in Cancer entered the ark of Khepra, the beetle, to cross through the three northern water-signs, the Moon rose at full in the Sea-goat to cross the three southern water-signs, and as they made their passage together, although on opposite sides, it was said to be made in the double ship or double-seated bark.[435]

In this case, the reference to Venus as star of the ship is not to its heliacal rising with Sirius in the month of July; rather to its solar proximity in or near the sign of Capricorn in mid-January. In either case, whether this Sun-Venus correlation is considered from the point of view of the summer months of inundation [Figure 24], or the winter months of subsidence [Figure 25], it is only possible for the phenomenon to occur in intervals of 243 years.

This Sun-Venus cycle is not to be confused with the so-called Venus transit, which also occurs every 243 years, wherein the planet can be seen to

---

[433] Massey. *Ancient Egypt, the Light of the World.* Book 5. Page 295.
[434] Budge, E. A. Wallis. *The Gods of the Egyptians.* Chapter V.
[435] Massey. *The Natural Genesis.* Book 11. Page 200.

move across the face of the Sun over a course of several hours. The two phenomena are related, but the respective time intervals are shifted, as can be seen in [Table 1].

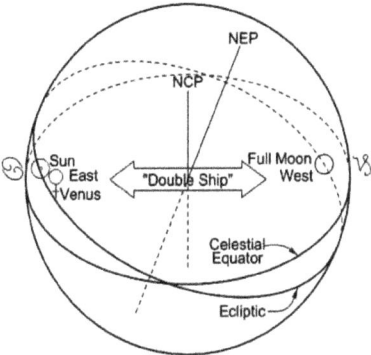

[Figure 24: The luni-solar "double-ship" of the ancient Egyptians, during the summer.]

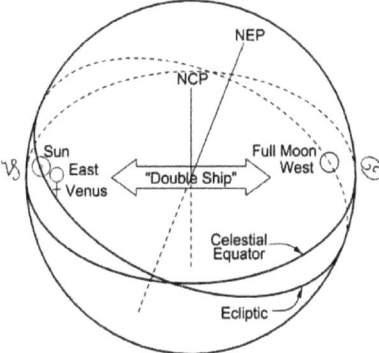

[Figure 25: The luni-solar "double-ship" of the ancient Egyptians, during the winter.]

Venus' heliacal rising w/ Sirius @ Summer solstice	Venus' transit to Sun
6-21-1593	7-24-1593
6-21-1836	7-26-1836
6-21-2079	7-28-2079

[Table 1: Comparison of time intervals for the heliacal rising of Venus and Sirius with the transit of Venus to the Sun.]

The connection between Sirius, Venus, and the phoenix, as evidenced by the aforementioned considerations in conjunction with the fact that the Bennu is the Egyptian phoenix naturally begs the question as to whether this

XV THE DEVIL

period of 243 years may in any way be associated with the lifespan of this mythical creature.

According to Massey:

> The phoenix was one of the great sayers or logoi...There were various phoenix cycles of time...There was a phoenix of the year, of 400, and of 500 years...The Sothiac cycle of 1,461 years was also a phoenix period. And because the bird was but a type, it was further continued as a figure of the Great Year, which enables us to understand the Talmudic legend of the bird over which the angel of death had no power, and which was fabled to have had no fall, because it refused to eat of the forbidden fruit when offered by Eve.
>
> Mr. R. S. Poole identifies the phoenix, the bird of birds, with the bennu, a periodic visitant that alighted in the Nile valley as a herald of the inundation. The Bennu constellation contained the most conspicuous star in heaven, that of Sirius or Sothis, the Dog-star. The bird was set above as the phoenix of the year related to the inundation, and the type was afterwards extended to the Sothiac cycle of 1,461 years, and finally to the Great Year of 25,868 years...Solinus affirms it as a fact well-known to all the world that the Great Year terminates at the same time as the life of the phoenix. Pliny tells us the revolution of the Great Year corresponds to the life of this bird in which (year) the seasons and stars return to their first places.
>
> Cicero had learned that the Great Year of the world extended to the length of 12,954 years. According to Solinus this is the exact period assigned by others to the life of the phoenix; and 12,954 years make one half the cycle of precession calculated with more than common closeness, as it differs only twenty years from Delambre's tables. Here the phoenix of the half cycle in the year of precession can be identified with the bennu or phoenix of Osiris, which marked two points of the Egyptian year six months apart by its heliacal and evening risings. Six months after it opened the year at Sunrise, "the bird" rose in the evening midway in the circle of the year, and this length of time in the lesser year corresponds to 12,934 years, or one half the cycle of precession.
>
> In the Egyptian fixed year founded on the primary four quarters represented zodiacally by the Lion, Scorpion, Waterer, and Bull, the initial point was marked by the heliacal rising of Sothis, with the Sun in the sign of Leo. When Berosus applies the signs of the zodiac to the double ending of the Great Year, he speaks of the conflagration occurring when the planetary conjunction took place in the sign of Cancer, and the flood when the same conjunction occurs in the sign of Capricorn. This shifts the initial point (not, however, as a mere fact in precession) from the Lion to the Crab, or the equinoctial point from the sign of the Bull to the Ram, but, as in the reckoning by the phoenix of 12,954 years, it also recognizes the midway reckoning of the Great Year according to the zodiacal signs of the lesser year and the Two Truths of fire and water, or summer and winter.[436]

---

[436] Massey. *The Natural Genesis*. Book 12. Page 340.

This last part, in which Massey refers to Berosus' conception of the "double ending" of the Great Year reveals several points worth mentioning, the first of which is that the ancients apparently conceived of the precessional cycle as being divided into two halves separated by 12-13,000 years, each of which was punctuated by some major cataclysm. Secondly, as the first worldwide disaster of flood has already been recorded universally to have occurred sometime in the distant past, we must take Berosus' mention of the associated planetary conjunction of Capricorn to have occurred when the Sun was in Cancer [Figure 26], thus placing the future date of the apocalyptic conflagration sometime during the Age of Capricorn, when the aforementioned planetary conjunction is opposite the Sun, in the sign of Cancer [Figure 27]. Lastly, we must recognize the fact that either intentionally, or due to incomplete knowledge regarding such matters, the ancients never gave a specific date as to when either of these diametric points of cataclysm could be localized, with the range of variability stretching into the order of 2,000 years or more, as is evidenced by Massey's comment regarding the shifting of the "initial point" (of the Great Year) from Leo to Cancer, "or the equinoctial point from the sign of the Bull to the Ram."

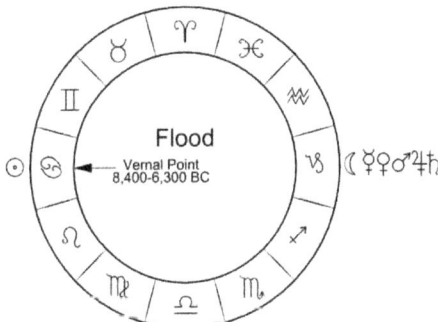

[Figure 26: The astronomical conditions of the "Great Flood" as described by Berosus.]

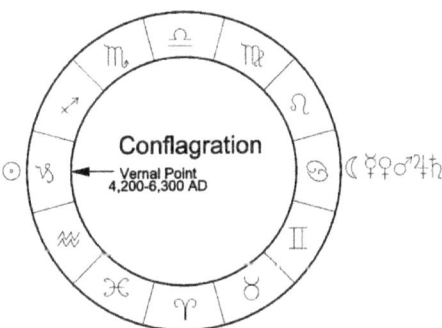

[Figure 27: The astronomical conditions of the "Great Conflagration" as described by Berosus.]

XV THE DEVIL

The fact that Massey mentions the phoenix[437] as associated with at least 5 different scales of time cycles (1, 400-500, 1,461, 12,954, and 25,868 years), as well as the inextricable connections of this mythic bird to Sirius and Venus suggests that whatever timescale is being considered, these two luminaries should always be given the utmost attention when ferreting out the various levels of lost or hidden gnosis buried within the mythic language of the ancients.

Another such cycle which comes to mind, and is directly relevant to the trump under consideration is the 8 year period comprised of 5 separate conjunctions amongst Venus, Earth, and the Sun, with respect to some fixed zodiacal point. This cycle – which repeats every 8 years – forms the familiar pentagram, a symbol around which has accreted multiple layers of associations over the millennia.

The association of the inverted pentagram with the devil, and the dark forces of perdition can be seen to have its fundamental roots in the 8 year pentagonal conjunction cycle of Venus as seen in relation to the so-called "Two Truths" concept, and its application to the precessional cycle of the Great Year. This is due to the fact that throughout the course of this 24-26,000 year period, as the vernal point slowly shifts in a retrograde fashion from sign to sign in the zodiac, so too can the Venusian pentagram of conjunction be seen to shift accordingly. In a 25,920 year precessional cycle, there are a total of 3,240 such 8 year "Venus conjunctions," and it therefore follows that halfway through, at conjunction cycle 1,620, the orientation of the pentagram will have gradually shifted to the point where what once was the apex of the star has become inverted to occupy the space between the legs of the original pentagram referenced some 12,960 years prior [Figure 28]. The inverted pentagram depicted above the head of the Baphometic figure in this trump can thus be seen as a reference to the nature of this portion of the Great Year, which is associated with a spiritual "winter" spanning several thousand years.

To summarize a few key points:

---

[437] Another suggestive example of the phoenix cycle is given by Plutarch, who cites 972 years for its lifespan in "The Cessation of Oracles" (see Plutarch's *Moralia*). This figure is arrived at by solving the following riddle: *"Nine generations lives the noisy Crow / Of lusty men: four times the crow the Stag. / Three stags outlives the Raven: but the Phoenix / Nine times the raven: ten phenices we / The long-haired Nymphs, daughters of mighty Jove."* As Plutarch gives a value of one year for the term "generation," the result is 972 years for the life of the Phoenix, and 9,720 years for the life of the Nymphs. This corresponds to exactly 4 Venus/Sirius heliacal cycles, and each 243-year portion can thus be considered to represent a stage of its life, akin to the 4 seasonal divisions of the year, albeit at a larger timescale. The last such cycle occurred in 1836 CE, and the next will take place in 2079 CE. Where in the lifespan of the current "Phoenix" this puts us is a question that merits further investigation.

[1] The ancients used allegory (which has become mythologized) in order to represent what to them were the natural truths pertaining to multiple cyclical levels of time; the four main cycles being the day, the month, the year, and the Great Year.

[2] Typhon represented an adversarial force of Nature, associated at various temporal levels with cometary intruders, the yearly typhoon season, and the periodic phenomenon of eclipse – both lunar and solar.

[3] Amphitrite was associated with both Venus and Sirius, specifically in relation to the 243-year Sothic cycle of that planet's heliacal rising during the season of inundation. This cycle is also connected to the so-called phoenix cycle.

[4] The pentagram was associated with Venus by virtue of the pentagonal geometry resulting from the conjunction amongst Earth, Venus, Sun, and any fixed zodiacal point over the course of an 8-year period.

[5] The pentagram's association with Sirius is explained by the heliacal rising of Venus with Sirius at the beginning of each 243-year phoenix cycle.

[6] The bidirectional representation of this pentagram refers to diametrically opposed periods during the cycle of the Great Year, as what is considered the "apex" of the pentagram is observed to shift in relation to what is used as the fixed zodiacal reference point.

[7] The inversion of the pentagram alludes to the "original" vernal point of the Great Year being in the sign of Cancer, which corresponds to the circumpolar astronome of Corona Borealis.

[8] This inverted pentagram and its placement above the head of the Devil, or adversary, refers to an Age of "darkness," as opposed to an Age of "enlightenment."

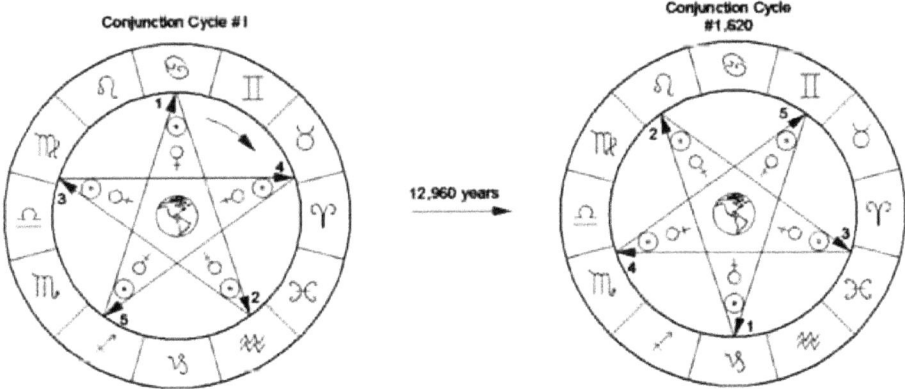

[Figure 28: The inversion of the Venusian "pentagram of conjunction" in the course of precession.]

XVI The Tower

WAITE CITES AN alternate title to this card as the "Tower of Babel," and indicates that the two figures falling from it are Nimrod and his minister [Figure 1].[438] There is an Arabic tradition that considers Nimrod to have been reincarnated as "Kai-Kaus," another name for the constellation Cepheus.[439] A corollary to the Arabic assignation of Nimrod to Cepheus can be seen in the Hebraic legends, which depict Pegasus as the horse of Nimrod.[440] Referencing a celestial sphere will show that any rider of this astral horse would fall within the longitudinal constraints of Cepheus [Figure 2].

[Figure 1: The Rider-Waite Tower trump.]

Nimrod is generally considered to be the chief architect of the Tower of Babel, and according to Massey, the tower in the legend refers to the celestial pole, and its destruction to the shifting of polestars throughout the course of

---
[438] Waite. *The Pictorial Key to the Tarot.* Part I. Section 2.
[439] Elliot, H. M. *The History of India as Told by its Own Historians – The Muhammadan Period. Posthumous Papers of H. M. Elliot.* Edited by J. Dawson. Volume VI. London, 1875. Appendix 5.1; 677.14
[440] Allen, Richard Hinckley. *Star Names: Their Lore and Meaning.*

precession.⁴⁴¹ The confusion of languages may refer to the necessity of readjusting the astronomical relationships of the planisphere according to the new arrangement resulting from the precessional shift, or as a result of northward migration.⁴⁴² This task is taken up by Moses as lawgiver, who supersedes Nimrod as anthropomorph of the constellation Cepheus. One aspect of this re-establishment of celestial order may be mythologically rendered in the story of Moses raising his serpentine staff in the wilderness for the northward bound Israelites circa 1543-1292 BCE,⁴⁴³ as the celestial pole becomes more predominant as a figure of supreme heavenly equipoise the further north (or south) one migrates [Figure 3].

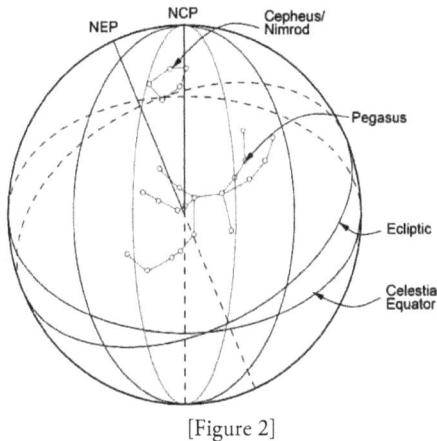

[Figure 2]

There is a strange allusion to leprosy in the story of Moses and the serpentine rod that may be explained as a reference to the stellar cult of Sut-Typhon, which gave predominance to the circumpolar genetrix Typhon (Ursa Major) and her son Sut (Ursa Minor) in their reckoning of time. These "worshippers" of Sut-Typhon can be traced back to the Age of Taurus, circa 2,300 BCE,⁴⁴⁴ and they were later to become associated with outcasts and most notably *lepers* during the period of the Exodus, when the mode of

---

⁴⁴¹ Massey. *Ancient Egypt, the Light of the World.* Book 9. Page 587: "The Tower of Babel was a symbol of the pole which had been overthrown or shifted by the waters of the deluge. To build the tower, then, was to replace the pole. The tower was the Babylonian Bab-illu, which the Hebrew writer has turned into the tower of 'babble' and confusion. The story itself is found on an Assyrian tablet in the British Museum, with this difference: In the older legend the structure is a mound, whereas in the Hebrew version it is a tower built of brick."

⁴⁴² Massey. *A Book of the Beginnings.* Section 20. Page 504: "The dispersion of language will be shown to represent figuratively the naming of places in the planisphere, which entered a second phase under Kepheus or Nimrod. There can be no difficulty in identifying Nimrod as Shu or Kepheus, Egyptian Kafi."

⁴⁴³ The Exodus is said to have occurred occured during 18th Dynasty. See Massey. *A Book of the Beginnings.* Section 18. Page 412.

⁴⁴⁴ Massey. *A Book of the Beginnings.* Section 18. Page 372.

chronometry was becoming primarily luni-solar, as opposed to luni-stellar. When Jehovah tells Moses to place his hand into his bosom and it alternately becomes leprous and whole again, we can read this as a reference to the changing position of the pole. During the Taurean Age, preeminence was ascribed to the circumpolar region by the typhonian cult, and as the feminine and stellar eschatology became anathemized, many of the negative aspects of life at the time – such as rampant leprosy – became associated with the followers of the Genetrix. When Moses' hand becomes unblemished again, the new polar station is established, along with the new system of temporal reckoning in which the mode is primarily luni-solar and patriarchal.

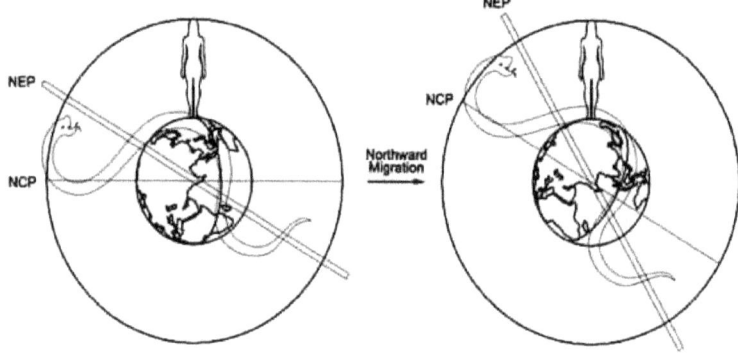

[Figure 3]

During this period, the sign of the inundation was transiting from Leo to Cancer, and the acronychal regulus Cepheus began to sink down below the horizon; thus the Tower of Nimrod fell [Figure 4]. Descending from the upper vault of heaven to take the new place as solstitial marker is Zeus' eagle, Aquila. Above this is the constellation Hercules – son of Zeus, who we have previously identified with the NEP. This Hellenized version of the regulus of inundation is a result of the shifting of cultural predominance from Egypt to Greece concomitant with the changing of the pole station from the regions of Ursa Major to those of Ursa Minor. Zeus then becomes the new regulus for the timing of the solstice, and the fact that this god is master of storms, thunder, and lightning is undoubtedly connected at some level with the knowledge that the yearly Egyptian inundation resulted from the summer storms of Equatoria.

There is a certain illustration depicting Jupiter and the Tower of Belus, after the late 16th/early 17th century Dutch engraver Crispijn de Passe the elder,[445] that may in fact illustrate such a concept [Figure 5]. This may be

---

[445] Original engraving after Dutch draftsman, Maarten de Vos.

interpreted as a pictorial representation of the acronychal rising of Aquila, Cygnus, and Lyra during the period of inundation when the solstitial point had moved from Leo to Cancer. Here the three great birds that draw the chariot of Zeus correspond to the three aforementioned constellations, associated with the eagle, swan, and vulture, respectively. The cloud upon which the company rides answers to the Milky Way, and the tower below may represent the constellation Aquarius (or a portion of Aquarius and Pegasus) as the ecliptic equivalent to Cepheus [Figure 6].

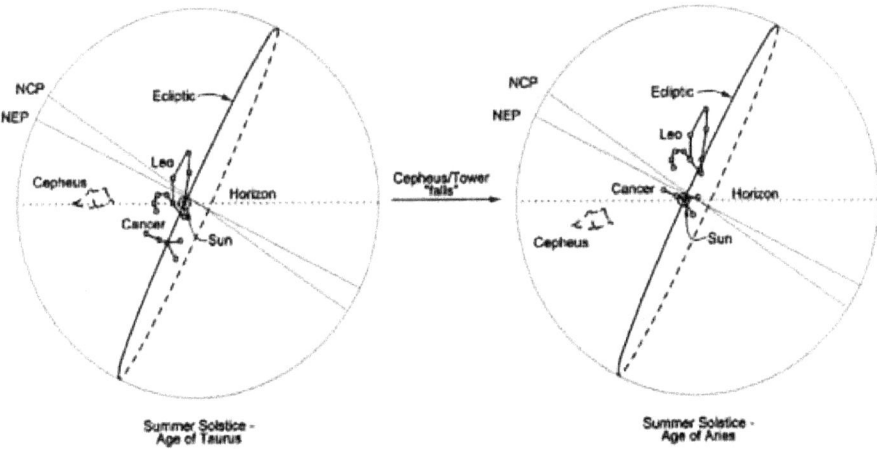

[Figure 4: The fall of the Tower of Nimrod = Cepheus.]

Now the obvious symbolism in this picture is in reference to Jupiter as astrological ruler of both Pisces and Sagittarius, however there is considerably more that can be read into the imagery. The tower depicted in the center is clearly a reference to the ziggurat of Marduk, which was said to have been erected within the sacred precinct of Zeus Belus in Babylon.[446] One of Zeus' roles was as a chronometric functionary of the horizon; that is, as a timekeeper for the celestial cycles. He was an anthropomorph of the planet Jupiter on either horizon – a form of the "star of the double horizon."[447] Belus is identified as "the inventor of sidereal science," and personification of the Dog-Star, Sirius.[448] Marduk fulfills a similar function as Zeus, as the Har-Makhu of the double horizon, albeit in the guise of the planet Mars. Furthermore, Marduk is considered as another appellation of Nimrod.[449]

Considering these significant points, it is evident that the destruction of the Tower of Belus refers to the ending of some cycle involving the

---

[446] Herodotus. *History*. Book 1. Verse 181.
[447] Massey. *The Natural Genesis*. Book 8. Page 522.
[448] Massey. *A Book of the Beginnings*. Section 15. Pages 225-226.
[449] Ibid. Section 20. Pages 489-507.

appearance of Sirius on the horizon, in conjunction with (or opposition to) both Jupiter and Mars. In the present instance, it is suggested that this astromythological story refers to the submergence of Sirius below the horizon during the summer solstitial flood season, as the vernal point shifted from Taurus, to Aries, to Pisces.

[Figure 5: Jupiter and the Tower of Belus.]

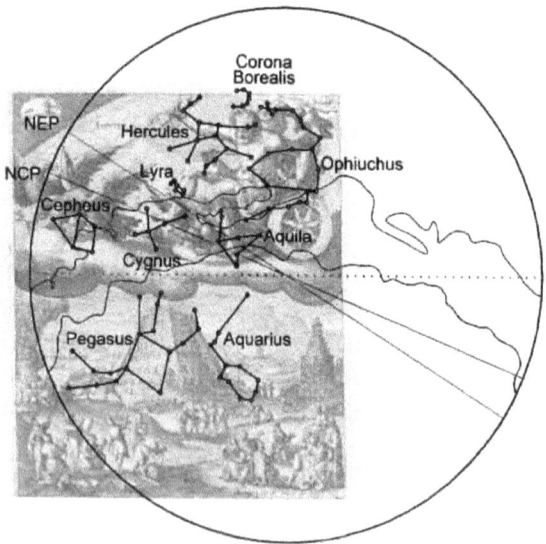

[Figure 6]

Thus, During the Age of Taurus, Cepheus was the acronychal circumpolar regulus, along with its ecliptic counterpart, Cor Leonis, each "governing" or "announcing" the commencement of the inundation during the summer months: the heart of the lion rising heliacally and the ape Kafi (Cepheus) venturing forth nocturnally, as the lion enters Amenta. As the vernal point shifted from Taurus to Aries, the sign of the inundation was transiting from Leo to Cancer, and Cepheus, the Tower of Nimrod, began to sink down below the horizon [Figure 4].

Yet another interpretation of this imagery is the depiction of Jupiter as savior of humanity from the nadir of the Great Year, with its concomitant "confusion of tongues" – the extents of which are bounded by the Ages of Pisces and Sagittarius.

If we consider Leo, or the juncture of Leo/Cancer to mark the "height" of the Great Year, the corresponding nadir is to be found in the Saturnian depths of Aquarius/Capricorn. This back-to-back rulership by Saturn of these two successive Ages may be referenced in the myth of Saturn eating his children (which can be said to occur as early as the Age of Aries, due to the association of Sevekh-Kronus with Saturn, instead of Mars). After this interminable period of leaden authoritarianism, greedy Saturn is finally overcome by the ascending Jovial influences of the Sagittarian Age; disgorging his children after defeat by Jupiter. In this sense, the degenerative influences of the Saturnine valley of darkness – associated with a general state of confusion, strife, and the insatiable drive towards conquest – correlate to the ultimate futility of any attempts made by humanity to unite harmoniously during this Winter of the Great Year.

According to Waite, this trump can be considered to "…signify…the end of a dispensation…"[450] and as has been indicated in the Wheel of Fortune chapter, the most fitting value for this trump is that of the Age "destroyed," prior to the beginning of a new Age. In our present time period, as we draw to the close of the Age of Jupiter-Pisces, and transition into the Saturn-Aquarian Age, the corresponding planetary attribution to this trump would therefore be Jupiter.

De Gébelin connects Herodotus' account of the story of Rhampsinitus to this card.[451] Rhampsinitus was said to be an ancient Egyptian king who amassed a treasure so vast he erected a tower around it for storage and safekeeping. He alone possessed a key to the single entrance into the tower. However, the tower's architect had covertly designed a revolving stone in the base of it, whereby one might secretly enter. This knowledge was never

---

[450] Waite. *The Pictorial Key to the Tarot.* Part II. Section 2; The Tower.
[451] De Gébelin. *Monde Primitif.* Volume 8, Book 1: The Game of Tarots.

XVI THE TOWER

disclosed to anyone, until on his deathbed, he revealed the workings of the door to his sons. Upon their father's death, they stealthily enter the tower, and systematically begin plundering its wealth. The king notices the diminishing treasures, and contrives a trap, which ensnares one of the brothers on their subsequent return. The ensnared brother resigns himself to his fate of death, and suggests his sibling cut his head off, thereby removing any possibility of identifying him. Upon finding the headless body, the king orders it to be hung up outside the palace walls in an attempt to provoke reaction by the deceased's family, thereby ensnaring them. The dead thief's mother learns of this, and persuades the surviving brother to go and retrieve the body. This he does by coming to the palace gates, driving asses before him laden with goatskins of wine. The wine is used to inebriate the guards at the gate, enabling the brother to obtain the body and return it to his mother.[452]

These motifs are similar to a Buddhist account of Indra (i.e., the Sun) who pursued Râhu, the dragon of eclipse, with his thunderbolt. Râhu had "smuggled himself into the presence of the gods," and drank of their amrit nectar of immortality. Indra beheaded him, but was unable to kill him. He subsequently placed the two halves of Râhu's body in the heavens as "time-symbols to represent the ascending and descending nodes of the moon on which the lunar eclipses depend."[453]

In the story of Rhampsinitus, the two brothers can likewise be seen as referencing the Sun, Moon, and lunar nodes. The gradual diminishing of the tower's treasure refers to the same "eating of the light" that the dragon of eclipse has been reproached for Age after Age. The ensnared brother can be seen as a symbol of eclipse, wherein the Sun is in one of the lunar nodes, and the Moon is in the other. The Solar brother is persuaded to return for his sibling's body, and rides to the place where it was hanging (i.e., the opposite lunar node). The asses and goatskins of wine are clearly references to the constellations Cancer (the ass) and Capricorn (the goat). The whole story in some way appears to be connected to a lunar eclipse where the North node of the Moon (the dragon's head) was in one of these two signs, and the South node in the other. In this case, the diminishing treasure may refer both to the darkening of the Moon during a lunar eclipse, as well as the decrease in the amount of daylight in the winter months. For example, as the days grew shorter (the treasure diminished) the Sun was making its way into Capricorn. Upon reaching this sign (which happened to be coincident with the South lunar node) the lunar brother "loses his head"; i.e., the Moon is eclipsed. 6 months later, the Solar brother comes driving asses before him (i.e., moves

---
[452] Cf. Herodotus' *History*. Book 2. Chapters 121-124.
[453] Massey. *The Natural Genesis*. Book 6. Page 343.

into the sign of Cancer), carrying the Dionysian wine in goatskins to commemorate the forthcoming condition of the visible fullness of the Moon in the opposite sign.[454] The Moon's body (its fullness) is returned at the place where the Mother (i.e., Sirius) resides.

There are other possible interpretations of this story – the point is that some astromythological connotation involving the constellations Cancer and Capricorn, the Sun, Moon, and Sirius was likely intended. It would not be surprising if this tale is yet another example of a derivation of the inundation myth that has as its basis the heliacal rising of Sirius in the Age of Aries, as during this period the solstices were localized in Cancer and Capricorn. According to de Gébelin:

> This tale brings back another fact which proves ... that statues of giants that appear in various festivals almost always designate the seasons. ... Rhampsinitus ... caused to be raised in the north and south of the temple of Vulcan two statues of twenty-five cubits, one titled Summer and the other Winter: they adored the one and sacrificed, on the contrary, to the other.[455]

Case suggests the Tower to be representative of the force that turns the Wheel of Fortune.[456] This is in keeping with the present interpretation of the Tower as the force of precessional change that brings about new modes of living, peculiar to each successive Age.

This force of change is active at the individual level via the epiphanic influence of the subconscious mind: one receives a brilliant flash of understanding "out of nowhere"; the lightning strikes, knocking the "crown" of the established mode of thinking from its lofty perch.[457]

The figures toppling from the structure are said to correspond to the chained lovers in the Devil trump.[458] These have previously been identified with the ecliptic and celestial axes, but at the level of the individual human they are said to represent the conscious and subconscious minds, or the Purusha and Prakriti.[459] The female is clad primarily in blue (subconscious) and the male has a red cape (conscious), but each figure wears both colors, signifying that the conscious mind has elements of the subconscious, and vice versa. Where once the subconscious and conscious minds were concealed

---

[454] Since the Moon's nodes retrograde through the zodiac every 18 years, the condition of total lunar eclipse can only occur twice in the same year at any particular ecliptic locus. In this scenario, since the 1st total eclipse occurred when the Sun was in Capricorn, the 2nd eclipse marks the return of the "body" of the Moon, as from thenceforth the opposition of the Sun and Moon at the same 2 ecliptic loci does not result in total eclipse.

[455] De Gébelin. *Monde Primitif.* Volume 8, Book 1: The Game of Tarots. Ellipses added.

[456] Case. *The Tarot: A Key to the Wisdom of the Ages.* Chapter 20.

[457] Cf. Ibid.

[458] Ibid.

[459] Ibid.

from each other, in the awakened state, their connection is revealed: "True knowledge makes Prakriti disappear, first as containing Purusha (the I AM), and then as separate from Purusha."[460]

The falling woman is said to be crowned because she represents the subconscious emotions, by which we are unwittingly ruled. Upon realization of this, a new understanding of Will arises. In the normal waking state of consciousness, the tendency is to see Will as an embodiment of individualism suggesting a separation of the part from the whole, but the new conception sees it as an instrument that works in harmony with the larger macrocosm, no longer a force to be "set against the impulse originating in the cosmic Purpose."[461]

---

[460] Ibid.
[461] Ibid.

XVII The Star

THE VISCONTI-SFORZA VERSION of this trump [Figure 1] depicts a woman holding a distinctive eight-pointed star, which bears a striking resemblance to the headdress of the Egyptian goddess Seshat [Figure 2],[462] "mistress of writings."[463] Sheshat was associated with the "stretching-of-the-cord" ceremony of the ancient Egyptians, in which the circumpolar stars were used in order to orient temples prior to their construction. According to Fagan:

> One of the most important ceremonies in the foundation of Egyptian temples was known as Pedjeshes (Pedj – "to stretch," Shes – "a cord") and it forms the subject of one of the chief monumental ornaments in the temples of Abydos, Heliopolis, Denderah, and Edfu. The reigning pharaoh and a priestess personifying Seshat, the goddess of writing, proceeded to the site, each armed with a golden mallet and a peg connected by a cord to another peg. Seshat having driven her peg home at the previously prepared spot, the king directed his gaze to the constellation of the Bull's Foreleg (this constellation is identical with Ursa Major, "Great Bear," and the "hoof" star is Benetnasch, Eta Ursae Majoris). Having aligned the cord to the "hoof" and Spica as seen through the visor formed by Seshat's curious headdress, he raised his mallet and drove the peg home, thus marking the position of the axis of the future temple.[464]

Special significance may have been ascribed to Spica and Eta Ursa Majoris during the Age of Taurus (from whence this ceremony derives) due to the fact that if extended upwards into the circumpolar heptanomis, the arcsegment formed between these two stars terminates at Thuban, in the tail of Draco. As Thuban was the station of the pole during the Age of Taurus, the Pedjeshes may have represented, in part, an ancient method of axially orienting structures to the NCP [Figure 3].

---

[462] Seshat is cognate with the feminine form of Sefekh (earlier Khevekh), consort of Taht (Thoth), and "goddess of the 7 stars." See Massey. *A Book of the Beginnings*. Section 17. Page 344.
[463] Massey. *A Book of the Beginnings*. Section 2. Page 286.
[464] Fagan, Cyril. *Zodiacs Old and New*.

[Figure 1: The Visconti-Sforza Star trump. Author's rendering.]

[Figure 2: depiction of Seshat from the north wall of the Temple of Seti. Abydos, Egypt. Author's rendering.]

The figure in the Rider-Waite version of this trump [Figure 4] is associated with the "Great Mother" by Waite.[465] The Egyptian goddess Ta-urt is a form of the Great Mother. According to Massey:

> The Great Mother, variously named Tiamat, Zikum, Nin-Ki-Gal, or Nana, was not originally evil. She represented source in perfect correspondence to Apt, Ta-Urt, or Rannut in the Egyptian representation of the Great Mother, who, howsoever hideous, was not bad or inimical to man; the "mother and

---

[465] Waite. *The Pictorial Key to the Tarot.* Part II. Section 2; The Star.

XVII THE STAR

nurse of all," the "mother of gods and men," who was the renewer and bringer forth of life in earth and water.[466]

Various other appellations of this mother goddess include "mistress of the horizon," "she who removes water," and "mistress of the water."[467] The ancient Egyptians sometimes depicted this goddess as a hippopotamus; a zootype used to represent the circumpolar region of Draco and Cepheus, as can be seen in the circular zodiac of Denderah [Figure 5].

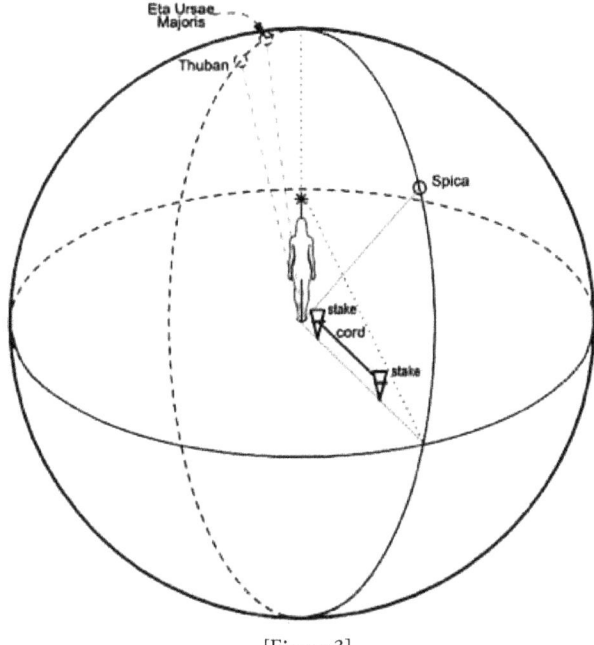

[Figure 3]

This zodiac – dated to August 19th, 16 CE[468] by Cyril Fagan – depicts Ta-urt with her "mooring post," representing the North ecliptic pole. The device resembling the "plow" hieroglyph ("mr," or "hb" – see [Figure 6]), upon which the jackal rests, signifies the celestial pole centered on alpha Ursa Minoris [Figure 7].[469] In this depiction, the head of Ta-urt corresponds to Cepheus, and her torso to a portion of Draco.

---

[466] Massey. *Ancient Egypt, the Light of the World.* Book 5. Page 272.

[467] Jennifer Houser-Wegner. From *The Ancient Gods Speak : A Guide to Egyptian Religion.* Pages 351-352.

[468] The Denderah zodiacs actually contain multiple zodiacs arranged compositely within the same image. According to Fagan: "…these Denderah sky charts are quite composite and international affairs. For example, the larger zodiac shows the ideograms of no less than five distinct New Year Days." Fagan. *Astrological Origins.* Page 92.

[469] Use of the plow hieroglyph should not be construed to reference Ursa Major, as that constellation is depicted as the "thigh" or "haunch" immediately adjacent.

CELESTIAL ARCANA

[Figure 4: "The Star" – Rider-Waite version. Author's rendering.]

[Figure 5: Lithograph of circular zodiac of Denderah, by Karl Landgraf, 1824. From *La Pierre Zodiacale du Temple de Dendérah*.]

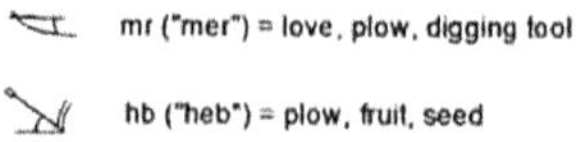

mr ("mer") = love, plow, digging tool

hb ("heb") = plow, fruit, seed

[Figure 6: Hieroglyphs for "plow." Redrawn by author from *An Egyptian Hieroglyphic Dictionary*. E. A. Wallis Budge.]

XVII THE STAR

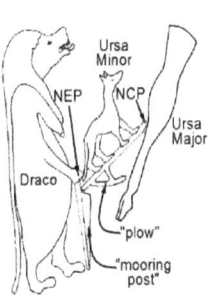

[Figure 7: Circumpolar region of circular zodiac of Denderah. Author's rendering.]

Ta-Urt was later refigured in the constellation Cassiopeia, as consort to Cepheus, regulus of the inundation.[470] When the floods had commenced and the Sun was in Leo, Cassiopeia and Cepheus – Queen and King of the inundation – rose acronychally at sunset, along with Aquarius and Pisces Australis [Figure 8]. The latter two constellations were seen as the "abyss of source" for the floodwaters,[471] which reveals one of the original connections between the Great Mother (Ta-Urt as Cassiopeia) and Aquarius. The Egyptians depicted Aquarius as Hapi, a hermaphroditic figure who poured the floodwaters forth from two ewers [Figure 9], and here the similarities between the Rider-Waite trump and Egyptian deity are evident: the pouring forth of the waters of life onto the earth and into the water echoes the Egyptian conception of Ta-Urt as the "renewer and bringer forth of life in earth and water."[472]

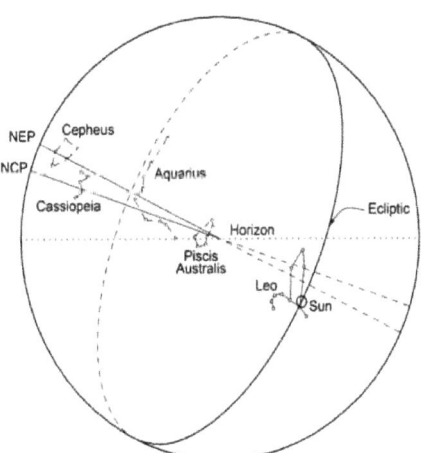

[Figure 8: The rising of the nocturnal Queen and King of the inundation during the Age of Taurus]

---

[470] Massey. *A Book of the Beginnings*. Section 16. Page 273.
[471] Massey. *Ancient Egypt, the Light of the World*. Book 5. Pages 282-284.
[472] Ibid. Pages 271-272.

255

CELESTIAL ARCANA

[Figure 9: Hapi/Aquarius as depicted in the rectangular zodiac of Denderah. Author's rendering.]

Ta-urt appears again in the rectangular zodiac from the temple of Hathor at Denderah [Figures 10 & 11]. Fagan dates this zodiac to April 16, 17 CE, based on the various positions of the planets hieroglyphically represented therein. Note the depiction of the initial waxing crescent of the New Moon over the Taurean bull [Figure 12]. The New Moon is formed when the Sun and Moon begin to disconjunct longitudinally, and the Moon becomes visible as a thin sliver during sunset. This phenomenon marks the beginning of each month and year in the Hebraic calendar,[473] and according to Fagan, the

> longitudes of the planets…for New Year's Day of the Lunar Year, 1st Nissan, which commenced at sunset of April 16, 17 AD at the first appearance of the crescent Moon near the Pleiades…tally perfectly with the position of the planets shown on the rectangular zodiac.[474]

The next detail concerns the representation of both solstitial points within the zodiac. A crowned falcon, representing the summer solstice,[475] is depicted between the pictograms for Sirius (the reclining cow in the "Barque of Isis") and Orion [Figure 13], placing the summer solstice essentially in Gemini, as would be expected from a zodiac from 17 CE. Logically, the antipodal solstice should be indicated somewhere in the region of Sagittarius, and the ideogram consisting of Ta-urt attached by a rope to the "bull's haunch" appears to fulfill this role, as the symbol for Sagittarius is adjacent [Figures 10 & 11].

If the "haunch of the bull" is considered to represent Ursa Major, this ideogram appears to signify the locus of the NCP during the Age of Taurus. The bull/haunch composite figure thus simultaneously represents Ursa Major as well as Taurus, and the spear-wielding Horus can be interpreted to signify the vernal point of the Sun during this Age. The seven stars recall Corona Borealis, which rose acronychally with the setting of the Taurean vernal point.[476] This ideogram may thus represent a kind of moveable marker, that

---

[473] Bromberg. *Moon and the Molad of the Hebrew Calendar*.
[474] Fagan. *Astrological Origins*. Page 92.
[475] Ibid. Page 88.
[476] Another possible reading would equate the seven stars with the Pleiades, which are effectively part of the constellation Taurus.

XVII THE STAR

shifts according to the precessional motion: what was once a symbol for the equinox during the Age of Taurus had come to represent the solstice, some 90 precessional degrees later, circa 17 CE.[477]

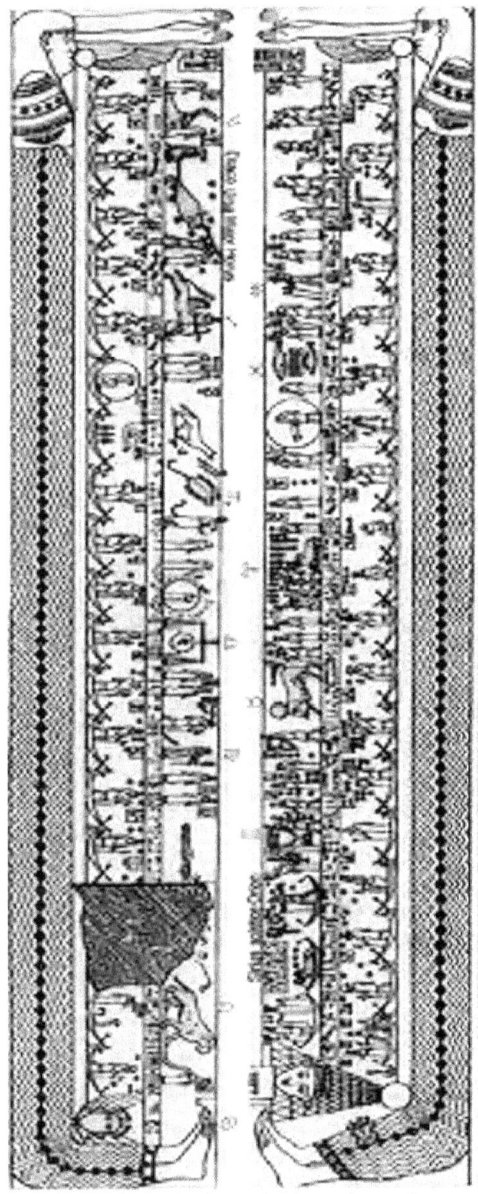

[Figure 10: Rectangular zodiac of Denderah. Author's rendering from J. Bentley's *A Historical View of Hindu Astronomy* (1825), Plate VII.]

---

[477] In the year 17 CE, Corona Borealis could be observed to rise heliacally with the Sun in Sagittarius on the winter solstice.

CELESTIAL ARCANA

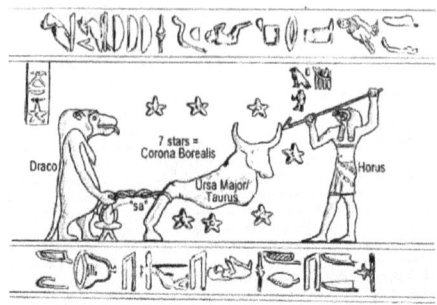

[Figure 11: Ta-urt in the rectangular zodiac of Denderah. Author's rendering.]

[Figure 12: The initial waxing crescent of the New Moon in the sign of Taurus.]

[Figure 13: The crowned falcon of Horus; representative of the summer solstice in Gemini/Orion.]

Examination of the Rider-Waite version of this trump shows that the arrangement of the seven stars around the larger central star very closely resembles that of the Egyptian ideogram, albeit rotated some 90° [Figure 14].

[Figure 14: Comparison of a portion of the rectangular Zodiac of Denderah with the Rider-Waite Star trump. Author's rendering.]

XVII THE STAR

Evidence supporting the hypothesized equivalency between Corona Borealis and the septenary represented in each of these pictograms may be found in Kircher's "Egyptian Planisphere" as presented by Massey in the frontispiece to *The Natural Genesis* [Figure 15].

[Figure 15: Athanasius Kircher's "Egyptian Planisphere," as seen in the frontispiece to Massey's *The Natural Genesis*.]

This planisphere represents a view of the northern celestial hemisphere, centered on the NEP, as seen from "outside" the celestial sphere. The hemisphere is divided up into 12 equal wedges, the outermost portions of which comprise the zodiacal signs (referred to as the various "Regnums," or "kingdoms"), whereas the inner portions towards the center correspond to the circumpolar astronomes, designated as "stanos" – archaic Italian for "digging out," "carving," or by inference drilling – perhaps in reference to the rotation

259

of the Celestial Pole. The most relevant portion of this planisphere in relation to the Rider-Waite trump are the circumpolar regions of sections 9, 10, and 11: Stano [undecipherable] Isidis (pole of Isis), Stano Numinum [undecipherable] ("divine" pole, with the connotation of a "nod of the head"), and Stano Typhonia (pole of Typhon). These three astronomes correspond to the circumpolar regions of Ursa Major (Stano Isidis), Corona Borealis (Stano Numinum, represented here by the horns of the Taurean bull, which are acronychal to Corona Borealis), and Hercules (Stano Typhonia); see [Figure 16].

[Figure 16: Enlarged portion of Kircher planisphere that corresponds to the Rider-Waite Star trump (relative positions reversed)]

It is postulated that the peculiar bird in the tree of the Rider-Waite trump answers to the same motif in the Stano Isidis portion (section 9) of the Kircher planisphere, although in a reversed orientation. The trump's crescent of 7 stars is interpreted to correspond to the adjacent section (section 10), where the arc of the cow's horns provides a different image and connotation for the same constellation: Corona Borealis. As the relative positions of these two adjacent sections of the planisphere are reversed in the Rider-Waite trump, this suggests the perspective of the card to be from "within" the celestial sphere.

The "bird in the tree" motif references the NGP, wherein lies the constellation Coma Berenices – the head of the Empress of trump III. The other tree to the left in the figure of the Rider-Waite Star trump can be seen

XVII THE STAR

to represent the region of Serpenes – a portion of the constellation Ophiuchus, also known as Typhon, thus agreeing with the (reversed) format of the Kircher Egyptian planisphere. If we consider the female figure to represent a version of the Egyptian Hapi, we thus have references to the following asterisms in this single card: Draco, Corona Borealis, Serpenes/Ophiuchus, Aquarius, Coma Berenices, and Sirius, all of which contain stars which are part of the so-called Sirius Supercluster, otherwise known as the Ursa Major Stream, of which "…the star alpha canis major is a well known member, and the cluster Ursae Majoris forms its core."[478] This system was discovered in the late 19th century by R.A. Proctor, who concluded from observations of stellar drift that the stars of the Ursa Major Stream "…form a distinct system moving closely around the center of gravity of a much larger system to which they belong."[479]

Waite associates Sirius and the Dog-Star with the eight-pointed star in this trump,[480] but also states ambivalently "…as regards the seventeenth card, it is the star Sirius or another, as predisposition pleases."[481] Thus, while not specifically assigning this trump to Sirius, he seems to be indicating that the gestalt of the card is in some way fundamentally related to it.

There is a considerable degree of symbolic conflation where the eight-pointed star is concerned. It was used originally in the stellar phase of chronometry to represent Sirius as the star of Ishtar.[482] Later, in the solar phase, it signified the Sun as Horus,[483] finally coming to represent the Star of Bethlehem as symbol of the gnostic Christ.[484] In the present case, it is suggested that its solar connotations be seen to represent the ecliptic pole/axis, residing in the empty space within the convolutions of Draco: the womb, so to speak, of Ta-urt. We see a similar representation of the ecliptic pole in the Sun trump, and in this connection, it should be noted that Waite specifically refers to the symbol of the Sun in earlier versions of that card [Figure 17] as the Dog-Star:

Beneath the dog-star there is a wall suggesting an enclosure-as it might be, a walled garden-wherein are two children, either naked or lightly clothed, facing a water, and gambolling, or running hand in hand.[485]

---

[478] Paulos, J.; Hauck, B. "The Sirius Supercluster." *Astronomy and Astrophysics* 162, Pages 54-61 (1986).
[479] Ibid.
[480] Waite. *The Pictorial Key to the Tarot*. Part I. Section 2; The Star.
[481] Ibid. Part I: The Veil and its Symbols. Section 4 – The Tarot in history.
[482] Massey. *The Natural Genesis*. Book 9. Page 80.
[483] Massey. *Ancient Egypt, the Light of the World*. Book 11. Pages 708-709.
[484] Ibid. Page 709.
[485] Waite. *The Pictorial Key to the Tarot*. Part I. Section 2; The Sun.

CELESTIAL ARCANA

Furthermore, in the Star trump from the *Ercole I d'Este* Tarot Deck (circa 1473 CE), it is evident that the ecliptic and celestial axes are being referenced in the crossed forearms of the pointing figures [Figure 18].

[Figure 17: The *Tarot de Marseille* Sun trump ("Le Soleil"). Image source: Wikipedia]

[Figure 18: The *Ercole I d'Este* Star trump, 15th century CE. Author's rendering.]

XVII THE STAR

Previously, the connection has been made between Osiris and the Sun, as well as the "power of the pole," i.e. the nocturnal facilitator of the passage of the Sun through the underworld.[486] The polestar as nocturnal Sun makes its 24-26,000 year circuit around the central axis of the NEP, and it is proposed that this mechanism serves in part as the basis for the concept of the "Sun behind the Sun."[487] Here the central and invisible/spiritual Sun is manifest as the immutable point about which the visible/physical station of the polestar revolves throughout the course of the Great Year. Whether or not Sirius is a central Sun to our own, Waite's ascription of the eight-pointed star[488] and the Sun (of the Sun trump) to Sirius[489] certainly invites the interpretation.

---

[486] See chapter on The Enneagram and the Tree of Life.
[487] Adams, Marshall. *The Book of the Master.* Pages 141-142.
[488] That is, the eight-pointed star of the Rider-Waite Star trump. See *The Pictorial Key to the Tarot.* Part I. Section 2; The Star.
[489] As per Waite's equation of the Sun in earlier Sun trumps with the "Dog-Star."

XVIII The Moon

THE "MOON" DEPICTED at the top of the Rider-Waite version of this card actually represents the conjunction of the Sun and the Moon; the reference being to the "New Moon" and/or eclipse. In the Hebraic calendric system, the beginning of each month is observed at the first visible sliver of the Moon as these two luminaries become disjunct. Similarly, the Hebraic New Year is considered to begin at the first New Moon of spring.[490] According to Waite, the Moon is depicted as waxing,[491] but Crowley sees it as waning[492] and "guarded by Tabu," which is "uncleanliness and sorcery."[493] This may refer to the observance of the monthly "taboo-time" in ancient Egypt, a period of 6 days that commenced as the waning Moon became invisible for 5 days (the "New Moon"), and was associated with menstruation.[494]

De Gébelin indicates that the dogs in this trump refer to the ancient Egyptian conception of the two tropics. These were seen as two dogs, which guarded the path of the ecliptic, so that none of the planets should deviate from their courses too far to the north or south.[495] This double-dog motif may derive from a period during which the jackal was considered a figure of the pole. In the past, when the ancestors of the Egyptians presumably dwelt closer to the equator, both poles were visible, and the two jackals as a pair served as the "openers of the roads in heaven."[496] The baying hounds are suggestive of the ancient Peruvian custom of beating dogs during eclipses to make them howl. Similarly, during eclipses in Greenland, the women would pinch the ears of dogs: if they howled this was considered a good omen, portending that the end of the world had not yet arrived.[497]

---

[490] Bromberg. *Moon and the Molad of the Hebrew Calendar.*
[491] i.e., moving out of conjunction/eclipse.
[492] i.e., moving into conjunction/eclipse.
[493] Crowley. *The Book of Thoth.* Page 112.
[494] Massey. *The Natural Genesis.* Section 12. Page 285.
[495] De Gébelin. *The Game of Tarots.* 1781. (Essay from *Le Monde Primitif*, translated by Donald Tyson.)
[496] Massey. *Ancient Egypt, the Light of the World.* Book 5. Page 275.
[497] Massey. *The Natural Genesis.* Book 1. Pages 46-47.

Crowley ascribes Pisces to this trump, and indicates that this sign "represents the last stage of winter."[498] Clearly, this is not the case if this statement is taken at face value, since the Sun currently enters Pisces after the spring equinox.[499] Crowley may be referring to the tropical zodiac, since he refers to Pisces as the last of the signs,[500] from which it follows that Aries must be the first; i.e., the location of the vernal point. This seems odd, since he appears to have considered it necessary to update the trumps to reflect precessional changes,[501] and presumably gave consideration to the sidereal zodiac in such modifications. One solution to this quandary is to attribute Crowley's reference to winter as applying to the Great Year, in which case the last stages of this "meta-season" would correspond to Pisces.

This card can be interpreted as representing the eastern view of the heavens at the equator, the vantage point associated with the ancient Egyptian mount of the vernal equinox. From this perspective, the two towers correspond to the northern (left) and southern (right) portions of the sky [Figure 1]. If we consider our current vernal point from this orientation, the dogs are suggestive of the hounds of Ganymede (Aquarius).[502] The crayfish would thus represent Pisces, with the waters below corresponding to "the abyss" beneath the horizon, the golden path leading upwards signifying the ecliptic. Waite refers to the crayfish as "the hideous tendency which is lower than the savage beast," that "strives to attain manifestation, symbolized by crawling from the abyss of water."[503] In the present context, this can be seen as a reference to the sinking of Pisces and the abysmal monster Ketos (Cetus)[504] back into the depths of Amenta, in the course of precession. The symbolic references to both Pisces and Aquarius can thus be explained by the fact that the current sidereal vernal point is to be found between the two constellations. It has been suggested that the crayfish represents the "retrograde functioning of the Moon,"[505] but there is another retrocession which is perhaps more relevant to the present reading of the trump, and that is the backwards-drifting motion of precession.

This trump is said to represent midnight,[506] which may be understood within the context of the Great Year, where vernal point Pisces represents the nadir or Dark Age of this cycle. Notable in this connection is the fact that

---

[498] Crowley. *The Book of Thoth*. Page 111.
[499] This is a necessary precondition if we are to consider ourselves as currently existing within the Aeon of Horus, i.e. the Age of Aquarius.
[500] Crowley. *The Book of Thoth*. Page 112.
[501] As an example, see Crowley's treatment of Trump XX (The Aeon) in *The Book of Thoth*.
[502] In addition to the aforementioned tropical and polar connotations.
[503] Waite. *The Pictorial Key to the Tarot*. Part II. Section 2; The Moon.
[504] Cetus was known as the "monster of the deep" (i.e., the abyss). See Massey: *Ancient Egypt, the Light of the World*. Book 5. Pages 278-279.
[505] De Gébelin. *The Game of Tarots*. Presumably, de Gébelin is referring to the retrocession of the lunar nodes of eclipse. Also see Manly P. Hall. *The Secret Teachings of All Ages*. Page 422.
[506] Crowley. *The Book of Thoth*. Page 112.

XVIII THE MOON

Crowley associates Anubis[507] – with whom we have previously identified our current polestar – with this card. As the vernal point shifted from Taurus to Aries to Pisces, the NCP was traveling *pari passu* from the tail of Draco to Ursa Minor (Anubis): the circumpolar locus of the Great Year's "winter," or "midnight."

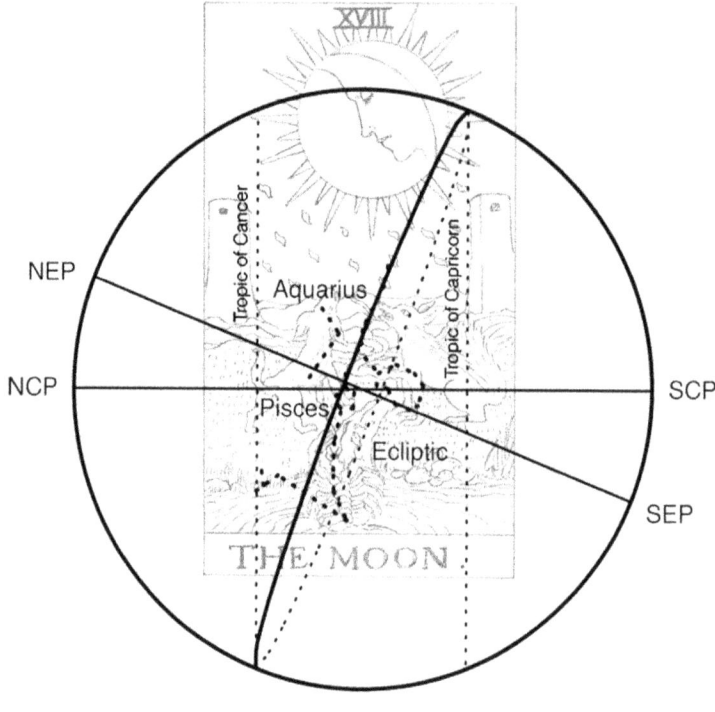

[Figure 1]

The above considerations are essentially macrocosmic. At the microcosmic scale – i.e., that of the individual – the conjunction of the Sun and Moon may be considered to represent the balancing of the rational and intuitive minds through certain meditative or yogic practices.

This card is commonly ascribed the Hebraic letter *Kuf* (*Qoph*), which reputedly means "back of the head,"[508] although there are several alternative meanings, such as "monkey," "to surround," "to touch," "great strength," and "eye of the needle."[509] At the macrocosmic level, the "eye of the needle" refers to the infinite contraction of the Light of God to a single point in order to

---

[507] Ibid.
[508] See Crowley's *The Book of Thoth*, and Case's *The Tarot: A Key to the Wisdom of the Ages*.
[509] Ginsburgh. *The Hebrew Letters*. Pages 280-293.

create the infinite void, thereby allowing the universe to grow and develop within. This void is called the *tzimtzum* in Kabbalistic doctrine.⁵¹⁰ In certain respects the *tzimtzum* is analogous to Prakriti as the feminine aspect of some universal biune deity.

Microcosmically, Prakriti refers to the subconscious. Case associates "back of the head" with the medulla and cerebellum regions of the brainstem. He suggests these control the more atavistic elements of the personality, as well as some of the autonomic physiological processes, and notes that the medulla remains awake when the rest of the brain is asleep.⁵¹¹

Since *Kuf* is anterior to *Reish*, which symbolizes the head proper, it is considered to represent the higher mental states prior to true enlightenment. The subconscious atavistic impulses still have dominance in some circumstances, but in others, cognizance of the subconscious is fully integrated into the waking consciousness, resulting in a higher state of perception. The Solar Force enters the brain, and is reflected by the subconscious. At first, the effects are "automatic" and take place without the conscious awareness of the individual. Later, however, they realize their connection to the macrocosmic Source, and "share consciously in the work."⁵¹²

Kuf (ק) is said to be composed of a *Zayin* (ז) and a *Reish* (ר). The *Reish* portion of the *Kuf* is imagined to hover above the *Zayin*, like the Spirit of God coaxing the individual soul to incarnate upon Earth for further spiritual development. Initially, the soul does not wish to comply out of fear of the lower realms, but the descending "foot" of the *Zayin* indicates the eventual acceptance of the Divine necessity. This vertical element of the *Zayin* is also considered a symbol of the connection to Divine realms the soul maintains while incarnate on Earth.⁵¹³

Case associates the lobster or shellfish in the imagery with Kephra, the beetle-headed Egyptian god, who is a therioceph of Cancer.⁵¹⁴ Crowley includes a scarab beetle in his version of this trump, noting that it is the beetle (Cancer) that "bears the Sun in his Silence through the darkness of Night and the bitterness of Winter."⁵¹⁵ As previously suggested, this can be interpreted to reference the precessional movement of the Sun through the darkest Ages of the Great Year. In the divisioning system suggested by Berosus' account of the Great Conflagration and Great Flood, the Golden Age may be considered

---

[510] Ibid.
[511] Case. *The Tarot: A Key to the Wisdom of the Ages.* Chapter 22.
[512] Ibid.
[513] Ginsburgh. *The Hebrew Letters.* Pages 280-293.
[514] Case. *The Tarot: A Key to the Wisdom of the Ages.* Chapter 22.
[515] Crowley. *The Book of Thoth.* Page 112.

XVIII THE MOON

as either at its height, or just ending when the vernal Sun arrives in Cancer.[516] Conversely, the Dark Ages may be at their nadir (or alternatively, just ending) when it passes into Capricorn. It is at this point that Cancer (or Kephra) is seen rising nocturnally as the vernal Sun in Capricorn sets. In this sense, the beetle is the acronychal counterpart to the Sun, and forms one half of the luni-solar "double ship" that gives the vernal Sun safe passage through the underworld of night at the level of the year. In addition, at the level of the Great Year, the same beetle may be considered as providing safe passage of the vernal Sun through the precessional Winter.

To further entangle matters, due to the tropical fixing of the solstitial points in Cancer and Capricorn, Crowley's statement can be read as referring to the present-day acronychal rising of tropical Cancer (actually Gemini/Taurus) during the winter months when the Sun is in tropical Capricorn (actually Sagittarius/Scorpio).

There is, however, another layer to all of this, which may help explain the apparent conflation between the sign of Cancer and a card that is supposedly connected to Pisces. Massey postulates that *Kuf* developed from the Egyptian hieroglyph *kha*.[517] This was the sign of the womb, figured celestially as the *meskhen*, or birthplace, constellated in the "hinder" or northern celestial regions as Cassiopeia.[518] This symbolism may have arisen as far back as the Age of Gemini. During that period, Cassiopeia and its paranatellons Pisces and Cetus, the monsters of the deep, were seen to rise acronychally to Virgo in the months of inundation. This is the very zodiacal arrangement suggested by the imagery of this trump [Figure 1].

Furthermore, there appears to be an etymological connection between Khepra ("the beetle") and Khep ("the enclosure"). The beetle was considered to be the "clasper," or "encloser," of the Sothic year during the Age of Aries.[519] Prior to this – at least as far back as the Age of Taurus, and probably much earlier – the "encloser" was feminine and circumpolar: she was Khept, the Great Mother whose symbol was the hinder thigh, or *kha*,[520] whether figured as Cassiopeia (during the Ages of Cancer and Gemini), or later as Ursa Major (in the Age of Taurus).

The Khep was the great Pool of the Mother Goddess; the circumpolar well from which the waters of the inundation were drawn.[521,522] In the Ages of

---

[516] See chapter on The Devil.
[517] Massey. *A Book of the* Beginnings. Section 13. Page 122.
[518] Massey. *Ancient Egypt, the Light of the World.* Book 6. Page 394.
[519] Massey. *A Book of the Beginnings.* Section 20. Page 464.
[520] Massey. *Ancient Egypt, the Light of the World.* Book 5. Page 303.
[521] Massey. *Lecture #5.*
[522] It is unclear whether Case was aware of this symbolism of Khep as the great pool. He indicates that the pool from which the shellfish emerges represents "the 'great deep' of cosmic

CELESTIAL ARCANA

Cancer and Gemini, this was represented by Cassiopeia, which rose acronychally during the months of inundation. When the vernal Sun had moved into Taurus, the Khep was constellated as Shu-Anhar, or Cepheus, symbolized by Moses in the later eschatology as the conjurer of water from the stone (the NCP), i.e., the circumpolar acronychal sign of the Sun during the flood season.[523] As the vernal point progressed from Taurus to its current locus (between Pisces and Aquarius), the great dragon-horse and her son (Draco and Ursa Minor) became ever more prominent as the dominant acronychal circumpolar constellations during the summer season. This fact may shed some light on the predominance of the "Great Harlot" (Draco) and Red Dragon with 7 heads (Ursa Minor) in the latest canonical eschatological documents of the West: the Revelation of St. John the Divine.[524]

---

mind stuff," comparing it to the subconscious mind. See *The Tarot: A Key to the Wisdom of the Ages*. Chapter 22.

[523] Massey. *Ancient Egypt, the Light of the World*. Book 10. Pages 666-667. Note that Massey considers the "stone," or "cleft in the rock" to be a figure of the "mount of sunrise" in the East. Whether the "rock" or "stone" is considered as a figure of the pole or the mount of sunrise, the two interpretations signify different aspects of the same thing: the astronomical conditions accompanying the arrival of the season of inundation as viewed from the perspective of the precessional cycle.

[524] See chapter on The Emperor.

XIX The Sun

THE POLAR STATION of the heptanomis corresponding to this most brilliant and potent stellar influence is Hercules. During the 3,300 years when the polestar was localized within this constellation, the vernal Sun was traversing the ecliptic through Leo, the only zodiacal sign ruled by the Sun.

The cherubic figure of the Visconti-Sforza (V-S) version of this trump echoes the shape of this circumpolar asterism, as seen from "outside" the celestial sphere. The Sun held aloft is suggestive of both the solar connotations of the NEP, as well as the NCP during the Age of Leo [Figure 1].

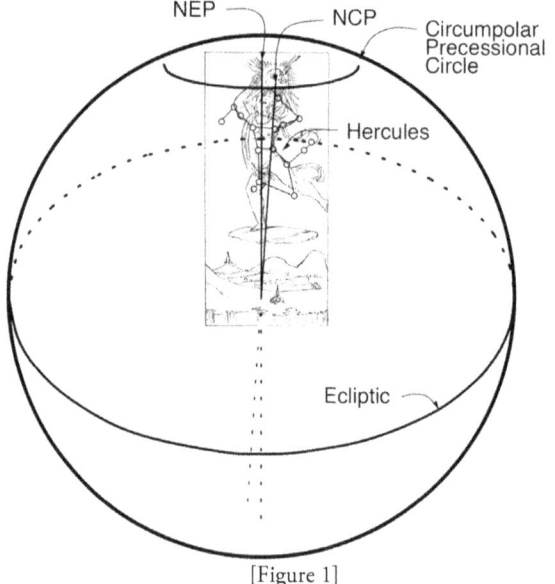

[Figure 1]

The *Tarot de Marseille* (TdM) version of this card depicts twin youths interacting somewhat ambiguously; whether amicably, contentiously, or otherwise, is unclear. Notably, Hercules was himself brother to a twin, Iphicles. As in the V-S version, the composition of the TdM Sun trump echoes the spatial relationship between the ecliptic pole/precessional circle, and the constellation Hercules [Figure 2]. The twin symbolism likely has its roots in the Egyptian concept of the battle between the two solstices or equinoxes as anthropomorphized in Sut and Horus.

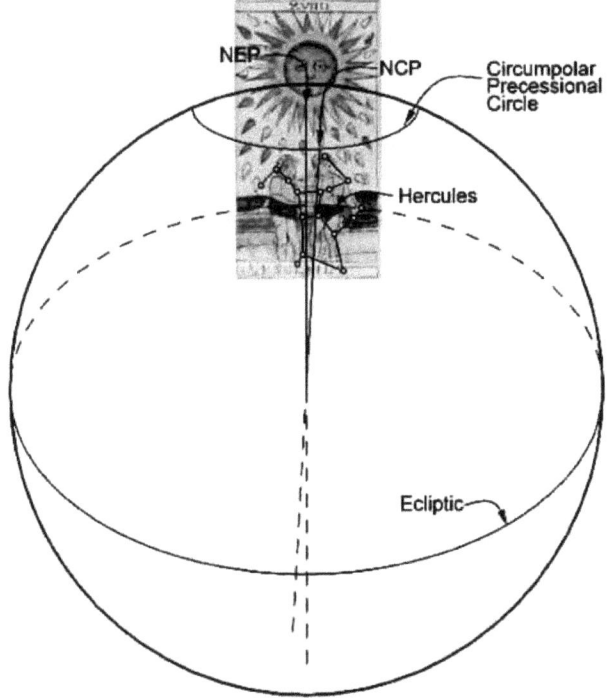

[Figure 2]

Manly P. Hall attributes the twins in this trump to Gemini.[525] When the foundations of our current astronomical symbolism were laid during the Age of Taurus, Gemini was acronychal to Hercules. This may explain a later tendency towards the conflation of certain symbols in the depictions of Gemini in the Medieval Era, such as the various representations pairing the symbols of the sickle and lyre, or the club and lyre [Figures 3 & 4].

The club and sickle are typically associated with Hercules; the former being a reference to the weapon the solar hero used to stun the Nemean Lion, and the latter signifying the device subsequently used to slay the Lernaean Hydra. The lyre equates to the constellation adjacent to Hercules (Lyra), and is associated with Apollo and Hermes/Mercury, the latter of the two being said to have invented it, after which it was catasterized by the former. Note also that in the Flamsteed illustration, one of the twins wields an arrow,

---

[525] Hall. *The Secret Teachings of All Ages*. Chapter XXIX – An Analysis of the Tarot Cards.

perhaps as an allusion to the arrow of Sagittarius, acronychal counterpart to Gemini.

[Figure 3: Medieval Era Image of Gemini from the *Deutsche Fotothek*.]

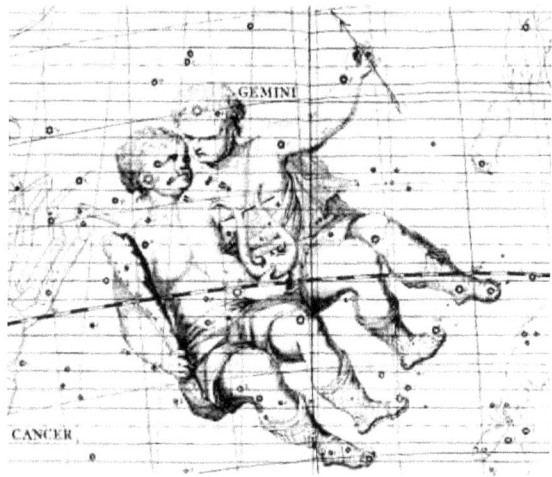

[Figure 4: 18th Century Depiction of Gemini from the *Atlas Coelestis* of John Flamsteed – 1729.]

The conflation of Gemini and Hercules is suggested in the following passage in the section entitled "Twins" by Hyginus in his *Astronomica*:

> These stars many astronomers have called Castor and Pollux. They say that of all brothers they were the most affectionate, not striving in rivalry for the leadership, nor acting without previous consultation. ... Others have called them Hercules and Apollo.[526]

---

[526] Hyginus. *Astronomica*. 2.22: "Twins". Ellipses added.

CELESTIAL ARCANA

Herakles translates to "the glory of Hera." One of the appellations of Hera was "cow-faced," which recalls the cow-headed Hathor of the ancient Egyptians. Herakles is the child Horus emerging as the vernal Sun in Taurus, or in the figure of Khunsu as Full Moon in the antipodal sign of Scorpio. This is undoubtedly due to the fact that Hercules was the circumpolar acronychal counterpart to Taurus. Thus, the Full Moon in Scorpio/Hercules marked the nocturnal passage of the Sun out of the annual darkness of Amenta, represented by the winter constellations. Hercules was the solar glory of Hera/Hathor by virtue of having overcome the drought of winter as he fully illuminated the sign of Taurus, which was the "hat-hor," or "house of Horus."[527] The Egyptian dearth of winter was not associated with the vicissitudes of freezing temperatures, as is the case for more northerly climes. Instead, the hardship came from the subsidance of the Nile; the consequent lack of water each year threatening famine, until the Child Horus came conquering at the vernal equinox. Here the balance was tipped in favor of the forces of light, as the days became longer and Horus grew in strength to his full power as the crowned and conquering king of the summer inundation.

Khunsu is an Egyptian Hercules, and represents the luni-solar form of the conquering child. He is a form of Harpocrates, and the two gods are both depicted with the so-called "infantine lock" of hair. "Khun-su" is "the brave child," or "victor son,"[528] a reference most likely to the birth of the year at the vernal equinox, as Khunsu was associated with the Full Moon of Easter.[529] In the transition from the Age of Taurus to Aries, this annual beginning occurred when the Sun was in the sign of the Ram, with the Full Moon in Libra. In the circular zodiac of Denderah, the child Horus is shown in the disk of the Full Moon, placed between the pans of the balance or scale. This is a figure of Harpocrates.[530]

Crowley touches upon this luni-solar method of demarcating the vernal inception of the year when he refers to the Sun trump as a representation of Heru-ra-ha.[531] This is a dual god comprised of the "extraverted" "Horus of the Horizon" (Ra-Hoor-Kuit) and the "introverted" Harpocrates (Hoor-pa-kraat).[532] Here the "extraversion" applies to the solar aspect of the god as vernal locus of the Sun, and the "introversion" to his lunar aspect as representative of the "nocturnal Sun" in the form of the Full Moon in the opposite sign.

---

[527] Budge, E. A. Wallis. *The Gods of the Egyptians, Volume 2.* Chapter XV – Set, or Suti, and Nephthys.
[528] Massey. *A Book of the Beginnings.* Section 6. Page 219.
[529] Ibid. Section 17. Pages 313-314.
[530] Cole, John. *Treatise on the Circular Zodiac of Tentyra, in Egypt.* 1824. Page 38.
[531] Crowley. *The Book of Thoth.* Page 113.
[532] Ibid. Page 115.

XIX THE SUN

The rationale behind the association of Khunsu/Harpocrates with Hercules and the Moon becomes evident when we consider the fact that during the Age of Taurus, the fullness of the Easter Moon was seen to rise on the eastern horizon in Scorpio, accompanied by the paranatellons of Ophiuchus and Hercules, the Kneeler. On the western horizon, the Sun was sinking below the horizon in the sign of Taurus, or between Gemini and Taurus at the beginning of the Age of the Bull, circa 4,500 BCE. Once again we can identify the celestial mechanism for the previously mentioned conflation between Gemini and Hercules, as the two signs are mutual antipodes, and are therefore cognates to one another, according to the luni-solar mode of chronometry.

With regard to the Rider-Waite version of this trump, if we posit the floral wreath of the child to correspond to the Wreath of Immortality (and hence Corona Borealis), a suggestive series of correlations emerge [Figure 5].

Inverting the celestial sphere such that the north pole of the vertical ecliptic axis points downwards, and taking our vantage point as looking from "within" the sphere, Corona Borealis is seen to fit precisely into the shape of the Child-Horus' headdress. The horse and standard are filled out by Draco (cf. the "dragon-horse" of Tiamat/Ta-Urt[533]), Hercules, and Ophiuchus. The position and angular relationship of the standard's pole and central vertical axis of the Sun overhead are analogous to those of the celestial and ecliptic axes in our inverted sphere. The tip of the standard points to the Galactic Center – between Scorpio and Sagittarius, wherein resides the Southern Crown (Corona Australis). Both the Northen and Southern Crowns can be seen as a complementary polar pair and the two are especially linked in this trump, for during the Age of Leo – when the NCP was in the region of Hercules and the Northern Crown – the SCP was localized in the Southern Crown. High above all of these is Ara, the Altar. This constellation was considered by the ancient Egyptians to represent the "altar of the equinox" during the Age of Taurus.[534] This is explained through the luni-solar conception of the double mount of the equinox, where the Full Moon in the decan of Ara bore witness to the ascent of Horus from Amenta, when in the course of the year the Sun had entered Taurus. The Egyptians viewed Ara as a figure of Mount Hetep,[535] which represented the horizon in the solar mythos,[536] and the circumpolar heavens in the stellar mythos.[537] In particular, the circumpolar Hetep was conceived of as a garden of paradise.[538]

---

[533] Massey. *Ancient Egypt, the Light of the World*. Book 5. Pages 277-278.
[534] Massey. *The Natural Genesis*. Book 11. Pages 201-204.
[535] Massey. *Ancient Egypt, the Light of the World*. Book 12. Pages 786-787.
[536] Ibid. Page 837.
[537] Ibid. Page 325, 349.
[538] Ibid. Page 375.

As noted in the discussion of The Star, Wait's reference to the Sun in this trump as the "Dog-Star" is interpreted as an allusion to the esoteric concept of precession as a function of our Sun's revolution about Sirius within a period of 24,000-26,000 years. This is manifest from a geocentric perspective by the retrograde motion of the NCP about the central NEP, functionally equating the NCP with the Sun, and the NEP with Sirius. Thus, when Waite subsequently refers to this trump as "the destiny of the Supernatural East and the great and holy light that goes before the endless procession of humanity,"[539] it is interpreted as a reference to the ongoing transition of the eastern vernal point from the Piscean to the Aquarian Age, driven by the "holy light" of Sirius, which "goes before" the Sun; i.e., leads the Sun as its lesser satellite throughout the endless cycles of precession, which themselves serve as the matrix for the endless procession of ever-reincarnating humanity.

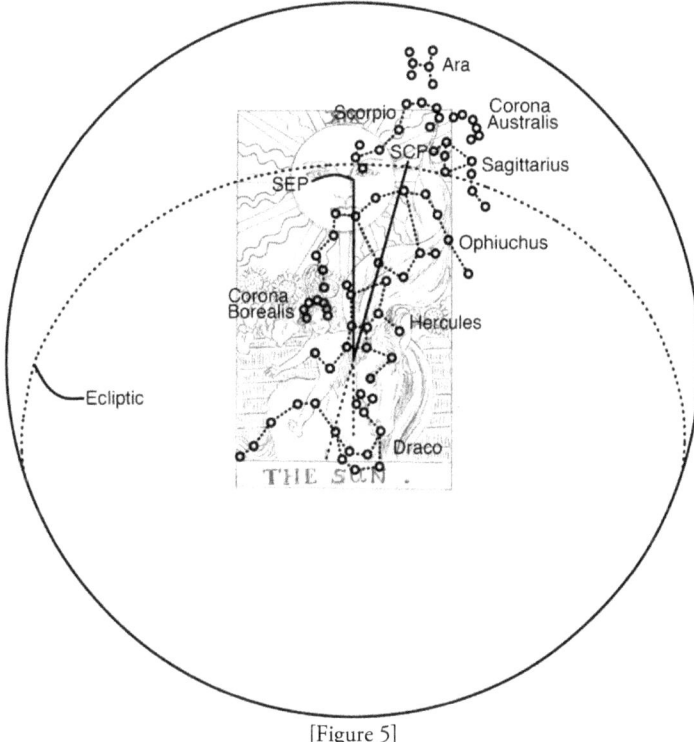

[Figure 5]

In the older versions of this card, there is a wall below the Sun/Dog-Star, which according to Waite is suggestive of an enclosure or walled garden.[540] This is hypothesized to represent the circumpolar garden of Hetep. Crowley

---

[539] Waite. *The Pictorial Key to the Tarot*. Part II. Section 2; The Sun.
[540] Ibid. Part I. Section 2; The Sun.

XIX THE SUN

depicts this as a red wall encircling a green mount/mound, which taken together are a symbol of the Rose Cross,[541] once it has

> completed the fire-change into "something rich and strange"; for the mound is green, where one would expect it to be red, and the wall red, where one would expect it to be green or blue.[542]

The blue is the sky/heaven; the red is the Sun on the horizon/earth.[543] Green is renewal;[544] red is the lower Sun, nether regions, feminine source, and menstrual phase.[545] Another pairing is white as upper, and red as lower, hence the white crown of Horus (of the inundation) and red crown of Sut (of the subsidence).[546] The Rose Cross, or Mystic Rose has been identified as representative of the "lotus of immensity"; i.e., the inverted conical shape formed by the precessional rotation of the NCP around the NEP.[547]

The wall atop the mound is in one sense a figure of the circumpolar "city of the white wall,"[548] where the wall is to be understood as both the outermost limits of the circumpolar region of the stars that never set within any particular Aeon, as well as the circumference of the precessional circuit which rings the central ecliptic pole. The pairing of colors is a reflection of the doctrine of the "two truths," whether considered from an equinoctial or solstitial perspective. In general, we can associate the "upper" with the spring and summer, and the "lower" with autumn and winter. The concept also extends to the "seasons" of the Great Year, and in this context, the implied reversal of "upper" (in this case, green) and "lower" (red) is interpreted to represent the idea that we are currently in the Winter phase of the Great Year.

Crowley mentions Hua (also spelled Hva) in connection with this trump, as representative of the number 12, and the most holy ancient ones.[549] The number 12 is suggestive of the zodiac, ecliptic, and ecliptic axis/pole. The "most holy ancient ones" may be a reference to the ancient Egyptian "*Ali*" (Hebrew "*Elohim*") – the circumpolar "watchers" associated with the keeping of time in the stellar phase of chronometry.[550] In Liber LVIII, Hua (Hva) is said to be cognate with *Keter* and the 12 zodiacal signs,[551] and here the solar references to the NEP and ecliptic are evident. *Keter* is also said to be represented in one form as Corona Borealis, the Kabbalistic "Crown of

---

[541] Crowley. *The Book of Thoth*. Page 114.
[542] Ibid.
[543] Massey. *The Natural Genesis*. Book 3. Page 178.
[544] Massey. *A Book of the Beginnings*. Section 3. Page 95.
[545] Ibid. Section 22. Page 582.
[546] Massey. *Ancient Egypt, the Light of the World*. Book 7. Page 419.
[547] See chapter on Death.
[548] Massey. *Ancient Egypt, the Light of the World*. Book 6. Pages 375-376.
[549] Crowley. *The Book of Thoth*. Page 114.
[550] Massey. *Lecture #5*. Page 116, 124.
[551] Crowley. Liber LVIII, Section IV – 12.

Crowns."552 These references all suggest a particular association of this trump with the solar aspects of the ecliptic and its axis, with special emphasis being placed on the circumpolar region of Corona Borealis.

Hua appears to be a version of the Celtic Hu, solar god of the horizon,553 and the youthful form of the Egyptian Atum.554 This suggests a vernal connotation, and it is notable that in ancient Briton, the Druids would unfurl the red dragon flag of Aeddon (Hu) in celebration of the vernal equinox.555 It is postulated that the red standard in the Rider-Waite trump is a reference to this ancient ceremony of the *Magnum Sublatum* ("Great Upraising").

Case and Crowley attribute the Hebraic letter *Reish* to this trump, and so it is with the currently proposed system. The numerical value of *Reish* is 200, which is related to the Hebraic phrase "the Sun is charity." The Gematrial value of "*Tzadik-Dalet-Kuf-Hei*" (צדקה), or "charity," is 199. According to Jewish law, a man is considered eligible for charity so long as he possesses no more than 199 *zuz*. Once he acquires 200 *zuz* (i.e., the numerical value of *Reish*), the once "poor man" becomes a "*Reish*," or "head" – that is to say, he is in charge of his own affairs.556

According to Hassidic interpretation, *Reish* and *Dalet* are related to each other through "the secret of the *Dalet*." *Dalet* (ד) has an extra "point" in the upper right, which *Reish* (ר) lacks. This point is said to represent the condition of *bitul*, or selflessness, a state that results from the recognition of and participation in the source of spiritual sustenance. The state represented by *Reish* does not have this direct connection to the infinite. Moreover, there are two alternate phonetic spellings for the letter *Reish*: ראש = "*Reish-Alef-Shin*" = "head/beginning," and רש = "*Resih-Shin*" = "poor man." It is said that the *Reish* must internalize the *Alef*, so as to acquire the selfless connection to the Divine – symbolized by the *Dalet* and thereby transcend the state of "poor man," becoming autonomous; i.e., the "head."557

This trump may therefore be seen to represent an intermediate phase of spiritual development, advanced beyond that denoted by the previous card, but not yet completely dissociated from the false ego arising from the misapprehension of separateness. According to Case, this trump represents the state of conscious identification with the "One Life," although not the final grade:

---

552 Massey. *Ancient Egypt, the Light of the World*. Book 9. Page 602.
553 Massey. *A Book of the Beginnings*. Section 8. Page 332
554 Ibid. Page 341.
555 Ibid. Page 342.
556 Ginsburgh. *The Hebrew Letters*. Page 305.
557 Ibid. Page 302.

XIX THE SUN

For though it is a stage wherein all physical forces are under the control of the adept, who, having himself become childlike, realizes in his own person the fulfillment of the promise, "A little child shall lead them" – yet a person who has reached this grade still feels himself to be a separate, or at least a distinct entity. This is not full liberation, though it is a higher stage than any of those preceding it. It is, in particular, the stage in which all physical forces are dominated by the will of the adept, because he is an unobstructed vehicle for the One Will which always has ruled those forces, since the beginning.[558]

This One Will is connected to the *Ruach*, or Spirit of God that is said to originate in the Primal Will, expressing itself in Wisdom. *Ruach* is Kabbalistically related to *Alef*, which Case ascribes to the Fool. In the present system, *Alef* is attributed to the Magician, for he represents the mediating force of the precessional triad, binding the fiery *Ruach* (*Shin* = NEP) to the watery *Mem* (=NCP). We must remember that there is no canon of attributions, and even if there were, it would be a false one, leading to stagnation. It is impossible to fix the universal forces represented by the trumps into some rigid structure, as these forces are dynamic and constantly changing their mutual relationships. Theoretically, there are innumerable possible attribution schemes, but the test for each is how well all of the elements are integrated into a dynamic and harmonious whole.

In one of the preliminary versions of *The Pictorial Key to the Tarot*, Waite indicates the child in this trump to be similar to the Fool, noting that the feathers of both figures represent the same thing.[559] The feather is that of an ostrich, which represents the celestial point of equipoise, both as a figure of the rotational axis of the celestial pole, as well as the equinox.[560] As will be further elucidated in the chapter on the Fool, this system ascribes *Mem* to that card. This is because the Fool is regarded as the "passive" force of the precessional triad, representing the station of the pole throughout the precessional Ages. Thus, in this system, the feathers of both the Fool and Child of the Sun represent different aspects of the same thing: in the case of the Fool, the feather represents the rotational point of equipoise; in the case of the Child, it signifies the equinox.

Waite also connects this trump to the Hanged Man: "For the stage of unfoldment represented by the Sun is the expression of the law the Hanged Man symbolizes."[561] Now, Crowley, Waite, and Case all attribute *Mem* to the

---

[558] Case. *The Tarot: A Key to the Wisdom of the Ages*. Chapter 23.
[559] Ibid.
[560] The ostrich feather is a symbol of equipoise between opposing forces. It was said by Horapollo to have been adopted as such, since the wing feathers of an ostrich are of equal length. It symbolizes the quinox, and is sacred to Maat, anthropomorph of the equinox, and celestial pole. See Massey. *A Book of the Beginnings*. Section 12. Page 53; Ibid. Section 23. Pages 635-636. Also, Massey. *Ancient Egypt, the Light of the World*. Book 9, page 602.
[561] Case. *The Tarot: A Key to the Wisdom of the Ages*. Chapter 23.

Hanged Man, but it should be noted that in the proposed system, *Dalet* is ascribed to that trump.[562] As has just been demonstrated, *Dalet* and *Reish* both bear an intimate Kabbalistic relation to each other. It thus follows that according to this system the relationship between *Dalet* and *Reish* should be demonstrably similar to that of the Hanged Man and the Sun. Previously, we have identified the Hanged Man with a precessional marker, wherein the feet represent the Dark Ages, and the head the Golden Age. But this inverted configuration also represents the state of *bitul*, in that the sacrifice of the Hanged Man is in reality an infinite blessing he receives from the dissolution of false ego.

The *Alef* (Magician) that the *Reish* (Sun) must sublimate to achieve the *Dalet's* (Hanged Man's) selfless connection to the Divine may be compared to the Azoth, or generative substance, energized by the Solar Force. This connection is symbolized by the *Dalet's* additional "point," which can be seen as an appended *Yud*.[563] Additionally, *Alef* is said to be composed of a lower and a higher *Yud* (The Hermit), linked together by a *Vav* (The Hierophant).[564] The Hermit can be seen as the upper *Yud* in this case, symbolizing the transmission of higher gnosis from the Age of Virgo, down into the lower *Yud* to be added to the *Reish* (Sun's vernal point) of the current Piscean Age, in order to potentiate its transmutation into the *Dalet* of the same Age, thereby transforming the inchoate energies of a Dark Age of separation into the harmonious and infinite Light.

---

[562] See chapters entitled "A Proposed System," "The Hermit," and "Judgment."
[563] Ginsburgh. *The Hebrew Letters*. Page 302.
[564] Ibid. Page 29.

XX Judgment

IN THE VISCONTI-SFORZA version of this trump [Figure 1], the top is dominated by a kingly figure, whose overall shape calls to mind the constellation Cepheus, with Gamma Cephei ("the shepherd") corresponding to the head of the king. Continuing this analogy, the crossed trumpets in the middle of the card are seen to represent the constellation Cygnus. The lowest portion – which depicts three figures in a coffin – would thus symbolize Delphinus, as that constellation bears the epithet "Job's coffin."[565] In this arrangement, the king's sword and *globus cruciger* answer to the NCP and NEP, respectively, if our vantage point is taken to be "within" the celestial sphere – with the sword's hilt representing the star Deneb in Cygnus, polestar during the Age of Sagittarius/Scorpio [Figure 2].

[Figure 1: The Visconti-Sforza Judgment trump. Author's rendering.]

---

[565] Allen, Richard Hinckley. *Star Names: Their Lore and Meaning.* (see chapter entitled "The Constellations"; "Delphinus, the Dolphin" entry.

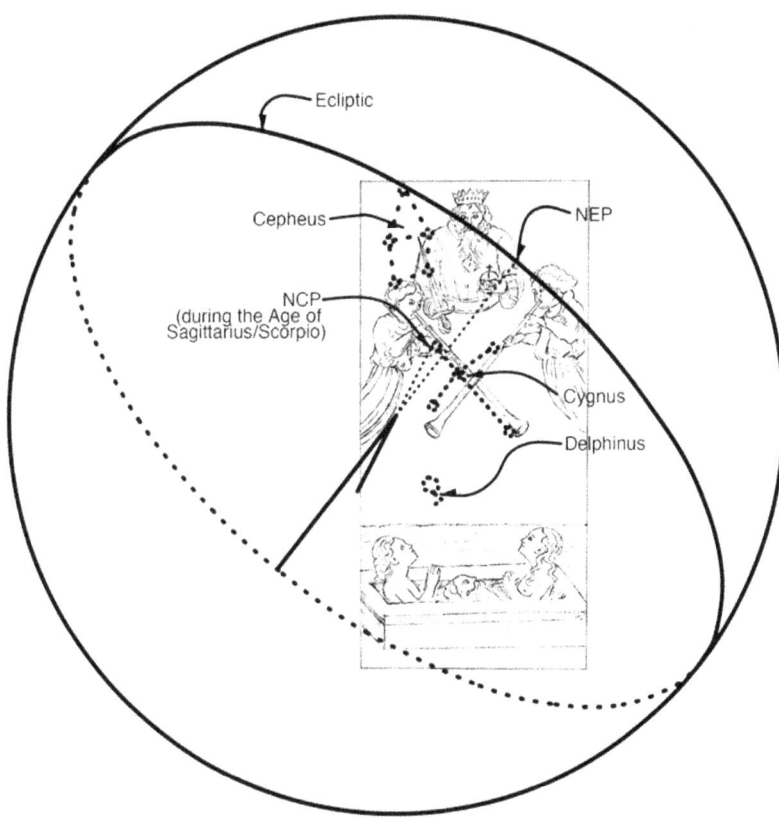

[Figure 2: Proposed celestial symbolism of the Visconti-Sforza Judgment trump.]

In the Rider-Waite version [Figure 3], much of the Visconti-Sforza symbolism remains intact. However, Cepheus no longer appears in the tableau; the composition is split between the angel (Cygnus), and the dead/resurrected (Delphinus). Instead of the more "spatially accurate" method of depicting the constellations used in the Visconti-Sforza version, there are more symbolic allusions: the swan-like wings referencing Orpheus/Cygnus; the Greek cross an allusion to the equinox, as well as the cruciform shape of Cygnus. The angel sounds its trump, signaling the beginning of a new Aeon.

Job's Coffin can be seen as an analog to the coffin of Osiris. In the Osirian mythos, this coffin was associated with the entrance of the Sun into the nether regions of Amenta.[566] During the Age of Taurus, this occurred when the Sun entered the Scorpion, the sign of Sut, who betrays Osiris, shutting him up in the coffin. Massey posits that Job's sufferings equate with the sufferings of Osiris in Amenta.[567] As Delphinus is a paranatellon of Aquarius and the

---

[566] Massey. *Ancient Egypt, the Light of the World.* Book 11. Page 706
[567] Ibid. Book 8. Page 493.

XX JUDGMENT

Southern Fish – which marked the winter solstice, and the darkest depths of Amenta in the Osirian mythos – the equation is evident.

Frances Rolleston refers to the star Deneb as "the Judge or Lord who cometh, in Cycnus."[568] It is possible that this usage harkens back to sometime during the Age of Aries, when Deneb rose heliacally with the Sun at the winter solstice, in the decans of Capricorn [Figure 4].[569] Deneb as the judge, then, is analogous to Atum – judge of the dead in the underworld of winter.[570]

[Figure 3: The Rider-Waite Judgment trump. Author's rendering.]

At a higher scale, the symbolism of Cygnus as judge relates to the vernal equinox of the Great Year. Here Deneb occupies the station of the pole, and is therefore doubly associated with the Judgment Hall of Maat, as the pole was a figure of Maat in the stellar mythos, whereas in the solar phase, the *maat* – and judgment – became associated with not only the autumnal equinox, but the vernal one as well. This so-called "Easter Judgment" was presumably carried out in England as the March Assizes.[571]

Waite refers to this card as "The Last Judgment."[572] As one of the members of the circumpolar heptanomis, Cygnus may be correlated to one of the Angels of the Apocalypse. Since the word "apocalypse" from the Greek

---

[568] Rolleston, Frances. *Mazzaroth*. 1862.
[569] In fact, due to the *pari passu* motion of the NCP with the vernal point throughout the course of precession, Deneb – "the judge who cometh" – can still be seen as the heliacal forerunner of the Sun on the eastern horizon at the winter solstice.
[570] Massey. *A Book of the Beginnings*. Section 8. Page 328.
[571] Massey. *Ancient Egypt, the Light of the World*. Book 6. Page 344.
[572] Waite. *The Pictorial Key to the Tarot*. Part II. Section 2; The Last Judgment.

"apocálypsis" literally means "uncovering," the Age of Sagittarius/Scorpio – with its corresponding polestar Deneb – thus becomes associated with an Age of revelation, as well as judgment. Whatever form this may take in the future, the ultimate cause may be considered to be (at one level) astrological and precessional. According to this interpretation, the rising dead are seen as the Delphinian region – once associated with the Underworld of winter – shifting to occupy the vernal point with Cepheus, Cygnus, Sagittarius, and Scorpio [Figure 5]. What was once the Underworld and winter has become the rebirth of spring; what was once dead reemerges into life.

One of the unique qualities of the region of the celestial heptanomis occupied by Deneb and Cygnus is that it is the only portion residing directly within the Milky Way. This alone is a compelling reason to consider the area a key locus within the precessional cycle. Additionally, during the Age of Sagittarius/Scorpio, the vernal Sun is conjunct the Galactic Center, or Throne of Brahma,[573] in the Hindu system.

Much has been said concerning a similar phenomenon during our current Age, where the Sun is conjunct the Galactic Center during the winter solstice, and of all the associated effects to be expected due to the influx of electromagnetic energies from the center of the Galaxy. This being said, any such influx should be significantly less than the effect of a solar/galactic conjunction during the vernal equinox in the Age of Scorpio, and the effect may be the greatest when said conjunction occurs during the summer solstice, when the polestar corresponds to Hercules, in the Age of Leo. This is due to the variable angles of incidence of cosmic radiations coming from the direction of the Sun associated with the changing seasons.[574]

According to Crowley, this trump is connected to Hrumachis and "the fall of the Great Equinox."[575] Apparently, this "Great Equinox" is a reference to the "equinox" of the Great Year. Hrumachis is cognate with Har-Makhu, god of the double equinox and double horizon.[576] It is unclear why Crowley should associate "the Aeon which is to follow this present one…in about 2,000 years" with the Autumn of the Great Year. Perhaps his use of the word "fall" is not intended to suggest the autumnal connotation, but the passage of the locus of the mundane vernal equinox from the Autumn to the Winter of the Great Year (i.e., the declination, or "falling" of the Great Year's vernal point). In 2,000 years, the vernal Sun will be approaching Capricorn, the last of the 3 Winter signs of this 24-26,000 year cycle. The Autumnal Equinox of the Great Year has already occurred some 6,000 years ago, when the vernal

---

[573] Cf. Sri Yukteswar's *The Holy Science*.
[574] See discussion in chapter on the Domicile System and World Ages.
[575] Crowley. *The Book of* Thoth. Page 116.
[576] Massey. *Ancient Egypt, the Light of the World*. Book 6. Page 334.

XX JUDGMENT

Sun entered Taurus. The next Equinox will not occur until we have passed into the Age of Scorpio, some 6,000 years hence, during the "springtime" of the Great Year, and is therefore associated with the resurrection of the Sun from the underworld on a vast scale.

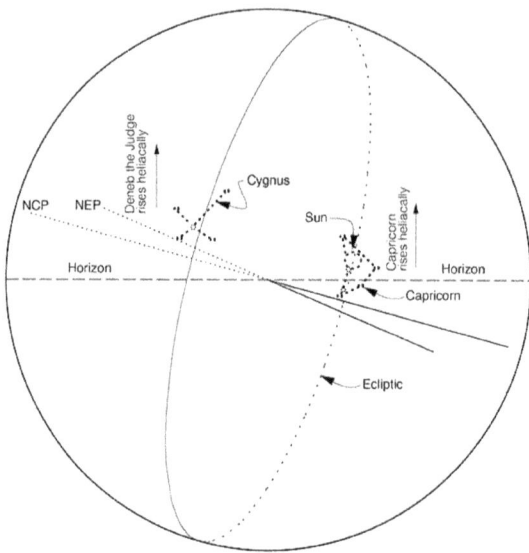

[Figure 4: The heliacal rising of Cygnus at the winter solstice, circa 1000 BCE in Egypt, looking east.]

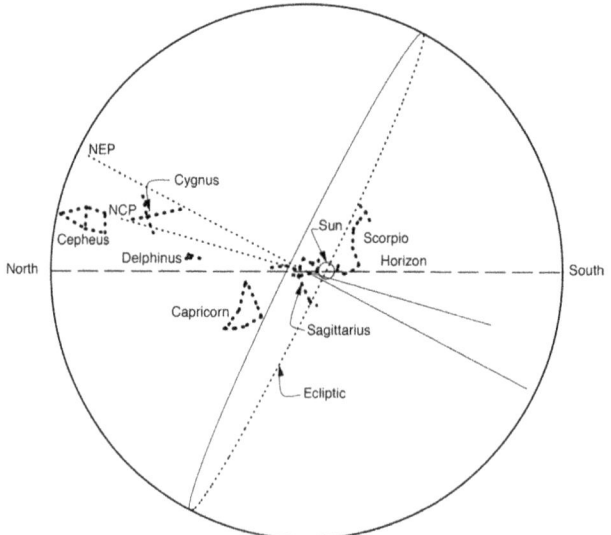

[Figure 5: The heliacal rising of Delphinus at the spring equinox during the Age of Sagittarius/Scorpio, circa 9,200 CE, as seen looking eastward in Egypt.]

Waite notes a definite connection between this trump and Temperance, suggesting that the correspondence is somehow related to the concept of eternal life. Within the precessional framework developed thus far, there is an implicit connection between these two cards that is analogous to the relationship between polestar and vernal equinoctial sign, similar to the previously discussed relationship between the Hanged Man and the Hermit, High Priestess and Chariot, The Sun and Strength. It would logically follow that Mars is the planet to assign to this trump, since the Age of Cygnus is a Sagittarian/Scorpion Age, and Jupiter has already been assigned to the Tower card.

In Roman and Greek mythology, Mors/Thanatos was minister to Pluto/Hades,[577] judge of the dead.[578] The connection between Hades and Mars as ruler of the 2nd house in ancient Egyptian astrology has already been discussed in the chapter on Death, and the association of Mors with Mars is attested to by Remigius of Auxerre in his work entitled *Martianum*.[579] Finally, strange as it may seem as a point of connection, the NCP of the planet Mars currently points to Deneb in Cygnus.[580]

This trump and its ecliptic-based partners Temperance and Death represent – in one aspect – the vernal equinox of the Great year. The darkest Ages of the Piscean Winter have past, and humanity as the collective Osiris emerges from Amenta into a new phase of life. With respect to the mundane year, this card represents the place of judgment in the lowest depths of Amenta, presided over by Tum, or Atum, the winter Sun. Whether the interpretation of this trump is viewed from the perspective of the mundane year, or the Great Year, judgment is carried out in the cycles of each. This is judgment in the sense of a weighing of the "balance of the year," and the determination of how much the equinoctial or solstitial points have shifted during the course of precession.

---

[577] Statius. *Thebaid*. 4.527 ff : "The void [of Haides] is flung open, the spacious shadows of the hidden region are rent, the groves and black rivers lie clear to view, and Acheron belches forth noisome mud. Smoky Phlegethon rolls down his streams of murky flame, and Styx interfluent sets a barrier to the sundered ghosts. Himself [lord Haides] I behold, all pale upon the throne, with Stygian Eumenides [Erinyes, Furies] ministering to his fell deeds about him, and the remorseless chambers and gloomy couch of Stygian Juno [Persephone]. Black Mors [Thanatos, Death] sits upon an eminence, and numbers the silent peoples for their lord; yet the greater part of the troop remains. The Gortynian judge [Minos] shakes them in his inexorable urn, demanding the truth with threats, and constrains them to speak out their whole lives' story and at last confess their extorted gains." Web. <theoi.com/Daimon/Thanatos.html>

[578] Cf. the Orphic hymn to Pluto: "O mighty daemon [Pluto], whose decision dread / The future fate determines of the dead." From *The Hymns of Orpheus*. Translated by Thomas Taylor. 1792.

[579] Chance. *Medieval Mythography: From Roman North Africa to the School of Chartres, A.D. 433-117*. Page 578; note 70: "Mars is called so as if *mors* (death)."

[580] Nadine G. Barlow. 2008. *Mars: An Introduction to its Interior, Surface, and Atmosphere*. Page 21.

XX JUDGMENT

Interestingly, it is predicted that around 2021-22 CE, a bright new star – visible to the naked eye – will suddenly appear in the constellation Cygnus. This phenomenon will occur via the merging of a closely orbiting binary system, resulting in a luminous red nova, that will be readily visible for several years afterwards.[581] Thus, around the winter solstice, or Christmas, this new star (KIC 9832227) will be seen to rise heliacally with Deneb – the "judge who cometh" – serving, in a sense, as an echo of the first Star of Bethlehem, whose appearance was said to usher in the birth of Christ [Figure 6].

If this were not enough fodder for the perpetual apocalypticists, we should also note that 2022 CE happens to be a year in which two full lunar eclipses and one partial solar eclipse will occur,[582] thereby fulfilling a couple of prophetic conditions concerning the end of the Age; namely, that the Sun will appear black, and the Moon will turn red, or appear "as blood":

> And I will shew wonders in heaven above, and signs in the earth beneath; blood, and fire, and vapour of smoke:
> The sun shall be turned into darkness, and the moon into blood, before the great and notable day of the Lord.[583]

> And I beheld when he had opened the sixth seal, and, lo, there was a great earthquake; and the sun became black as sackcloth of hair, and the moon became as blood.[584]

Thus, by the time the Sun enters sidereal Sagittarius/Scorpio on Christmas (or the "day of the Lord"), at least two of these prophetic conditions will have nominally been met. Moreover, this particular "day of the Lord" will have the additional connotation of judgment associated with it by virtue of the new star, or "wonder in heaven" appearing in Cygnus, the constellation of Judgment. Needless to say, it will be interesting to see how events play themselves out on the world stage during this period.

Although Case and Crowley ascribe *Shin* to this trump, in this system, the ascription is *Tav* (ת). This letter is said to be formed by the combination of *Dalet* and *Nun*, and the two components דנ (*Dalet-Nun*) connote "to judge."[585] As discussed in the chapter on The Sun, *Dalet* is in turn formed by the transformation of the Azoth (א=*Alef*) by the Solar Force (ר=*Resh*), which results in the state of *bitul*, or selflessness, represented by the Hanged Man. When this state is conjoined with the egoic dissolution that answers to *Nun*

---

[581] Molnar, Lawrence et al. "Prediction of a Red Nova Outburst in KIC 9832227." Submitted to *The Astrophysical Journal*, Januaray 4, 2017.
[582] The solar eclipse will occur October 25 (Sun and Moon in sidereal Virgo), and the lunar eclipses will occur May 16 (Sun in sidereal Pleiades; Moon in sidereal Libra) and November 8 (Sun in sidereal Libra; Moon in sidereal Pleiades).
[583] Holy Bible (KJV). Acts 2:19-20.
[584] Ibid. Revelation 6:12.
[585] Ginsburgh. *The Hebrew Letters*. Page 327.

(which is ascribed to Death), the resultant disappearance of the false ego in combination with the transformed Azoth yields the seal of judgment (*Tav*), which represents the application of correct discernment (i.e., justice), in accord with higher spiritual law.[586]

[Figure 6: The heliacal rising of KIC 9832227 at the winter solstice/Christmas, circa 2022 CE in Egypt, looking east.]

Tav is said to embody the secret of reincarnation. It symbolizes the effect of the "impression" or "stamp"[587] of the previous lifetimes upon the present one.[588] The specific "stamp," or "*Tav*," as a relictual echo from the past incarnations contains within it a "blemish" (*mum*) that is related to *Mem* (מ). As the sins of the past incarnation are rectified through the "conscious labors and intentional sufferings"[589] in this lifetime, the *mum* opens, issuing forth *Mem* – the living waters of regeneration and revelation. Thus, inheritance of one's unique *Tav* in this life serves as a basis for the further refinement of the soul. The Mem of the blemish and the *Tav* of the relictual impression combine to form מת = *Mem-Tav* = "death," which is rectified by "Divine Providence," or *Alef*, to produce אמת = *Alef-Mem-Tav* = "truth." This truth

---
[586] Cf. Ibid.
[587] Ibid. *Tav* represents a stamp, seal, or impression.
[588] Ginsburgh. *The Hebrew Letters*. Page 328.
[589] Cf. Gurdjieff. *Beelzebub's Tales to His Grandson*. Pages 773, 792, 808.

is the realization of regeneration, that is, the "resurrection of the dead" (i.e., dead parts of one's soul), which is said to be dependent on the consciousness of right action in the present.[590]

---

[590] Ginsburgh. *The Hebrew Letters*. Pages 328-329.

XXI The World

THE RIDER-WAITE VERSION of this trump [Figure 1] is essentially a view of the northern celestial hemisphere from "outside" the celestial sphere. The female figure with sinuous drapery is seen to represent the Great Mother, who is associated with the circumpolar regions and Draco;[591] her navel corresponds to the NEP. Looking down from this vantage point, one can see the four fixed zodiacal constellations – represented by the man/angel (Aquarius), eagle (Scorpio), lion (Leo), and bull (Taurus). The wreath thus loosely represents the ecliptic, with the red ties indicating the two solstitial loci. The batons the figure holds correspond to the equinoctial nodes [Figure 2].

[Figure 1: The World (Rider-Waite version). Author's rendering.]

The configuration of the Matriarch and her wands is also representative of the 4 primary seasons of the Great Year. From this perspective, her head

---

[591] See discussion of Ta-urt and the circular zodiac of Denderah in the chapter on The Devil.

corresponds to the head of Draco; station of the pole during the Golden Age – or Summer – of the Great Year. Her foot rests upon the star of Sut-An (Polaris), which marks the Iron Age – or Great Winter – of this vast cycle. The two extended wands represent the flanking Silver Ages. The right-hand wand (from our perspective) corresponds to the circumpolar region of Ursa Major, and the Age of Gemini/Taurus; the descending Silver Age – or Autumn – of the Great year. The opposite wand corresponds to the interstice between Cepheus and Cygnus, the station of the pole during the upward-trending (Vernal) Silver Age [Figure 3].

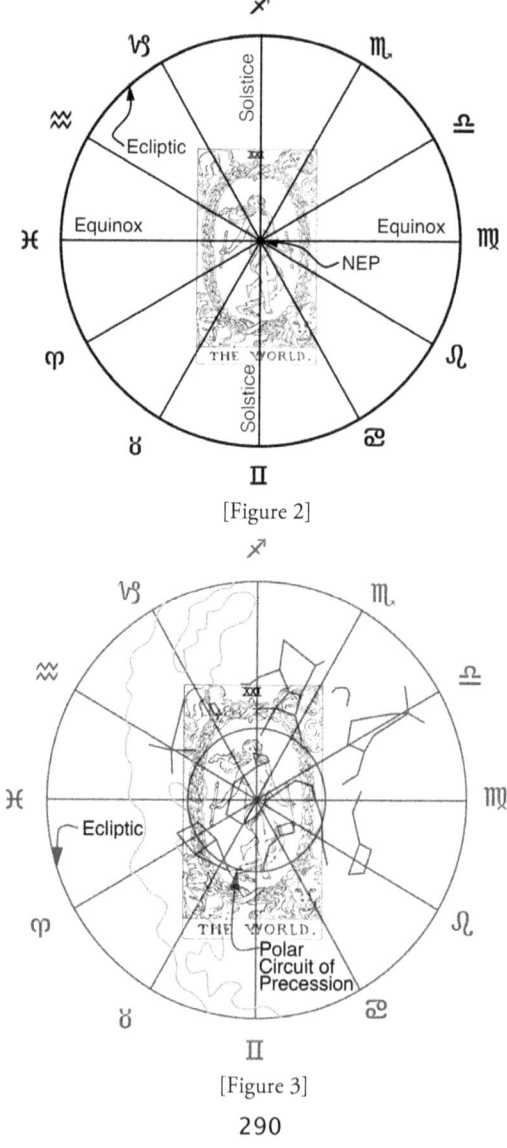

[Figure 2]

[Figure 3]

XXI THE WORLD

Case depicts these wands as spirals, turning in mutually opposite directions. The right-hand spiral is said to represent the involution of "Life-power," and the left hand its evolution.[592] "Involution" refers to the descent of spirit into matter, in order to learn and acquire new knowledge. The resultant experience is then transubstantiated into the material plane, thereby lifting it up to a higher evolutionary state. According to Blavatsky:

> The whole of antiquity was imbued with that philosophy which teaches the involution of spirit into matter, the progressive, downward cyclic descent, or active, self-conscious evolution… One and all, they allegorized and explained the FALL as *the desire to learn and acquire knowledge* – to KNOW. This is the natural sequence of mental evolution, the spiritual becoming transmuted into the material or physical. The same law of descent into materiality and re-ascent into spirituality asserted itself during the Christian era, the reaction having stopped only just now, in our own special *sub-race*.[593]

Within the precessional framework, involution may be seen as effected via the transmission of gnosis from the higher to the lower Ages during the Autumnal Equinox of the Great Year (Age of Gemini/Taurus). The ultimate assimilation of this gnosis takes place at the cycle's nadir, where the mystery of transubstantiation is played out in the Piscean eschatology of Death, Burial, and Resurrection. The process of evolution is carried out by the subsequent reassimilation of this transubstantiated gnosis during the Great Vernal Equinox (Age of Sagittarius/Scorpio).

The Sola-Busca trump 21 (Nabuchodenasor, i.e., Nebuchadnezzar) [Figure 4] appears to represent a vantage point from "inside" the celestial sphere, the constellation Draco being clearly identifiable. The circle around the dragon roughly corresponds to the precessional path of the north celestial pole around the ecliptic pole over a period of 24,000-26,000 years. The wand held by the man may represent the position of the polestar at some point within this precessional cycle. Any high degree of accuracy is impossible, due do the technical and symbolical methods employed in creating these decks. It is clear, however, that the tip of the wand points to a location near the tail of the dragon, indicating a date after 2,000 BCE, but before our current time [Figure 5].

Nebuchadnezzar I was king of Babylon from 1126-1103 BCE, which would correspond to the Age of Aries/Taurus. Interestingly, he claimed to be from a long line of kings from before the flood; i.e. a previous Aeon, associated with a different polestar.[594] Perhaps Nebuchadnezzar's crown is

---

[592] Case. *The Tarot: A Key to the Wisdom of the Ages*. Chapter 25.
[593] Blavatsky. *The Secret Doctrine*. Volume 1. Pages 416-417.
[594] Lambert, W. G. "Enmeduranki and Related Matters". *Journal of Cuneiform Studies*, 21. 1967.

intended to represent Corona Borealis, and the base of the wand shows from whence he purportedly descended; that is, from some "noble lineage" stretching back into the Golden Age.

[Figure 4: The Sola-Busca trump 21: Nabuchodenasor (Nebuchadnezzar). Author's rendering.]

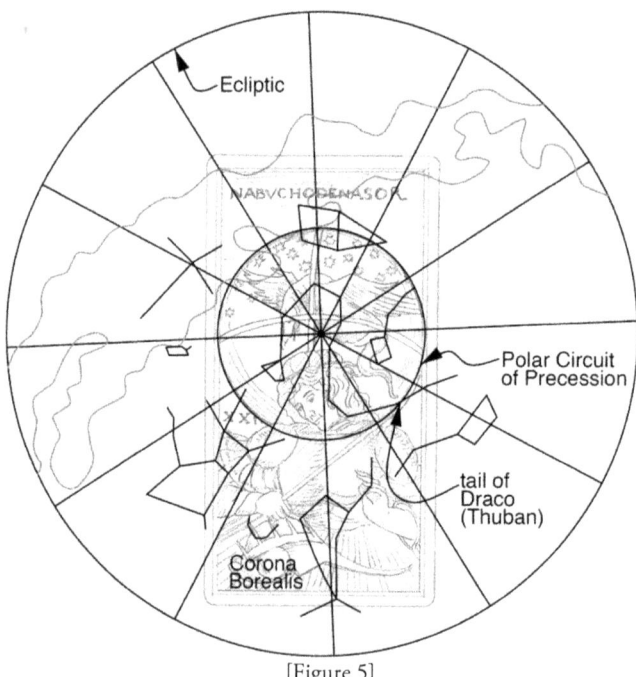

[Figure 5]

The Visconti-Sforza version of this trump [Figure 6] deviates significantly from the previously discussed versions in relation to the apparent vantage

XXI THE WORLD

point depicted in the card. Instead of a view of the northern celestial hemisphere, centrally justified with respect to the NEP, what appears to be represented is a view looking due east (or west) at the equator.

[Figure 6: The Visconti-Sforza World trump. Author's rendering.]

From this perspective, the circle may represent the celestial sphere, with the roiling waters below the island signifying the invisible stars beneath the horizon; the horizon being indicated by the flat slab of land upon which the castle is situated. Closer investigation of the spires of the castle reveals that they are arranged in such a way as to form a lemniscate. In this case, the lemniscate formed by the spires matches the conformation of the Sun's analemma when plotted throughout the course of one year [Figure 7].

CELESTIAL ARCANA

The two puti, or twins (Horus and Sut), likely correspond to the northern and southern celestial poles. In the center – where the celestial equator crosses the ecliptic – the perpetually warring twins are reconciled at the equinoxes, indicated by their mutual holding up of the heavens; a function relegated to Shu the Reconciler in the ancient Egyptian astro-mythology.[595]

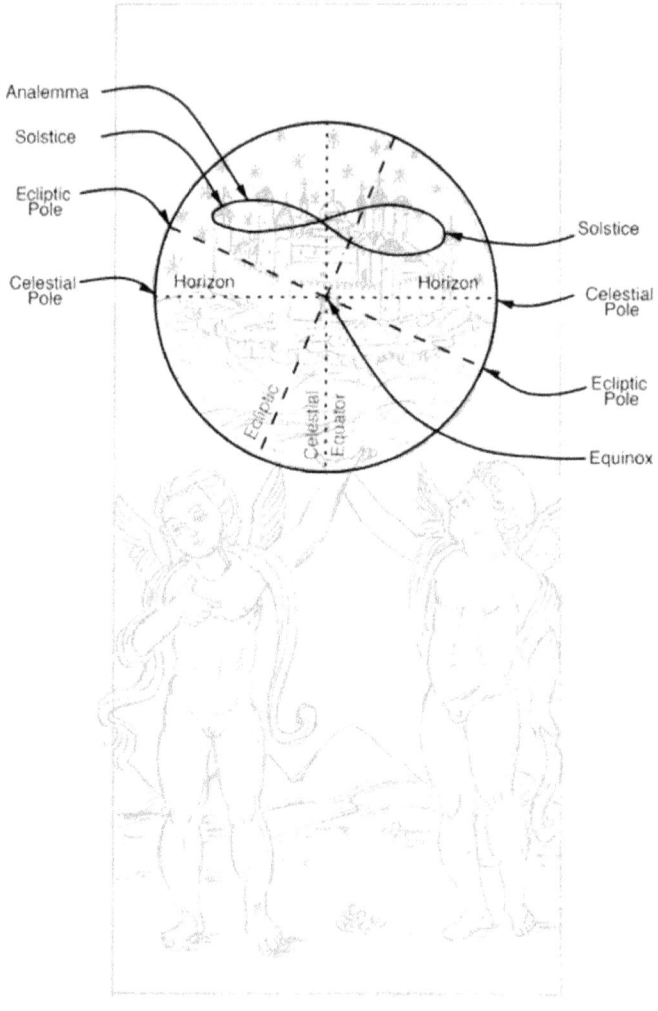

[Figure 7]

If we take this card to represent an eastern view of the celestial sphere, from the vantage point of Earth's equator, the left-hand side of the lemniscate formed by the spires corresponds to the summer portion of the year, whereas the right equates to the winter. In ancient Egypt, the flooding of the Nile

---

[595] Massey. *Ancient Egypt, the Light of the World.* Book 6. Page 326.

XXI THE WORLD

occurred during the summer, which allowed the crops to be planted; thus the association with the "good" twin Horus. The drought and pestilence of the winter season was ascribed to the "evil" brother Sut. It was at the equinox that crops were harvested, and these two antagonistic forces of flooding and drought – Horus and Sut – were reconciled.

Due to the solar connotations of the NEP and its centrality in the Rider-Waite and Sola-Busca versions, as well as the solar elements inherent in the Visconti-Sforza analog, the suggested Yetzirac attribution for this trump is the active principle of fiery *Shin*.

In a sense, *Shin* may be seen as comprising all 3 elements of the precessional triad: the NEP, NCP, and "binding force." This triple nature is reflected in the 3 "heads" of the letter, which correspond to the changeless Essence (i.e., NEP), the potential for change (i.e., the binding force), and the actualization of change (i.e., the NCP).[596] *Shin* is said to be comprised of an inner, unchanging flame – eternally at rest – and an outer flame that exists in a state of continual motion and change. *Shin*

> reflects its divine symmetry into the three lower worlds of Creation, Formation, and Action, in the secret Name *Tzevakot* ... This Divine Sign, present in all worlds as an axis around which the hosts of heavens and earth revolve, is the inner pillar of Immanent Light which "fills all (lower) worlds."[597]

It is said that the 3-headed *Shin* is a symbol of our present Age, but during the next phase of humanity's evolution, it will be transformed into a 4-headed *Shin*. This doctrine is reflected in the design of the head *tefilin*, or phylactery – a box containing scriptural parchments that is bound to the forehead during Jewish morning prayers:

> On the right side of the head-*tefilin* is engraved a three-headed *shin*, the *shin* of *chochmah*, the secret of the root of the souls of the three patriarchs. On its left side is engraved a four-headed *shin*, the *shin* of *binah*, the secret of the root of the souls of the four matriarchs. This is the *shin* of the World to Come: when the symmetry-axis of this world will itself split into two, and reveal that the "intermediate," the middle line, entails the roots of the two extremes which it serves to unite.[598]

At one level, this may be interpreted as an allusion to the entrance into the Aquarian Age proper, when the disharmonies between the matriarchal and patriarchal cosmologies finally begin to be resolved, thereby catalyzing the unification of gnosis from the Higher and Lower Ages into a balanced and harmonious whole.

---

[596] Cf. Ginsburgh. *The Hebrew Letters*. Page 310.
[597] Ibid. Page 315. Ellipses added.
[598] Ibid.

0 The Fool

THIS TRUMP HAS been alternately ascribed to *Shin* (Waite), and *Alef* (Case, Crowley), but within the context of our current model, the attribution is *Mem*. This is because the Fool embodies the passive element of the triad of forces governing the precessional mechanism.

Mem is attributed to water,[599] and here the reference is to the "waters of the deep"; in other words, the starry firmament. In the stellar mythos, Tef was the ancient Mother presiding over the waters at the celestial pole.[600] Subsequently, it was Osiris as the Lord of the Waters[601] who became associated with the "power of the pole."[602] Notably, the Hebrew word for Orion – constellation par excellence of Osiris – is *Kesil*; that is, "fool."[603,604]

Waite himself expressed dissatisfaction with the attribution of *Shin* with regard to this card:

> In later times the cards have been attributed to the letters of the Hebrew alphabet, and there has been apparently some difficulty about allocating the zero symbol satisfactorily in a sequence of letters all of which signify numbers. In the present reference of the card to the letter *Shin*, which corresponds to 200, the difficulty or the unreason remains. The truth is that the real arrangement of the cards has never transpired.[605]

The Rider-Waite version of this trump [Figure 1] depicts the Fool in the act of stepping off a cliff, his canine friend appearing to encourage him. The vast chasm awaiting him can be seen as the starless and foreboding interstitial regions of the heptanomis. One can imagine the ancients looking on in horror as they bore witness to the gradual spiraling away of their once stable polestar, not knowing whether some catastrophe was in fact looming behind such an ill omen.

---

[599] Massey. *A Book of the Beginnings*. Section 13. Page 121.
[600] Ibid. Section 14. Pages 131-132.
[601] Massey. *Ancient Egypt, the Light of the World*. Book 8. Page 483. Web.
[602] Ibid. Book 2. Page 574.
[603] Strong's Concordance. Entry 3684.
[604] Job 9:9; Job 38:31; Amos 5:8.
[605] Waite. *The Pictorial Key to the Tarot*. Part I. Section 2.

[Figure 1: The Rider-Waite Fool trump. Author's rendering.]

Of course, the Fool has nothing to fear, for the precessional binding force represented by the Magician will hold him aloft through these interstitial periods until once again, he alights upon a solid stellar ground:

> The edge which opens on the depth has no terror; it is as if angels were waiting to uphold him, if it came about that he leapt from the height … The Sun, which shines behind him knows whence he came, whither he is going, and how he will return by another path after many days.[606]

Case suggests this card to be a solar archetype, representative of Osiris, Christ, and the Solar Force.[607] He proposes the Fool's belt to be indicative of the year, but in light of the proposed system, it may be more fitting to enlarge the scale of this concept to the level of the Great Year. Here, the Fool represents the eternal development of the Solar Force via Humanity's evolution. This development is ideally one in which all suffering is transformed into bliss. Such is the task of the eternally returning soul[608] – represented by the Fool – throughout the precessional Ages.

This card is typically seen to encompass all the other trumps. In the present case, this is demonstrated by the fact that it represents the passage of

---

[606] Ibid. Part II. Section 2; The Fool. Ellipses added.
[607] Case. *The Secret Doctrine of the Tarot*. Chapter 2.
[608] Cf. Case's "Supreme Spirit."

0 THE FOOL

the pole throughout the 12 zodiacal trumps, which are ruled by the 7 planetary trumps, with said passage being governed by the 3 elemental trumps.

The feather in the Fool's cap has been equated with the Egyptian goddess Maat,[609] who – in one guise – represented the point of equipoise as figured in the Celestial Pole.[610] Some versions of this card portray the archetype as a bearded ancient,[611] which is in accordance with the proposed attribution of this trump to the Celestial Pole, as this is clearly a representation of the Ancient of Days – the Old Man of the Mountain; the patriarchal form of the Great Mother, ruler of the circumpolar regions in the ancient stellar phase of chronometry. Crowley suggests this card to be connected to the vulture form of Maat,[612] whose spiral-shaped neck can be interpreted to represent the shape described by the station of the pole as it cycles through the heptanomis every 24,000 or so years.

Some maintain that the Fool should be the 1st card, and therefore equated with *Alef,* since the numerical value of 0 precedes all other positive integers.[613,614] However, this is something of a spurious argument, since the first versions of this trump never had any numerical value, which is more in keeping with the notion of the Fool encompassing all the other trumps. In many ways, the Fool can be seen as the eternal "Wandering Jew," who may have anciently been portrayed as the ever-coming "Iu."[615] This "ever-coming" aspect refers to the cyclical nature of the Sun's rebirth every day, year, or Great Year – this last cycle representing the "highest" manifestation of Iu, as the continuously changing Avatar throughout the cycles of precession.

Iu is a form of the father-god Ptah, but in the guise of the son.[616] In this relationship, we can see a parallel with the celestial mechanism of precession, wherein Ptah represents the NEP, and Iu the NCP. In this case, Iu – the one who always returns at the end and beginning of each Age – represents an anthropomorphization of the Celestial Axis and Pole as they journey through the precessional cycles.

There is a distinction made between the open and closed *Mem.*[617] The open one is symbolic of an above-ground stream or fountain, whereas the

---

[609] Case. *The Secret Doctrine of the Tarot.* Chapter 2.
[610] Cf. Massey. *Ancient Egypt, the Light of the World.* Book 9.
[611] Case. *The Secret Doctrine of the Tarot.* Chapter 2.
[612] Crowley. *The Book of Thoth.* Page 53.
[613] Case. *The Secret Doctrine of the Tarot.* Chapter 1.
[614] Crowley. *The Book of Thoth.* Page 53.
[615] According to Massey, "Jew" is derived from "Iu," the ancient Egyptian personification of the ever-coming son (sun). Cf. Massey. *A Book of the Beginnings.* Section 18: The Egyptian Origin of the Jews Traced from the Monuments. Page 432.
[616] Massey. *Ancient Egypt, the Light of the World.* Book 7. Page 423.
[617] Ginsburgh, Yitzack. *The Hebrew Letters: Channels of Creative Consciousness.* Pages 193-206.

closed form connotes a subterranean one.[618] This letter is said to be a symbol for "mother,"[619] and here this is interpreted as the primordial Mother, who – before she became associated with any particular astronome of the heptanomis (i.e. Corona Borealis, or Ursa Major) – seems to have been associated with the central axis of the whole precessional cycle; that is, the NEP. But not only is she the NEP (which in the World trump is symbolized merely by her navel), she is also the heptanomal circle which encloses it, as the womb; the zero; the cipher; the chaotic matrix, from which all discrete manifestations in humanity's history arise.

The open *Mem* is said to represent the soul conscious of itself, whereas the closed *Mem* signifies the opposite.[620] Within the context of the present system, this may be taken to signify a distinction between the soul unaware of its higher-dimensional potentiality throughout multiple precessional cycles, and the soul that has progressed into a spiral representation of this once planar cycle. Another way of stating it is the closed *Mem* represents the soul of this lifetime, whereas the open *Mem* represents the soul of "all" lifetimes. When the circle is broken in the open *Mem*, this represents the movement into a perpendicular dimension, from whence issue the waters of life down to the present incarnation.[621] This may be seen as accessing the Supreme Spirit, which labors continuously towards the transmutation of suffering into bliss, since its fundamental nature is Love.[622] Notably, the gematrial addition of the Hebraic word for Love (אהבה) yields 13, the ordinal value of *Mem*.[623]

The upper (open) *Mem* flows downwards into the lower (closed) *Mem*, as the soul entering into the body. There are specific moments at which this process is naturally the most active; these are the moments of conception, development, birth, and other key phases when one's "root shines with special brilliance."[624]

This concept of the upper *Mem* is similar to Case's notion of the Fool as the Spirit prior to self-expression:

> The abyss at the Fool's feet is in contrast to the height on which he stands. It represents what Lao-Tze, the Chinese sage, called the "Mother Deep" in the Tao-Teh-King. At its bottom is the plain, which, as the scene of labor, constructive activity, struggle, competition, and a multiplicity of

---

[618] Ibid.
[619] Ibid.
[620] Ibid.
[621] Ibid: "The flowing stream, the source of wisdom, hints at the soul flowing down into the body from the source above. This takes place primarily at the moments of conception, formation, birth, and other crucial moments in life when one's root shines with special brilliance." Compare this with Case's notion of the Fool as the Spirit prior to self-expression.
[622] Case. *The Secret Doctrine of theTarot*. Chapter 1.
[623] Ginsburgh. *The Hebrew Letters: Channels of Creative Consciousness*. Pages 193-206.
[624] Ibid.

0 THE FOOL

manifestations, is the polar opposite of the perfection, singleness, and simplicity suggested by the mountain-top.

The Fool is on the verge of descending, because this picture shows Spirit as we think of it prior to self-expression. He is unafraid, for he knows nothing can harm him. No matter how far into the depths he plunges he will surely rise again. His purpose in descending is to find a path leading to the loftier height beyond.[625]

The closed *Mem* represents the womb in gestation, and the open one signifies the womb giving birth. At the diurnal level, this ties the closed *Mem* to the night, where the womb of the Great Mother can be seen active and churning in the northern circumpolar heavens, whereas the open *Mem* represents the opening of the day, as the Sun is birthed from the Mother's womb that has provided safe passage to it through the Underworld. The closed *Mem* represents the "subterranean source,"[626] i.e., the circumpolar fountain of the Waters of Life,[627] and the open *Mem* represents the "point of conscious insight,"[628] symbolized by the emergence of the Solar Child from the mysterious depths of Amenta.

The gematrial value of *Mem* is 40, and the Jews are said to have wandered in the wilderness for forty years. According to Massey, the wilderness represents the Underworld, or Amenta. He suggests that the forty years were a modification of the previously designated 40 days of the ancient Egyptian Lent, wherein Sut the adversary held predominance as the Sun traversed at this time through the driest portion of the year (i.e., the end of the Egyptian winter).[629] Today, this symbolism remains extant in the celebration of Easter, coinciding roughly with the spring equinox. In Egypt, during the Age of Taurus, these 40 days coincided with the passage of the Sun out from the abyss of Cetus/Pisces[630] into the newly emergent "bull of the mother," i.e., Taurus/Orion. Today, the transition occurs between the adjacent signs Aquarius and Pisces.

According to Gertrude Moakley, the 7 feathers in the headdress of the Visconti-Sforza Fool represent the 7 weeks (or 40 days) of Lent,[631] which fits well with the proposed connection of this trump with the letter *Mem*, but it should be noted that the 7 feathers could also represent the 7 astronomes that the Osirian "power of the pole" traverses throughout the precessional cycle.

---

[625] Case. *The Secret Doctrine of the Tarot*. Chapter 1.
[626] Ginsburgh. *The Hebrew Letters: Channels of Creative Consciousness*. Pages 193-206.
[627] Massey. *Ancient Egypt, the Light of the World*. Book 8. Page 483.
[628] Ginsburgh. *The Hebrew Letters: Channels of Creative Consciousness*. Pages 193-206.
[629] Massey. *Ancient Egypt, the Light of the World*. Book 10. Pages 642-643.
[630] Ibid. Book 5. Pages 279-280.
[631] Moakley, Gertrude. *The Tarot Cards Painted by Bonifacio Bembo for the Visconti-Sforza Family: An Iconographic and Historical Study*. 1966.

If we arrange the 7 heptanomal trumps (i.e., the planetary ones) in the order in which they occur throughout the precessional cycle, and organize their corresponding Hebrew letters analogously – starting with the High Priestess – we derive "בגף" = "Division" and "רדתך" = "Frequency," traits of the Fool as eternally returning Green Man of spring,[632] who marks out the pace of the Ages throughout the cycles of precession. The Green Man is Osiris, and more anciently, Ptah – both deific figures of the Pole; both known as the architects, dividers, and timekeepers of heaven.[633]

Whether the Fool is considered as representative of *Alef, Shin, Mem*, or any of the other Hebraic letters, one must recognize that ultimately, the Trumps can never be rigidly confined to any structure or sequence, and to insist on a particular arrangement as immutable is presumptuous, at best. This being said, once the individual elements have been defined within any given structure (Kabbalistic, or otherwise), one should expect the whole to function harmoniously and provide ample insight into the subject at hand, perhaps serving to reveal certain aspects which have been invisible according to other systems of classification. This phenomenon may well be explained by the fact that humans are above all pattern-seekers, and will therefore attempt to find structure and meaning in chaos. While this may be the case, it does not deprecate the act of willing the pattern into existence.

---

[632] Cf. Crowley. *The Book of Thoth*. Page 56.
[633] Cf. Massey. *A Book of the Beginnings*. Section 13. Also *Ancient Egypt, the Light of the World*.

The Minor Arcana

GIVEN THE LACK of a solid historical narrative concerning the origins of the Tarot, it should come as no surprise to learn of the uncertainty as to whether the Minor Arcana was originally intended to be included as part of the same deck of cards as the Major Arcana. Waite expresses his view on the subject in the following passage from *The Pictorial Key to the Tarot*:

> I observe that there has been a disposition among experts to think that the Trumps Major were not originally connected with the numbered suits. I do not wish to offer a personal view; I am not an expert in the history of games of chance, and I hate the *profanum vulgus* [i.e., "common rabble"] of divinatory devices; but I venture, under all reserves, to intimate that if later research should justify such a leaning, then – except for the good old art of fortune-telling and its tamperings with so-called destiny – it will be so much the better for the Greater Arcana.[634]

Although Waite offers no opinion regarding whether or not the Minor Arcana was originally intended as adjunctive to the Trumps Major, it is apparent that his inclination to see them as separate derives in no small part from his disdain for the use of the cards in the art of divination. He seems particularly ill-disposed towards the use of the Major Arcana for such purposes, stating:

> It will be seen that, except where there is an irresistible suggestion conveyed by the surface meaning, that which is extracted from the Trumps Major by the divinatory art is at once artificial and arbitrary, as it seems to me, in the highest degree. But of one order are the mysteries of light and of another are those of fantasy. The allocation of a fortune-telling aspect to these cards is the story of a prolonged impertinence.[635]

---

[634] Waite. *The Pictorial Key to the Tarot*. Part I. Section 4 – The Tarot in History. Bracketed comment added.
[635] Ibid. Part III. Section 3 – The Greater Arcana and Their Divinatory Meanings.

Nonetheless, Waite devotes a good portion of *The Pictorial Key to the Tarot* to the presentation of manifold divinatory attributions of both Arcana, as well as detailed instructions concerning various methods of "good old fortune-telling." One is led to surmise that he must have been skeptical concerning the majority of divinatory devices, but was still open to the possibility of obtaining some sort of useful results, given the proper methodology and abilities of the persons involved in the art:

> The records of the art are *ex hypothesi* the records of findings in the past based upon experience; as such, they are a guide to memory, and those who can master the elements may – still *ex hypothesi* – give interpretations on their basis. It is an official and automatic working. On the other hand, those who have gifts of intuition, of second sight, of clairvoyance – call it as we choose and may – will supplement the experience of the past by the findings of their own faculty, and will speak of that which they have seen in the pretexts of the oracles.[636]

> The value of intuitive and clairvoyant faculties is of course assumed in divination. Where these are naturally present or have been developed by the Diviner, the fortuitous arrangement of cards forms a link between his mind and the atmosphere of the subject of divination, and then the rest is simple. Where intuition fails, or is absent, concentration, intellectual observation and deduction must be used to the fullest extent to obtain a satisfactory result. But intuition, even if apparently dormant, may be cultivated by practice in these divinatory processes. If in doubt as to the exact meaning of a card in a particular connexion, the diviner is recommended, by those who are versed in the matter, to place his hand on it, try to refrain from thinking of what it ought to be, and note the impressions that arise in the mind. At the beginning this will probably resolve itself into mere guessing and may prove incorrect, but it becomes possible with practice to distinguish between a guess of the conscious mind and an impression arising from the mind which is sub-conscious.[637]

This aspect of guessing until one hits upon a plausible narrative – presumably arising through some mystical connection to the normally untapped reservoirs of the oracular psyche – is also suggested in Crowley's presentation of a certain method of divination in his *Book of Thoth*:

> 5. Tell the Querent what he has come for: if wrong, abandon the divination.
>
> 6. If right, spread out the pack containing the Significator, face upwards … Make a story of these cards…

---

[636] Ibid.
[637] Ibid. Section 8 – An Alternative Method of Reading the Tarot Cards.

THE MINOR ARCANA

8. If this story is not quite accurate, do not be discouraged. Perhaps the Querent himself does not know everything. But the main lines ought to be laid down firmly, with correctness, or the divination should be abandoned.[638]

The skeptic will easily recognize the problem with such methodologies: of course it is to be expected – due to the loose interpretive nature of card reading – that one will eventually hit upon coincidences which appear relevant to the situation at hand. The question remains as to whether such results are to be interpreted as mere coincidence, or as those of one tapping in to some "mystical dimension" – perhaps the subconscious.

An alternative perspective, which does not exclude the use of the Tarot in the divinatory arts, but frames the prognosticative aspects within an astrological context, may serve to shed some light on one of the originally intended uses of the cards.

According to the sidereal astrological system developed by Cyril Fagan through studying the ancient Babylonian and Egyptian systems, the 4 so-called "angles" of any particular horoscope (designated as the Ascendent, Midheaven, Descendent, and Antemeridian) are the loci at which the heavenly luminaries have their greatest effect [Figure 1].

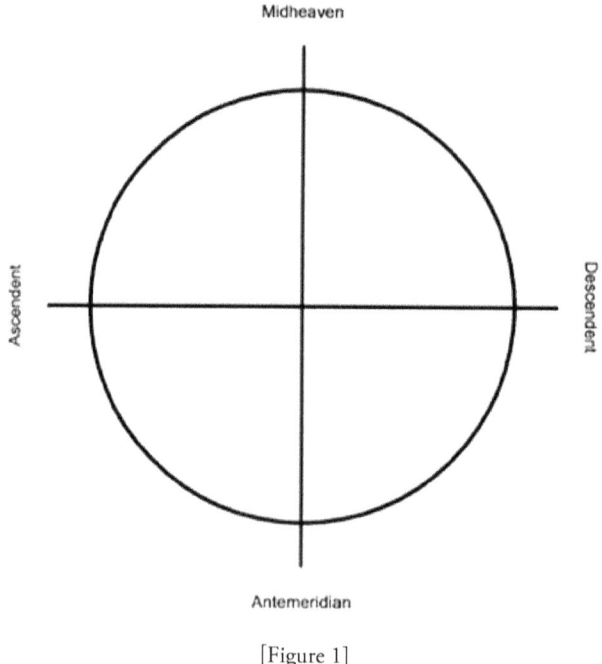

[Figure 1]

---

[638] Crowley. *The Book of Thoth*. Appendix A – The Behaviour of the Tarot. Pages 250-251. Ellipses added.

In this system, the 4 "angles" are of utmost importance. Any luminary within 7° of either side of them is said to be in the "foreground," where it exerts a maximum influence. Once this narrow margin has been exceeded, the planet, constellation, or star is said to reside in the "background," and to have the weakest effect [Figure 2].

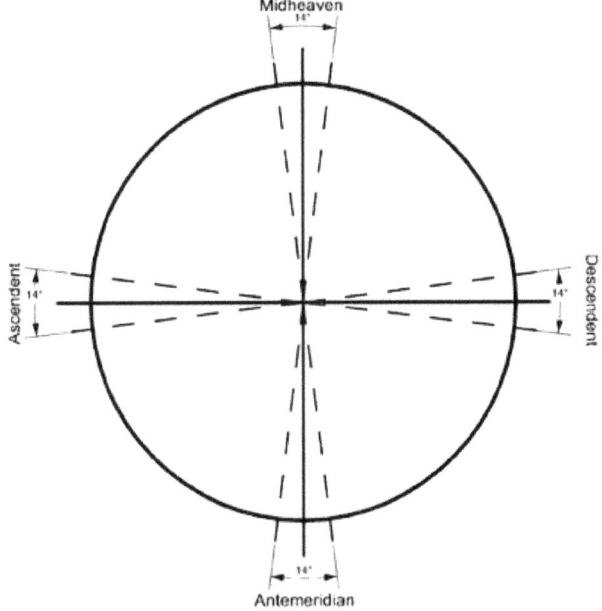

[Figure 2]

The angles themselves are arranged according to a descending potency in the following order: Midheaven ("Medium Coeli," or MC), Ascendent (eastern horizon), Descendent (western horizon), and Antemeridian (the "subterranean" locus 180° from the Midheaven) [Figure 3].

The connection of these 4 "angles" to the 4 suits of the Minor Arcana will be elucidated shortly; first, we examine the relationship between these suits and the 4 elements.

Nowhere in the *Pictorial Key to the Tarot* does Waite specifically equate the four suits to the same number of alchemical or astrological elements (except in the case of Cups), although one may infer a certain arrangement based on the considerations of the symbols used in the various court cards of each suit.

The King and Queen of Wands both employ lions, with the addition of a cat in the case of the Queen. This feline symbolism is suggestive of the constellation Leo (with the cat possibly indicating the inclusion of Leo Minor), and thus the element of fire. Bulls are prominently displayed in the King and Queen of Pentacles – the obvious correlation being Taurus, and thus the earthy element. The prevalence of water in the suit of Cups implies – according to Waite – "that the Sign of the Cup naturally refers to water."[639] Using the same rationale as employed with the suits of Wands and Pentacles, one might assume that Cups would naturally correspond with Scorpio, and Wands – which bear the symbol of the butterfly – would therefore be considered as representative of the airy element, as well as the constellation Aquarius. Whereas in the case of the elemental attributions these designations appear to obtain, it could reasonably be argued that regarding the fixed Signs, Scorpio is more appropriately to be attributed to Swords, and Aquarius to Cups.

In *The Natural Genesis*,[640] Massey gives the following arrangement based on an ancient Egyptian scheme, which pairs each cardinal direction with one of the four elements and a corresponding astrological zootype:

North = water (hippopotamus = Typhon)
East = wind (ape = Hapi)
South = fire (phoenix = Har)
West = earth (crocodile = Sevekh).

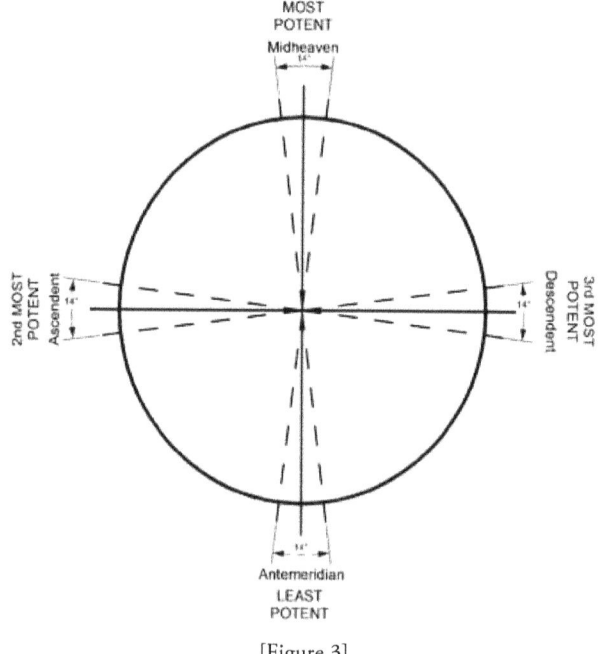

[Figure 3]

---

[639] Waite. *The Pictorial Key to the Tarot*. Part III. Section 2 – The Lesser Arcana, otherwise, the Four Suits of Tarot Cards.
[640] Massey. *The Natural Genesis*. Book 9. Page 32.

CELESTIAL ARCANA

Making reference to the Celestial Sphere [Figure 4], we consider the NCP to be centered in Ursa Major in deference to the postulate that much of the symbolism of this Egyptian arrangement derives from the Age of Taurus. Here, the North refers to the circumpolar stars of Ursa Major, Draco, and Bootes – each of which was associated with the hippopotamus and/or Typhon during the Taurean Age. Being only visible at night, they were identified with the waters of the abyss, or Amenta, which corresponded to the three nocturnal "water signs" of the inundation: Pisces, Aquarius, and Capricorn. Taking this into account, as well as the fact that the dolphin symbolism of the King of Cups can be read to signify the constellation Delphinus ("the dolphin" – antiscion of Capricorn and Aquarius), it seems reasonable to impute a similar reference in the Rider-Waite iconography to the waters of the abyss, and therefore ascribe Aquarius to the suit of Cups, rather than Scorpio.

Framing our attribution scheme from the perspective of the Egyptian arrangement; if we consider the suit of Swords to represent the airy element, Scorpio would be the corresponding fixed zodiacal sign, as it was considered an air sign[641] during the Age of Taurus.[642] The corresponding cardinal direction for this suit would thus be east.

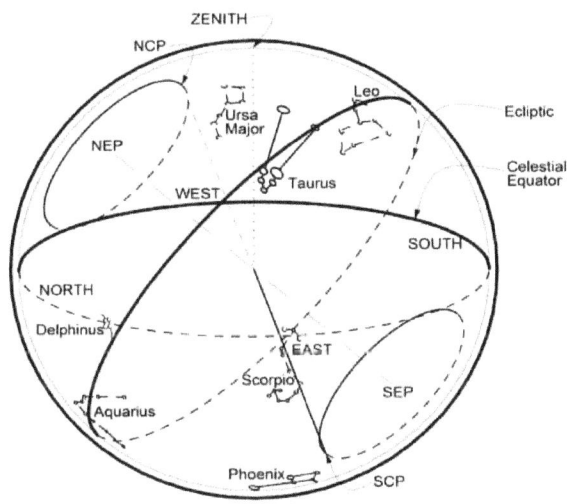

[Figure 4]

---

[641] Fagan. *The Solunars Handbook*. Page 12.
[642] Ibid. Page 10: "The evening rising of Scorpio during April (Sun in Taurus) ushered in the deadly Khamsin (Arabic = 50 days' wind), bringing with its pestilential hot sandstorms from the Sahara hordes of scorpions to infest the land. The demotic ideogram for Scorpio was a serpent and in Egyptian symbolism serpents always signified wind, storms and hurricanes." [Note that the condition of the Sun in Taurus during the month of April is met circa 3,000 BCE, which falls within the nominal timeframe of the Taurean Age.] Bracketed comment added.

THE MINOR ARCANA

The south, according to this arrangement, corresponds to fire and the phoenix zootype. As Egypt is some 30° north of the equator, the Sun is always seen to be south of the zenith, and this cardinal direction may have become associated with that luminary which exerted its most intense influence by virtue of its high angle of incidence during the summer months of inundation when in the sign of the lion. Furthermore, the phoenix zootype – associated by the ancient Egyptians with the south – corresponds to the constellation of the same name, where the South Celestial Pole was to be found during the Age of Taurus [Figure 4]. During this Age, the west embodied the earthy element, typified by the crocodile Sevekh, otherwise known as the "bull of the seven cows."[643] Here the seven cows are the seven stars of Ursa Major, and Sevekh as the bull of the seven cows may refer to the fact that when the vernal point was to be found in Taurus (4,000-2,000 BCE), the NCP was localized within the "seven cows" of Ursa Major.

Again, Fagan sheds some light on the ascription of element to sign as seen from the perspective of ancient Egyptians in the Taurean Age:

> Taurus the bull rose at sunset during the month of October (Sun in Scorpio). Not only was this the month in which the cattle were coupled, but the waters of the Nile had so far receded as to permit the oxen being yoked and the land ploughed.[644]

The plowing of the land during the acronychal rising of Taurus thus serves as rationale for the ascription of the element earth to the sign of the bull. It should be noted that the aforementioned Egyptian arrangement per Massey appears to be from a period during which the reckoning of time was still in the stellar phase, as opposed to solar, due to the fact that the elemental and directional ascriptions only make sense if read acronychally during the Taurean Age.

Based on these considerations, we generate the arrangement of Tarot suits, cardinal directions, and fixed signs as given in [Table 1].

However, we are currently some 5,000-6,000 years removed from the Taurean Age, and the elemental attributions for the zodiacal signs have changed, presumably based on some logic of seasonal connotations.

Thus, if we consider the Age of Aries to represent the beginning of a primarily luni-solar phase of timekeeping, as opposed to the earlier stellar paradigm, it would reasonably follow that during this Age, the cardinal direction of east (or west – depending on sighting convention) became associated with the vernal point of the Sun in Aries, as opposed to this constellation's acronychal counterpart, Libra. Here we have the element of fire

---
[643] Massey. *The Natural Genesis*. Book 9. Page 1.
[644] Fagan. *The Solunars Handbook*. Page 9.

signifying the rebirth of the Sun from the frozen earth of winter, represented by Capricorn. The Moon, as regulator of tides and ruler of Cancer naturally receives the elemental ascription of water, leaving the element of air to be applied to Libra, suggesting the dry breezes of autumn ridding the trees of their desiccated foliage.

TABLE 1			
CUPS	WATER	AQUARIUS	NORTH
SWORDS	AIR	SCORPIO	EAST
WANDS	FIRE	LEO	SOUTH
PENTACLES	EARTH	TAURUS	WEST

Clearly there is no empirical basis for such an elemental attribution scheme with respect to the zodiacal signs, and it can be argued that, indeed, there never was. This being said, the Egyptian arrangement seems to be based on a relatively sound footing, which at least can be related to the seasonal variations as described by Fagan.

The so-called Ptolemaic attribution scheme that relates elements to zodiacal signs actually appears to have originated much later than Ptolemy, via William Lilly's *Christian Astrology*, circa 1647 CE. There is no specific mention in the *Tetrabiblos* of any of the four elements having a zodiacal counterpart, so the reason for the association of this scheme with Ptolemy is unclear. Perhaps it is due to the fact that Lilly appears to have drawn much of his astrological source material from him. Whatever the case may be, Lilly proclaims what has now become the standard system of "elemental triplicities" in *Christian Astrology* without any explanatory rationale:

Fiery	♈, ♌, ♐.
Earthy	♉, ♍, ♑.
Airy	♊, ♎, ♒.
Watery	♋, ♏, ♓.[645]

Taking the Egyptian arrangement, and applying a "precessional shift" of one sign, we arrive at the following relationship of zodiacal signs to cardinal directions:

North	♑
East	♎
South	♋
West	♈

---

[645] Lilly. *Christian Astrology*. Chapter 1.

THE MINOR ARCANA

If we then apply the necessary acronychal inversions to convert these correlations from a stellar to a solar paradigm, we generate:

North   ♋
East   ♈
South   ♑
West   ♎

Adding Lilly's elemental attributions and our putative suit correlations, we have:

TABLE 2			
North	♋	WATER	CUPS
East	♈	FIRE	WANDS
South	♑	EARTH	PENTACLES
West	♎	AIR	SWORDS

Interestingly, of each of the trumps we have associated with these four cardinal signs, three of them unambiguously bear the accouterments of the respective suit in accordance with the attributions as given in [Table 2]. Thus, the figure in the Justice trump (corresponding to Libra) wields a sword; the Capricornian Devil has an inverted pentacle above its head; and the Aryan Emperor holds a tau cross, which can be seen as a form of wand. One might justifiably add that the yonic element on the shield depicted in the Chariot trump can be read as a Grail reference, thereby answering to the suit of Cups; and here the additional reference to the circumpolar astronome of Corona Borealis, and its central star Clavis Corona as the polestar during the Age of Cancer should not go unnoticed.

Reading through Waite's description of the denaries of the Minor Arcana, one cannot help being struck by the similarities in many of their descriptions with the various characteristics and qualities associated with the 12 zodiacal houses. The following tables collate various keywords from Waite's denaries in *The Pictorial Key to the Tarot* versus similar keywords found in Fagan's discussion of the ancient Egyptian house system in *The Solunars Handbook* [Tables 3-6].

Although at first glance the house and sign distribution appears uneven, if we center our analysis on the cardinal directions and ascribe 3 houses/signs to each direction, we arrive at the following percentages wherein each of the four quadrants has a clearly defined majority of denary suits associated with it [Figure 5].

TABLE 3 – CORRESPONDENCES BETWEEN FAGAN'S HOUSE DESCRIPTIONS AND WAITE'S DENARY KEYWORDS - *WANDS*

CARD	HOUSE (FAGAN)	HOUSE KEYWORDS AND CORRESPONDING SIGN (FAGAN)	SELECT DENARY KEYWORDS (WAITE)
2W	II	♏ (agony of death)	physical suffering
3W	VI, IX	♓ (traders), or ♊ (strength)	commerce, strength
4W	I	♎ (prosperity)	prosperity
5W	II	♏ (wars, violence, upheaval, rebellion)	mimic warfare, competition, struggle, battle of life
6W	XI	♌ (hopes, luckiest house)	hopes, victory
7W	VII	♈ (contention)	contention
8W	XII	♍ (commencement of work day, awakening from slumber)	activity in undertakings
9W	VII	♈ (battle, strife, oppositions)	strength in opposition, formidable antagonist
10W	VI	♓ (grief, persecution)	oppression

TABLE 4 – CORRESPONDENCES BETWEEN FAGAN'S HOUSE DESCRIPTIONS AND WAITE'S DENARY KEYWORDS - *PENTACLES*

CARD	HOUSE (FAGAN)	HOUSE KEYWORDS AND CORRESPONDING SIGN (FAGAN)	SELECT DENARY KEYWORDS (WAITE)
2P	IX	♊ (gaiety, writing)	gaiety, writing
3P	X	♋ (skilled trades)	skilled trades
4P	IV	♑ (inheritance)	inheritance
5P	V	♒ (marriage, mistress)	marriage, mistress
6P	X	♋ (merchant)	merchant
7P	X	♋ (trade, barter)	trade, barter
8P	X	♋ (all craft of which a man is master)	craft, tradesman
9P	IV	♑ (material possessions, land, garden, orchards, real estate, castles, houses)	plenty in all things, wide domain, garden, manorial house
10P	IV	♑ (wealth, place of residence)	riches, abode of family

TABLE 5 – CORRESPONDENCES BETWEEN FAGAN'S HOUSE DESCRIPTIONS AND WAITE'S DENARY KEYWORDS - *SWORDS*

CARD	HOUSE (FAGAN)	HOUSE KEYWORDS AND CORRESPONDING SIGN (FAGAN)	SELECT DENARY KEYWORDS (WAITE)
2S	VIII	♉ (passivity, peace)	conformity, accord
3S	VI	♓ (secret toils of the mind)	mental alienation, error, confusion
4S	VII	♈ (rest, sleep)	repose, tomb
5S	II	♏ (wars, death)	destruction, burial
6S	VII	♈ (relationship with marriage partner)	proposal of love
7S	VII	♈ (contentions)	quarreling
8S	VI	♓ (incarceration)	bondage
9S	VI	♓ (imprisonment)	imprisonment
10S	VI	♓ (assassination, grief)	what is intimated by the design [*i.e. murder*]*, pain, affliction [*author's interpretation]

TABLE 6 – CORRESPONDENCES BETWEEN FAGAN'S HOUSE DESCRIPTIONS AND WAITE'S DENARY KEYWORDS - *CUPS*

CARD	HOUSE (FAGAN)	HOUSE KEYWORDS AND CORRESPONDING SIGN (FAGAN)	SELECT DENARY KEYWORDS (WAITE)
2C	V	♒ (love affairs, sex ties)	love, interrelation between the sexes
3C	V	♒ (merrymaking)	merriment
4C	VI	♓ (secret toils of the mind)	imaginary vexations
5C	IV	♑ (inheritance, patrimony)	inheritance, father
6C	IV	♑ (old Age)	past memories
7C	IV	♑ (magic, occultism)	visions of the fantastic spirit, fairy favours, imagination
8C	III	♐ (spiritual aspirations)	rejection of what was once thought important
9C	VIII	♉ (sound of body)	physical well-being
10C	V	♒ (marriages, love affairs, sex ties)	married couple, love, friendship

CELESTIAL ARCANA

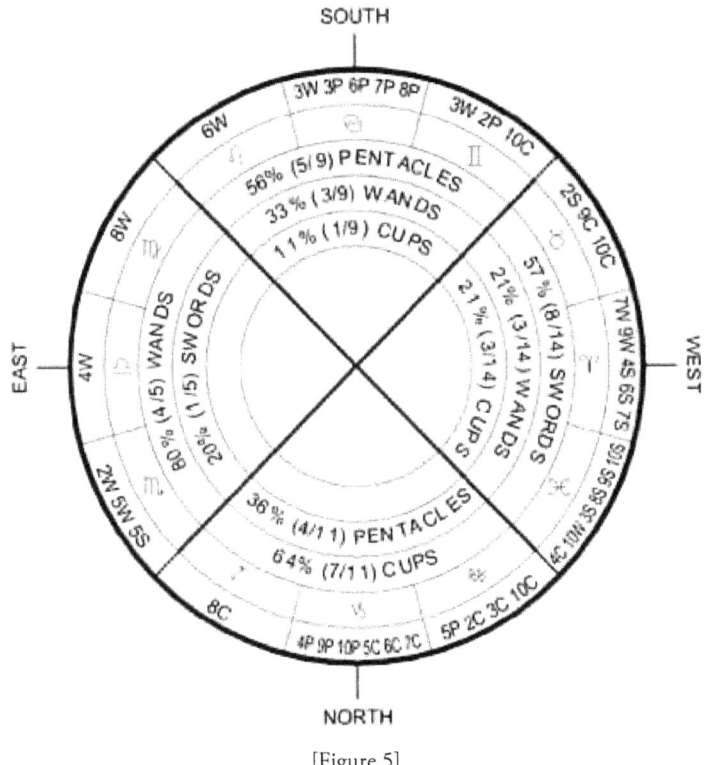

[Figure 5]

Considering the fact that Fagan's house system is acronychal with respect to the Ptolemaic system, applying the requisite inversions yields the following arrangement [Figure 6], which corresponds with that given in [Table 2].

If we consider a construct wherein the Aces of each suit represent the 4 horoscopic "angles" proper, it will be seen that select keywords from Waite's descriptions, and those of the astrological houses corresponding to the Ascendent, Midheaven, Descendent, and Antemeridian (as given by Fagan) are suggestive of the arrangement in [Figure 6] (see [Table 7]).

Regarding the uneven clustering of denaries per house, one possible explanation is that over time, the primary use of the small cards as a decan system wherein each astrological house was divided into three equal units became permutated into the colloquial divinatory descriptions we see today, along with various erroneous attributions creeping in throughout the course of time, either intentionally or accidentally. This does not preclude the notion of the original usage of the cards as having been for divinatory purposes, but a distinction needs to be made between the concept of prognostication via horoscopic analysis and divination by means of random card drawing in combination with some nebulous "mystical" or "magical" factor.

THE MINOR ARCANA

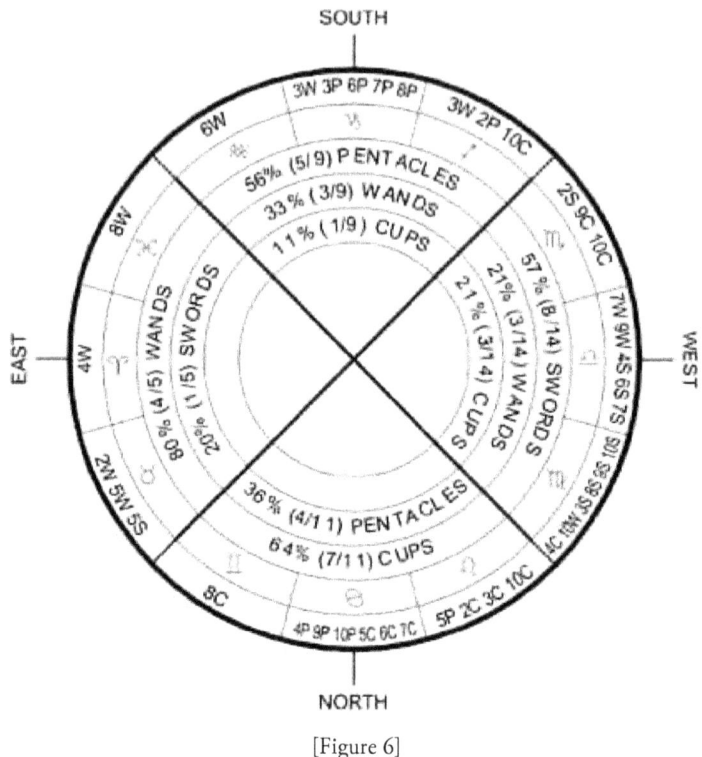

[Figure 6]

TABLE 7 – CORRESPONDENCES BETWEEN FAGAN'S HOUSE DESCRIPTIONS AND WAITE'S ACE KEYWORDS			
ACE	**HOUSE/ANGLE**	**WAITE**	**FAGAN**
Wands	I/Ascendent	beginning, source, birth	life, birth
Pentacles	X/Midheaven	gold, wealth, material conditions	credit, trade, merchants
Swords	VII/Descendent	triumph of force	war, battle, contentions, oppositions
Cups	IV/Antemeridian	abode, nourishment, fertility, abundance	houses, place of residence, womb of earth, prosperity

Notably, in the various methods of divination that Waite discusses, the terms "significator" and "querent" are frequently mentioned. It is quite possible that the use of these terms in popular "fortune-telling" arose from an earlier context of prognostication via strictly astrological methodologies. According to Lilly:

The *querent* is he or she that propounds the question and desires resolution; the quesited is he or she, or the thing sought and inquired after.

The *significator* is no more than that planet which rules the house that signifies the person or thing demanded; as if ♈ is ascending, ♂ being lord of ♈, shall be *significator* of the *querent*, viz. the sign ascending shall in part signify his corporature, body, or stature: the lord of the ascendant, according to the sign he is in, the ☽ and planet in the ascendant, equally mixed together, shall shew his quality or conditions; so that let any sign ascend, what planet is lord of that sign shall be called lord of the house, or *significator* of the person inquiring, &c.

So that, in the first place, when any question is propounded, the sign ascending and his lord are always given unto him or her that asks the question [*querent*].

Secondly: You must then consider the matter propounded, and see to which of the twelve houses it does properly belong: when you have found the house, consider the sign and lord of that sign, how, and in what sign and what part of heaven he is placed, how dignified, what aspect he has to the lord of the ascendant, who impedites your *significator*, who is a friend unto him, viz. what planet it is, and what house he is lord of, or in what house posited; from such a man or woman signified by that planet shall you be aided or hindered, or by one of such relation unto you as that planet signifies; if lord of such a house, such an enemy; if lord of a house that signifies enemies, then an enemy verily; if of a friendly house, a friend.

The whole natural key of astrology rests in the words preceding, rightly understood.[646]

---

[646] Lilly. *Christian Astrology*. Chapter 20. Italics and bracketed comment added.

A Proposed System

THE CORRESPONDENCES BETWEEN the Hebraic alphabet and Major Arcana, according to the considerations of the previous chapters, are given in [Table 1]. The alphabet has been left in its natural sequence and the order of trumps adjusted, as necessary, to match.[647] The reasons such adjustments are required are on the one hand structural: every card has been assigned some astro-planetary or elemental value, and such is the case for each letter of the Hebraic alphabet. The order of trumps is thus a function of which value is assigned to each element in both sets. But there are other reasons for each adjustment that provide additional justification beyond the structural necessities.

In the case of the Hanged Man, the rationale for the ascription of Mercury has already been given. The fact that it is a planetary trump (in this system) gives us leeway to ascribe it to any of the seven "doubles" via recourse to the "blind,"[648] as discussed in the Origins and Attributions chapter. *Dalet* has been chosen – in part – due to the fact that in older versions of this trump, the gallows were depicted in the form of a *Dalet*.[649] This letter signifies a door, and according to Crowley, it is the door of the tree upon which the Adept of Man (the Hanged Man) hangs.[650] Additionally, Mercury/Hermes is known as the gatekeeper of dreams,[651] and the "thief at the gates,"[652] which connects Mercury with "the door," and *Dalet*.

The Fool's ascription to water as the passive element of the precessional triad has been discussed. The placement of the "zero" card between Justice and Death is not an adjustment in the strict sense, as the Fool actually has no numerical value; in earlier decks, no zero is present on the card.

---

[647] See the highlighted rows.

[648] The possibility of different planetary ascriptions being the result of the fact that the planets are "the wanderers" par excellence needs to be addressed. Different orderings would correspond to different dates, and may therefore provide additional astromythological context.

[649] Crowley. *The Book of Thoth*. Page 98.

[650] Crowley. *John St. John – The Record of the Retirement of G. H. Frater, O∴M∴*.

[651] G. C. Lewis. *The Fables of Babrius*. Translation of Aesop's 6th century BCE. "The Sculptor and Mercury".

[652] See Homer's *Hymn to Hermes* in *The Homeric Hymns*, by Hugh G. Evelyn-White.

Hebraic Alphabet		Proposed attribution scheme
Alef	א	I Magician – Air - א
Beit	ב	II High Priestess - ☽ - ב
Gimel	ג	III Empress - ♀ - ג
Dalet	ד	**XII Hanged Man - ☿ – ד**
Hei	ה	IV Emperor - ♈ - ה
Vav	ו	V Hierophant - ♉ - ו
Zayin	ז	VI Lovers - ♊ - ז
Chet	ח	VII Chariot - ♋ - ח
Tet	ט	VIII Strength - ♌ - ט
Yud	י	IX Hermit - ♍ - י
Kaf	כ,ך	X Fortune - ♄- כ,ך
Lamed	ל	XI Justice - ♎ - ל
Mem	מ,ם	0 Fool – Water - מ,ם
Nun	נ,ן	XIII Death - ♏ - נ,ן
Samech	ס	XIV Temperance - ♐ - ס
Ayin	ע	XV Devil - ♑ - ע
Pei	פ,ף	XVI Tower - ♃ - פ,ף
Tzadik	צ,ץ	XVII Star - ♒ - צ,ץ
Kuf	ק	XVIII Moon - ♓ - ק
Reish	ר	XIX Sun - ☉ - ר
Shin	ש	**XXI World – Fire - ש**
Tav	ת	**XX Judgment - ♂ - ת**

[Table 1]

The rationale for the transposition of Judgment and the World is given as follows: the symbolism of the latter has been demonstrated to be fundamentally concerned with the NEP, which in this system represents the active force of the precessional triad, and is associated with the Sun; thus fire; thus *Shin*. The relegation of Judgment to the last card is fitting, in that judgment in the Egyptian, Hebraic, and Christian eschatologies comes at "the end," whether that end is of one's life, or of the Age. Furthermore, the letter *Tav* – to which the Judgment trump has been ascribed – is composed of a *Dalet* and a *Nun*,[653] which together spell "Dan."[654] In Hebrew, "dan" or "din" signifies the administration of judgment.[655] In the Old Testament, Rachel names her son Dan in commemoration of God's judgment upon her, as she

---
[653] Ginsburgh. *The Hebrew Letters*. Page 328.
[654] Ibid.
[655] Strong's Concordance. Entry 1777.

had previously been barren.[656] Dan bore a standard crowned with a serpent in the clutches of an eagle.[657] These are symbols of Scorpio, and Rolleston attests to the fact that the tribe of Dan was ascribed this sign of the zodiac.[658] The Cepheus/Cygnus region of the heptanomis corresponds to the station of the pole during the Age of Scorpio, and the Judgment trump has been linked to that polar station. Considering all of these factors, the transposition of Judgment and the World seems adequately accounted for.

The totality of elements contained within the Major and Minor Arcana are sufficient – when taken together and combined in a certain way – to construct or depict any number of horoscopes, progressions, returns, etc.; in short, anything in which the elements of the zodiac and planets are arranged relative to each other to depict their relationship and cumulative influence upon the terrestrial sphere at any given moment in time.

Regarding the planets outside the sphere of Saturn, little if anything was known during the timeframe in which the Tarot as we know it originated (i.e. the Medieval Ages).[659] Nonetheless, if it is desired to include the outer planets in a scheme utilizing the 22 trumps of the Major Arcana, the 3 "elemental" cards (represented by the Fool, Magician, and World cards in the currently proposed scheme) present themselves as an option. Another (and perhaps more fitting) possibility is the doubling of functionality for the cards ascribed to Mars, Jupiter, and Saturn, as the classical association of these planets to rulership of Scorpio, Sagittarius, and Capricorn has been superseded by the modern attributions of Pluto, Neptune, and Uranus, respectively.

Proceeding to the Court cards, if we take the Kings to represent the foreground angles of any particular sidereal horoscope, return, progression, etc., the Queens the next level of potency of influence, followed by the Knights who exert still less influence; finally arriving at the Pages which occupy the so-called "background" angles of the horoscope, we arrive at the totality of relationships as illustrated in [Figure 1].

If we consider the "denary" or "pip" cards (numbers 2-10) to represent an even finer subdivision of the court card division system, the following configuration presents itself, wherein a total of 72 discrete arcsegments of any particular horoscope may be referred to, each having its own set of properties – unique in and of themselves – with the Aces being attributed to the foreground angles proper, representing the most potent distillation of the properties ascribed to each [Figure 2].

---

[656] Genesis 30:6.
[657] Rolleston, Frances. *Mazzaroth*. (See note on verse 17 of "Genesis XLIX").
[658] Ibid. (See note on Jacob).
[659] According to Fagan, "the outer planets Uranus, Neptune, and Pluto…were unknown prior to the 18th century…". Fagan. *Astrological Origins*. Page 6.

CELESTIAL ARCANA

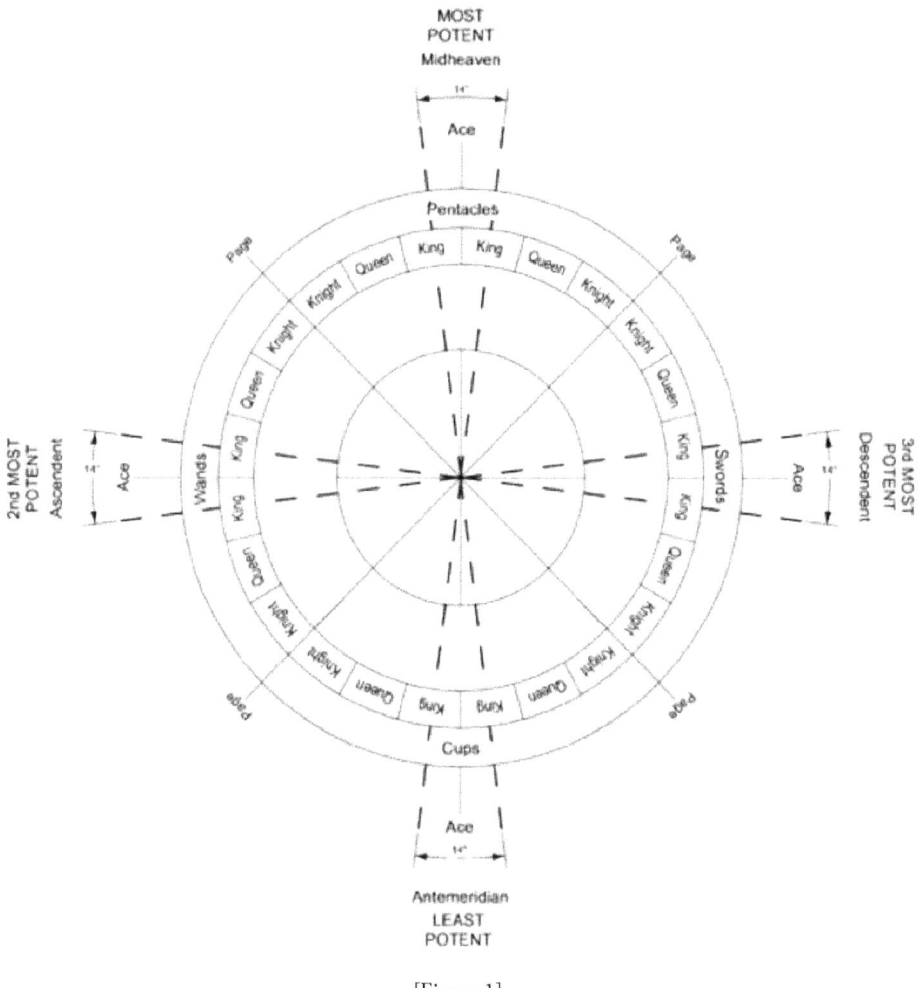

[Figure 1]

In this arrangement, there is a bilateral doubling of each suit of court cards and pips along the axes formed by the Aces. On the surface, it may seem more parsimonious to divide each suit's quadrant into 9 units (2-10), rather than the proposed 18, thereby eliminating the duplication and avoiding any misunderstanding as to which sector of the horoscope is being referred to. For example, in the proposed system, the 9 of Wands could refer to the 5° arcsegment to either side of the Ace of Wands, thus seeming to present an ambiguity.

This difficulty is resolved when we add the zodiacal trumps around the circle, their rotational justification being determined by which sign occupies the Ascendent, Midheaven, etc., for any particular moment in time.

A PROPOSED SYSTEM

Additional specificity can be achieved by the following convention of Minor Arcana card inversion: if the card is upright, it is to be read as referring to one side of the angle or Ace, and if inverted, to the opposite side [Figure 3].

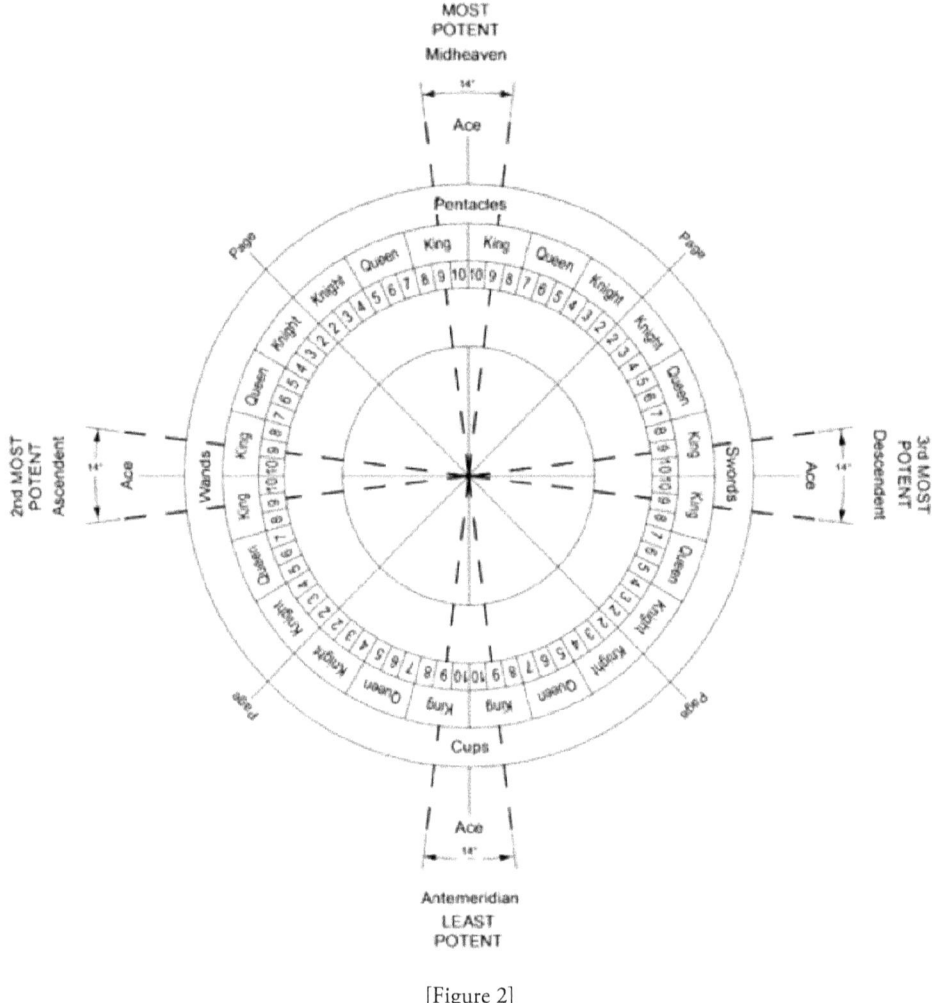

[Figure 2]

Another advantage of this arrangement is that the numerical value of the pips can be seen to correspond to the relative strength of the foreground and background angles. Adding the house divisions to this arrangement produces the configuration as illustrated in [Figure 4]. Here each house is subdivided into six 5° arcsegments, each of which corresponds to one of the denaries, exclusive of the Aces. Here it should be noted that houses II, V, VIII, and XI– which are "cadent," and according to Fagan, "sluggish" – are each ascribed to the lowest numbered denaries, which is in keeping with the aforementioned concept of angular potency as represented by numerical magnitude. [660]

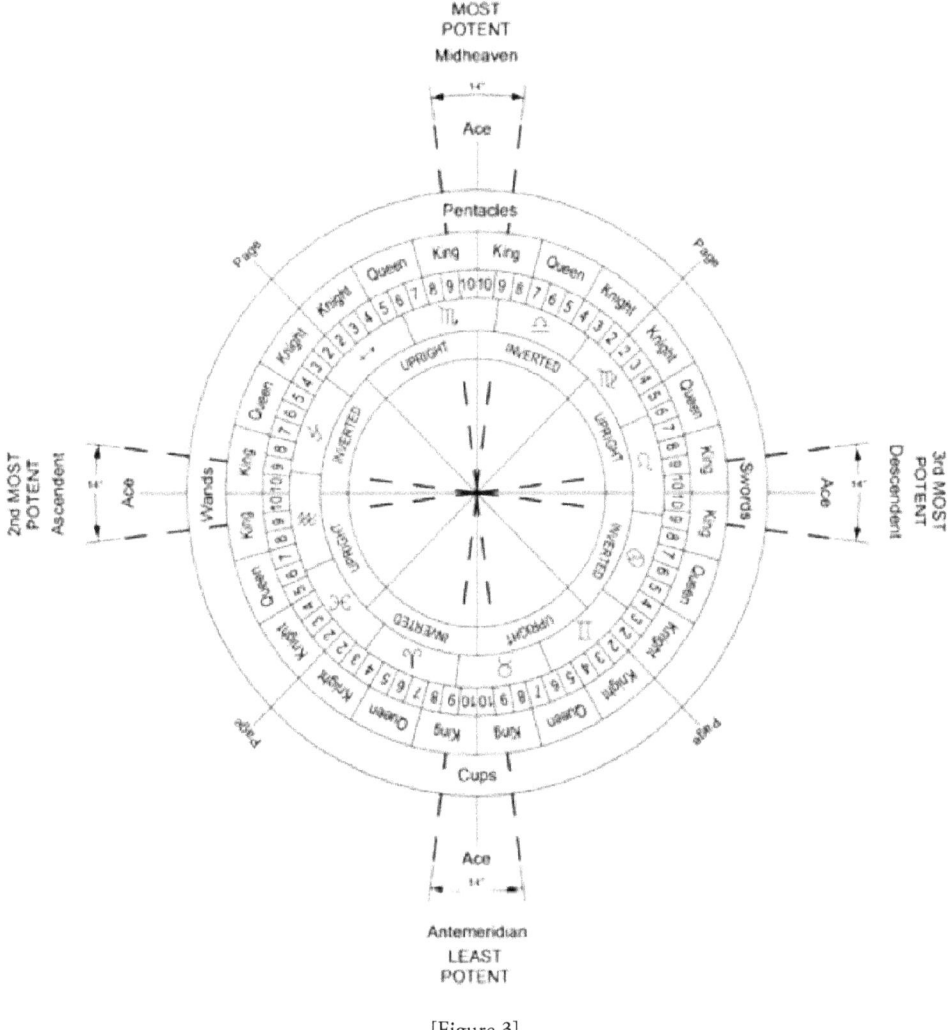

[Figure 3]

---

[660] Roman numerals for houses should not be confused with those used for the Trumps Major; there is no fixed connection between the two.

A PROPOSED SYSTEM

[Figure 4]

There is also the categorization of the four suits relative to the four "divisions of man"[661] – i.e. the four "centers," or "brains" – as developed by Ouspensky. These are the intellectual center, localized in the head, the emotional center, corresponding to the bundle of nerves in the solar plexus,

---

[661] Ouspensky. *A New Model of the Universe*: "The four principles or the four letters of the Name of God, or the four alchemical elements, or the four classes of spirits, or the four divisions of man (the four Apocalyptic beasts) correspond to the four suits of the Tarot: wands, cups, swords and pentacles. Each suit, each side of the square which as a whole is equal to the point, represents one of the elements, or governs one of the four classes of spirits. Wands are fire or elves, cups are water or water-sprites, swords are air or sylphs and pentacles are earth or gnomes. Moreover, in each suit the King stands for the first principle or fire, the Queen for the second principle or water, the Knight for the third principle or air, and the Page (Knave) for the fourth principle or earth."

and the moving and instinctive centers, generally considered to be governed by the nerves of the spinal column, although it should be noted that each system interpenetrates the other. This discrete, yet interpenetrating nature of the centers is suggested by the division of each one into intellectual, emotional, and moving (or "mechanical") parts, and the further subdivision of these into still finer gradations according to the same formula. Thus, if each suit corresponds to a particular "brain" of the human organism, the court cards represent the first level of subdivision, and the denaries the second [Figure 5].

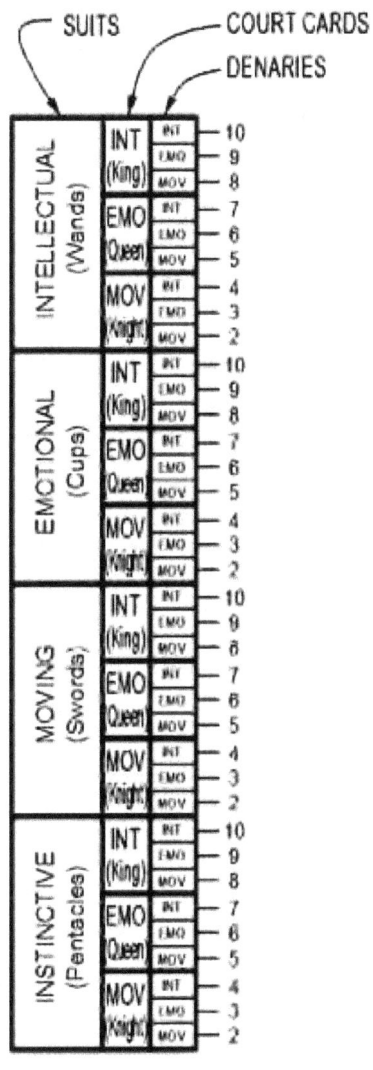

[Figure 5]

A PROPOSED SYSTEM

When the suits are considered from this perspective, yet another rationale is provided for the bilateral doubling of pips, for according to Ouspensky, in addition to the various "vertical" divisions of the centers, each one was also divided into two halves, one "positive," and the other "negative"[662] [Figure 6].

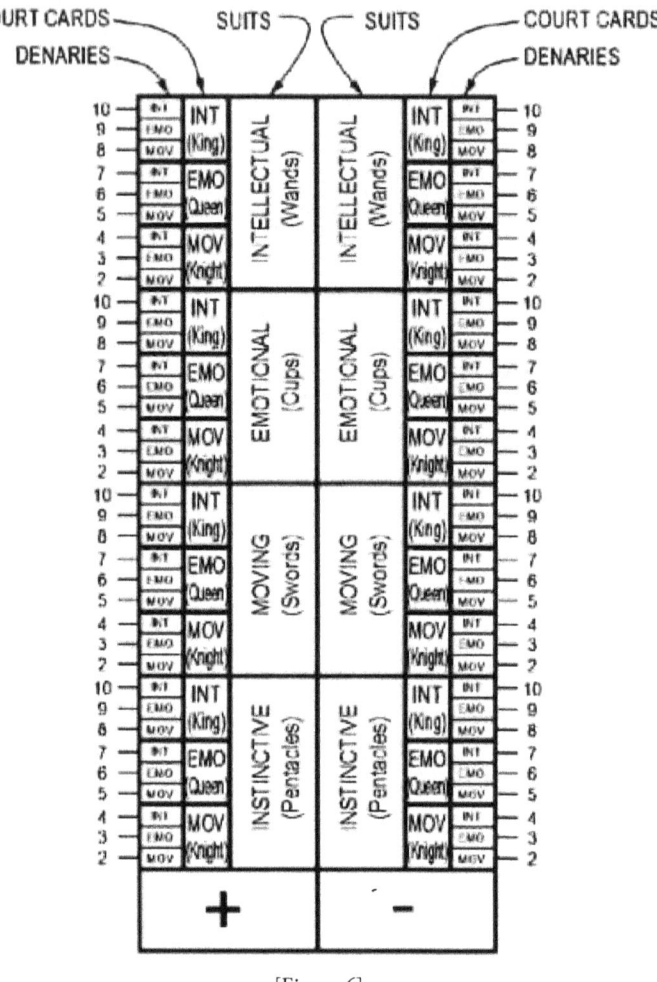

[Figure 6]

While it may be objected that the divisioning per [Figure 4] into what essentially amounts to a pentade system is at odds with the fact that there are 36 denaries (exclusive of the Aces), which suggests an intended decan system, Fagan has shown that ancient Egyptian astrology was based on a pentadic divisioning of the ecliptic into 5° units – giving a total of 36 pentades for the

---

[662] Ouspensky. *In Search of the Miraculous.* Page 257.

diurnal heavens, and a complementary 36 for the underworld.[663] This bilateral duplication of a set of 36 discrete units may have at some point given rise to the concept of card reversal, wherein a drawing of an inverted denary is said to indicate a diametrically opposed or complementary meaning with respect to its upright counterpart. This being said, the pentade system may prove too cumbersome for some operations, in which case a decan system is just as easily implemented. There are numerous possible arrangements for such a system, each potentially being the best suited for a particular application.

The question of which suit corresponds to which brain or center is ultimately one of convention. For a determination of the correspondences between the suits of the standard deck of playing cards and those of the Minor Arcana, we reference Waite's *Pictorial Key*, in which he hypothesizes the following equations: Diamonds=Wands, Cups=Hearts, Pentacles=Spades, and Swords=Clubs. There is, however, no canonical relationship between the suits of both types of deck. Since these correspondences have been presented *ex hypothesi* by Waite, and the airy element represented by Swords seems to better represent activity and movement more so than the earthy Pentacles, I have chosen to equate Swords with Spades and the moving center, and Pentacles with Clubs and the instinctive center in [Figures 5 and 6].

One should not make the mistake of considering any of the suits or centers to be fixedly associated with a particular zodiacal sign, in any direct sense. Let us take for example Ouspensky's indication that the four apocalyptic beasts (and the dove) were somehow connected to the concept of "centers."[664] For the sake of illustration, we consider the intellectual center to be associated with Aquarius, the emotional center with Leo, and the instinctive/moving (i.e. "gut") center with Taurus. The eagle, or Scorpio can then be said to be related to the Higher Self, which draws up the dove (which we have previously identified with Columba and/or Sirius) from the depths of the physical world to the higher spiritual world.[665]

It hardly seems reasonable to interpret these correspondences as indicating that the constellations of Aquarius, Leo, and Taurus in any way govern the intellectual, emotional, or instinctive/moving centers, respectively. It would make more sense to consider these relationships to be symbolic conventions, which indicate anthropomorphic or zoomorphic associations particular to

---

[663] Fagan. *Astrological Origins*. Chapters 7 & 8.
[664] Ouspensky. *In Search of the Miraculous*. Page 295.
[665] Cf. al-Qashani. *A Glossary of Sufi Technical Terms*. Page 86. Entry 393: "THE EAGLE: For Sufis this sometimes expresses the Primal Intellect, while at other times it expresses nature in its totality. This is because they refer to the Rational Soul as a 'dove,' which the Primal Intellect snatches up, like an eagle, from the depths of the lower, physical world, to the sublime world and sacred outer reaches of space. Alternatively it may be snatched and captured by nature, and fall down with it to the lower depths. That is why both have been called the Eagle: the difference between them lies in the context."

each center. Thus, the intellectual center and its frontal lobes are preeminently human in nature and characterized by the only human zodiacal sign. The lion can be seen as representing the desirable qualities of courage and strength in the emotional center, while also illustrating the potential dangers of emotions unchecked. The bull may represent the brutish "unthinking" nature of the instinctive/moving center – which, however, can be utilized to perform useful work. The dove might signify the latent soul, or the potential to develop an astral body that may resonate to the frequencies emanating from the Galactic Center, located in the direction of Scorpio (the eagle).

The problem of the somewhat misogynistic tinge to this arrangement – where the King is associated with the intellectual parts of centers, as well as the angular and most significant portions of the horoscope, is of course a function of the medieval culture which gave birth to the classes of suits as we know them today. Crowley's replacement of the King with the Knight (who presumably serves the Queen), and the Page with the Princess – who is elevated to the lofty heights of the circumpolar realms – in a sense can be seen as a suggested rectification of the lopsided gender roles in anticipation of the new Aeon.

The Princesses are described as representing four distinct "elemental" types of people who

> lack ... all sense of responsibility, [and] whose moral qualities seem to lack "bite." They are sub-divided according to planetary predominance. Such types have been repeatedly described in fiction. As Eliphas Levi wrote: "The love of the Magus for such creatures is insensate, and may destroy him."[666]

Here Crowley is quoting from Levi's *Magical Ritual of the Sanctum Regnum*, but in fact, the creatures being spoken of towards which the Magus is so profoundly attracted are not human types at all.[667] They are the spiritual manifestations, or aspects of consciousness, of the four elements – the so-called "elementals." According to Levi, these forces of elemental consciousness have an intense curiosity, are innocent of good and evil, and may be controlled by the Magus towards either end, although such manipulation entails enormous responsibility on the Magus' part; and if such conscious forces are abused, can lead to his or her undoing.[668]

In Crowley's arrangement of small cards, the order of planetary attributions for each sequential denary clearly follows the so-called "Chaldean Order" of the "faces" of planets as given by Lilly in *Christian Astrology* [Tables 2-6].[669]

---

[666] Crowley. *The Book of Thoth*. Page 150. Ellipses added.
[667] Additionally, there is no well-defined astrological basis for the notion of certain "types" being connected with the circumpolar regions.
[668] Levi. *Magical Ritual of the Sanctum Regnum*.
[669] Tables derived from Crowley's *Book of* Thoth (page 283), and Lilly's *Christian Astrology*, Chapter 18: "Table of the Essential Dignities of the Planets."

Table 2: Crowley's Titles and Attributions of the Wand Suit (Clubs)	Denary
The Root of the Powers of Fire (Ace of Wands)	1
♂ in ♈ = Dominion	2
☉ in ♈ = Virtue	3
♀ in ♈ = Completion	4
♄ in ♌ = Strife	5
♃ in ♌ = Victory	6
♂ in ♌ = Valour	7
☿ in ♐ = Swiftness	8
☽ in ♐ = Strength	9
♄ in ♐ = Oppression	10

Table 3: Crowley's Titles and Attributions of the Cup Suit (Hearts)	Denary
The Root of the Powers of Water (Ace of Cups)	1
♀ in ♋ = Love	2
☿ in ♋ = Abundance	3
☽ in ♋ = Luxury	4
♂ in ♏ = Disappointment	5
☉ in ♏ = Pleasure	6
♀ in ♏ = Debauch	7
♄ in ♓ = Indolence	8
♃ in ♓ = Happiness	9
♂ in ♓ = Satiety	10

Table 4: Crowley's Titles and Attributions of the Sword Suit (Spades)	Denary
The Root of the Powers of Air (Ace of Swords)	1
☽ in ♎ = Peace	2
♄ in ♎ = Sorrow	3
♃ in ♎ = Truce	4
♀ in ♒ = Defeat	5
☿ in ♒ = Science	6
☽ in ♒ = Futility	7
♃ in ♊ = Interference	8
♂ in ♊ = Cruelty	9
☉ in ♊ = Ruin	10

A PROPOSED SYSTEM

Table 5: Crowley's *Titles and Attributions of the Pentacle Suit (Diamonds)*	Denary
The Root of the Powers of Earth (Ace of Pentacles)	1
♃ in ♑ = Change	2
♂ in ♑ = Work	3
☉ in ♑ = Power	4
☿ in ♉ = Worry	5
☽ in ♉ = Success	6
♄ in ♉ = Failure	7
☉ in ♍ = Prudence	8
♀ in ♍ = Gain	9
☿ in ♍ = Wealth	10

Table 6: Lilly's System of Faces as extracted from his *Table of the Essential Dignities of the Planets*			
Zodiacal Sign	1st Decan of Sign = 1st FACE	2nd Decan of Sign = 2nd FACE	3rd Decan of Sign = 3rd FACE
♈	0°-10° ♈ similar to influence of ♂	10°-20° ♈ similar to influence of ☉	20°-30° ♈ similar to influence of ♀
♉	0°-10° ♉ similar to influence of ☿	10°-20° ♉ similar to influence of ☽	20°-30° ♉ similar to influence of ♄
♊	0°-10° ♊ similar to influence of ♃	10°-20° ♊ similar to influence of ♂	20°-30° ♊ similar to influence of ☉
♋	0°-10° ♋ similar to influence of ♀	10°-20° ♋ similar to influence of ☿	20°-30° ♋ similar to influence of ☽
♌	0°-10° ♌ similar to influence of ♄	10°-20° ♌ similar to influence of ♃	20°-30° ♌ similar to influence of ♂
♍	0°-10° ♍ similar to influence of ☉	10°-20° ♍ similar to influence of ♀	20°-30° ♍ similar to influence of ☿
♎	0°-10° ♎ similar to influence of ☽	10°-20° ♎ similar to influence of ♄	20°-30° ♎ similar to influence of ♃
♏	0°-10° ♏ similar to influence of ♂	10°-20° ♏ similar to influence of ☉	20°-30° ♏ similar to influence of ♀
♐	0°-10° ♐ similar to influence of ☿	10°-20° ♐ similar to influence of ☽	20°-30° ♐ similar to influence of ♄
♑	0°-10° ♑ similar to influence of ♃	10°-20° ♑ similar to influence of ♂	20°-30° ♑ similar to influence of ☉
♒	0°-10° ♒ similar to influence of ♀	10°-20° ♒ similar to influence of ☿	20°-30° ♒ similar to influence of ☽
♓	0°-10° ♓ similar to influence of ♄	10°-20° ♓ similar to influence of ♃	20°-30° ♓ similar to influence of ♂

It should be noted that this arrangement is quite artificial. One may be tempted to assume from such a configuration that the 1st decan of Aries is truly similar to the Martial influence – since, after all, Mars is the purported ruler of Aries. However, this logic breaks down as soon as we come to the 1st decan of Taurus – which we would expect to have qualities similar to Venus, as she is the supposed planetary ruler of that sign; and on and on in a similar fashion throughout the rest of the sequence, until we come to the 1st decan of Scorpio, which – according to this arrangement – has the qualities of Mars, which also happens to be the classical ruler of that sign. Subsequently, from Sagittarius to Pisces, said logic again fails to obtain. The fundamental rationale behind this dubious arrangement appears to be a hybridization of conventions, where *Mars has been ascribed to Aries by virtue of the tropical designation of that sign as the vernal opener of the year, and each face or decan thereafter – throughout the entire zodiac – is ascribed according to the Chaldean (or apparent geocentric) order.*[670] It must be admitted that such an arrangement has no basis in causality, but is in fact somewhat arbitrary.

It is unclear why Crowley would use such a system to organize his denaries, although one interesting detail may indicate a possibly satirical motive, as William Lilly himself had the following to say about such an arrangement:

> A planet being in his decante, or face, describes a man ready to be turned out of doors, having much to do to maintain himself in credit and reputation and in genealogies it represents a family at the last gasp, even as good as quite decayed, hardly able to support itself.[671]

It is difficult if not impossible to say where the face system espoused by Lilly originated. It could well be that such a concept ultimately derives from Ptolemy's *Tetrabiblos*, but as the following quote from Chapter 26 ("Faces, Chariots, and Other Similar Attributes of the Planets") of this work shows, Ptolemy's concept of face by no means implied a fixed ecliptic locus nor was it in any way organized according to the Chaldean planetary order:

> The familiarities existing between the planets and the signs are such as have been already particularised.
>
> There are also, however, further peculiarities ascribed to the planets. *Each is said to be in its proper face, when the aspect it holds to the Sun, or Moon, is similar to that which its own house bears to their houses: for example, Venus is in her proper face when making a sextile aspect to either luminary, provided she be*

---

[670] The explanation of the Chaldean order is given as follows: Earth was considered to be at the center of the cosmos, with the rest of the planets being arranged concentrically outwards according to their apparent orbital speed, giving the order of Moon, Mercury, Venus, Sun, Mars, Jupiter, and Saturn.
[671] Ibid. Chapter 17: "Of the Essential Dignities of the Planets".

A PROPOSED SYSTEM

occidental to the Sun, but oriental to the Moon, agreeably to the primary arrangement of her houses.[672]

It is possible that Lilly's system of faces arose from a conflation between Ptolemy's face concept and the so-called "disposition of terms" as given in chapter 23 of the *Tetrabiblos*. Essentially, a "term" is a fixed portion of each zodiacal sign (ranging anywhere from 3-12°), which has ascribed to it the characteristics of one of the 5 ancient planets (Saturn, Jupiter, Mars, Venus, Mercury). Each sign is divided somewhat evenly into 5 terms, resulting in 60 terms comprising the totality of the ecliptic.

Ptolemy gives two methods of term distribution in chapter 23 – one according to an ancient Egyptian system; the other to a Chaldean system, each of which he considers either "defective," or "highly imperfect."

Another interesting face system is that developed by William Salmon throughout the first 13 chapters of his *Horae Mathematicae* of 1679. This system divides each zodiacal sign into six pentades for a total of 72 faces, the planetary attributions of which are illustrated in [Figure 7].

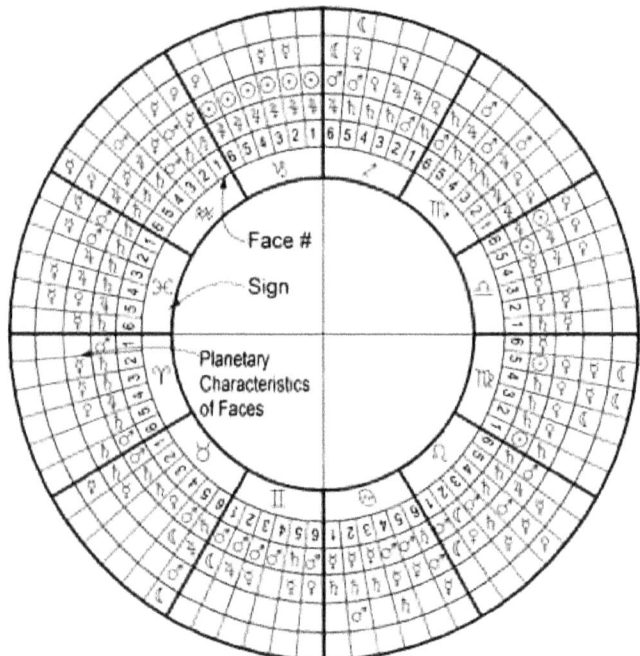

Diagram Derived from William Salmon's
Hora Mathematicae
(Chapters 2-13)

[Figure 7]

---

[672] Ptolemy. *Tetrabiblos*. Chapter 26. Italics added.

Salmon does not indicate the basis for his system, but it seems likely that it ultimately derives from some permutation of the original ancient Egyptian or Chaldean distribution of terms.

All of these systems are unsatisfactory in no small part due to the fact that the planets are constantly changing their positions in the zodiac, and therefore their influence can never be once-and-for-all ascribed to any particular ecliptic locus. As has been noted previously, even the Ptolemaic domicile system is suspect, as far as attribution of planetary qualities to entire zodiacal signs is concerned.

The question must be asked as to why none of the different face or term systems proposed by some of the most influential minds shaping Western astrology agree with each other. The answer appears to be that because at best they are based upon subjective interpretations of the astrological effects of the various faces, and at worst, the rationale underlying such attributions is completely erroneous.

The final Minor Arcana organizational scheme to be discussed (see [Figure 8]) is similar in many ways to the one illustrated in [Figure 2], except the Page court card has been eliminated and "replaced" with the Dame,[673] and the Aces have been renamed as Avatars. Strictly speaking, however, the Dame does not replace the Page in this arrangement, as she is to be seen as equal and complimentary to the Knight. Similarly, the King and Queen are here presented as complimentary equivalents. In this scheme, the Avatars represent the foreground angles, which are themselves inclusive of the denary cards 8, 9, and 10. The aforementioned issue of bilateral duplication is solved in the case of the court cards by the pairing of the King and Knight to one side of each angle, and the Queen and Dame to the other. The question regarding which side of the angle to which any particular Avatar may refer is answered via recourse to the convention of card-reversal. Because this scheme utilizes the same number of trump, court, and denary cards as standard tarot decks (i.e., the Rider-Waite, or the Crowley-Harris decks), it may also be used for any of the prognosticative operations and "card spreads" as any other 78-card set. Some popular spreads include the "10-card" or "Celtic cross" spread, the "name" spread, the "horseshoe" spread, the "royal" spread, the "7th-card" spread, and the so-called "gypsy" spread.[674]

In the event one wishes to utilize the pentadic arrangement for prognosticative operations involving the random drawing of cards, the following 22-card spread is recommended:

---

[673] "Dame" is the official title of a female Knight.
[674] See *The Encyclopedia of Tarot, Volume I*, pages 337-344 for further details on each of these spreads. Also, see *The Pictorial Key to the Tarot* (Part III, Sections 7-9) for a discussion of the Celtic cross spread, as well as an alternate gypsy spread, and a 35-card spread.

A PROPOSED SYSTEM

[1] Separate the entire deck into 3 stacks, consisting of:
    (a) 56 cards: denaries (36 cards); court cards (16 cards); Avatars (4 cards) – shuffle.
    (b) zodiacal trumps (12 cards) – shuffle.
    (c) planetary trumps (10 cards) – shuffle.[675]

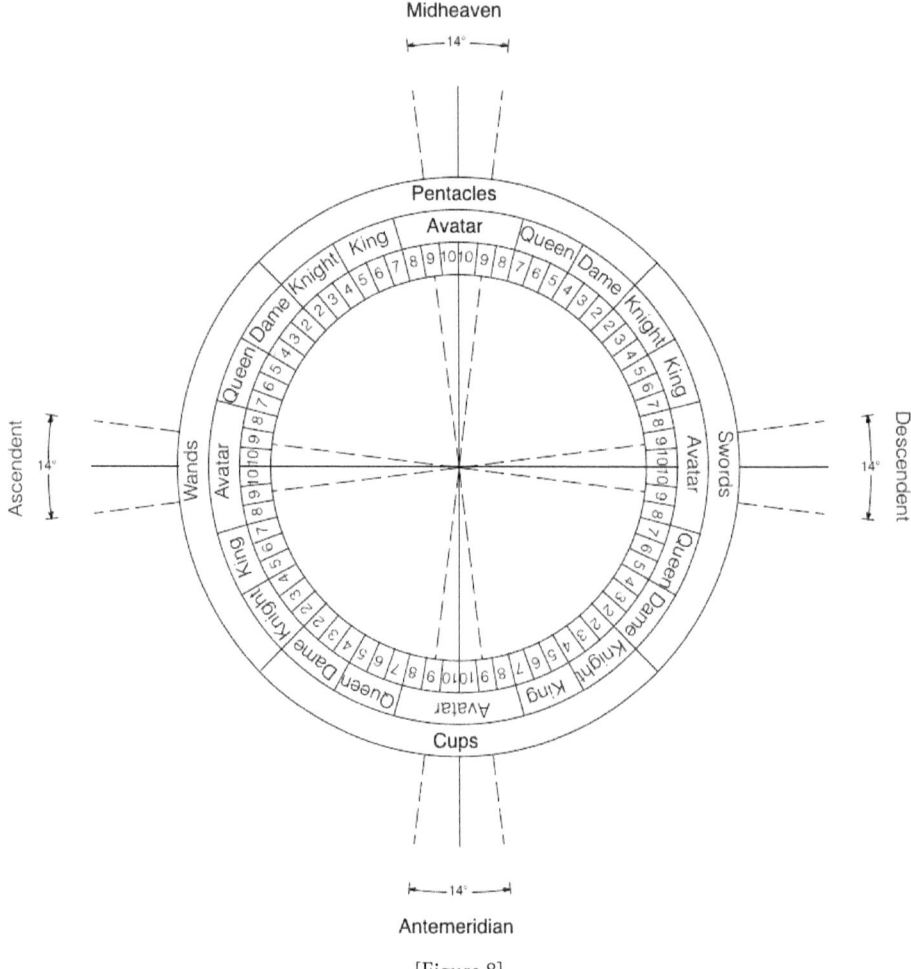

[Figure 8]

[2] Draw one card each from stack (a) and (b). This sets the rotational justification of the zodiac, and thus determines the signs occupying the Ascendent, Midheaven, Descendent, and Antemeridian. The exact

---

[675] According to the present system, the Fool is equated with Water and Neptune, the Magician with Air and Uranus, and the World with Fire and Pluto.

CELESTIAL ARCANA

positioning of the zodiacal trump is found by matching its 0° point to the 0° point of the denary, court card, or Avatar drawn in this step [Figure 9].

[3] Draw one card each from stacks (a) and (c) until all planetary trumps are paired with a specific denary, court card, or Avatar. This sets each planet in a specific house, sign, decan, etc.

[4] Read the resulting horoscope as though applied to the Querent.

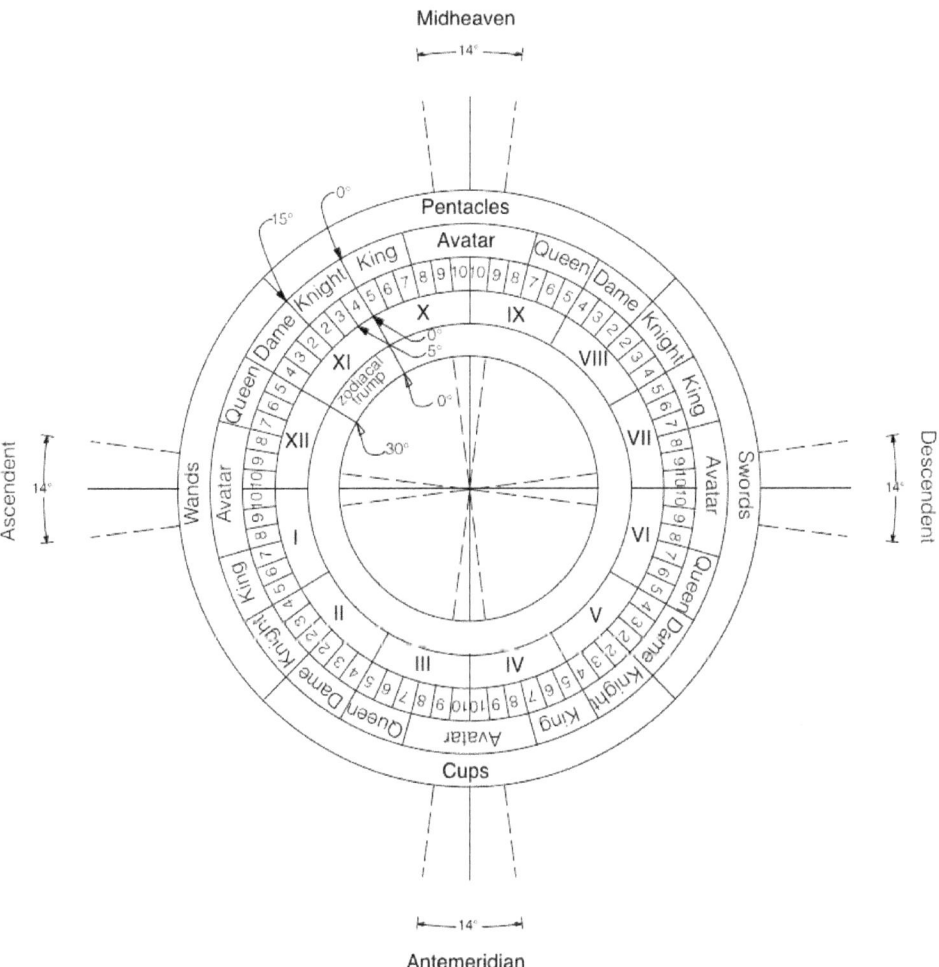

[Figure 9: Setting the rotational justification of the zodiac via matching the 0° point of the zodiacal trump drawn with that of the card drawn from stack (a). In this example, the card drawn from stack (a) could be the Knight of Pentacles, or an upright 4 of Pentacles.]

A PROPOSED SYSTEM

In the 22-card spread, the court and Avatar cards give the advantage to the Querent by enhancing angularity (increasing potency of effect) for benefic configurations, and mitigating it (decreasing potency of effect) for malefic ones. According to an "associative thinking" rationale, this may be considered to be the result of these cards having humanistic sympathies, as they are anthropomorphic. Structurally, this is possible due to the fact that the court cards and Avatars each comprise 15°, whereas the zodiacal trumps comprise 30°, and the planetary trumps 5°. Thus, the rotational justification of the zodiacal trumps can be varied by ±15°, and that of the planetary trumps by ±10° [Figure 10].

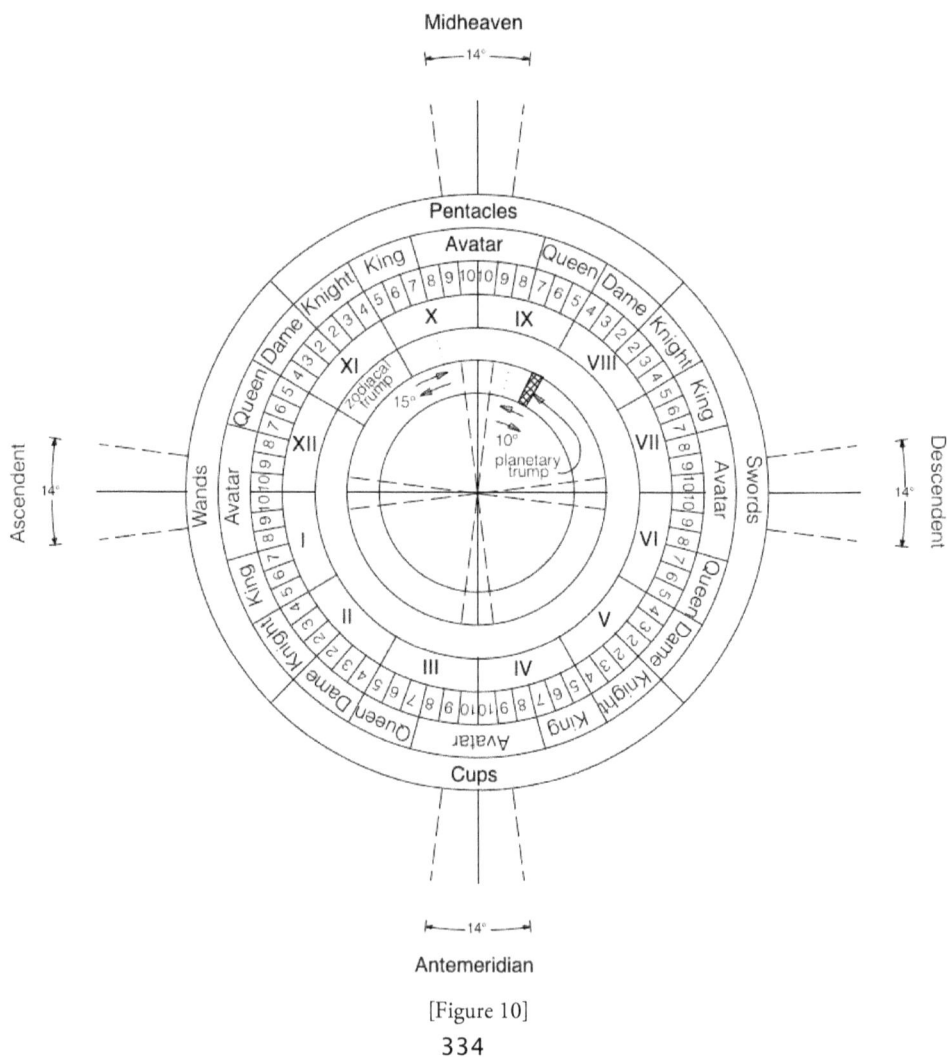

[Figure 10]

CELESTIAL ARCANA

The denary pairings are unable to be mitigated or enhanced. Associatively, this is because the denaries' effects derive from the world of numbers, which is abstract and impartial. Structurally, the reason is that each denary and planetary card comprise 5°, thereby leaving no room for adjustment.

The following sample reading will clarify some of these points:

[1] The 3 stacks are separated and thoroughly shuffled, making sure to randomize the directionality of the cards in stack (a).

[2] The first 2 cards are drawn: an upright 9 of Pentacles from stack (a), and Death from stack (b). This sets the Ascendent at 25° Capricorn [Figure 11].

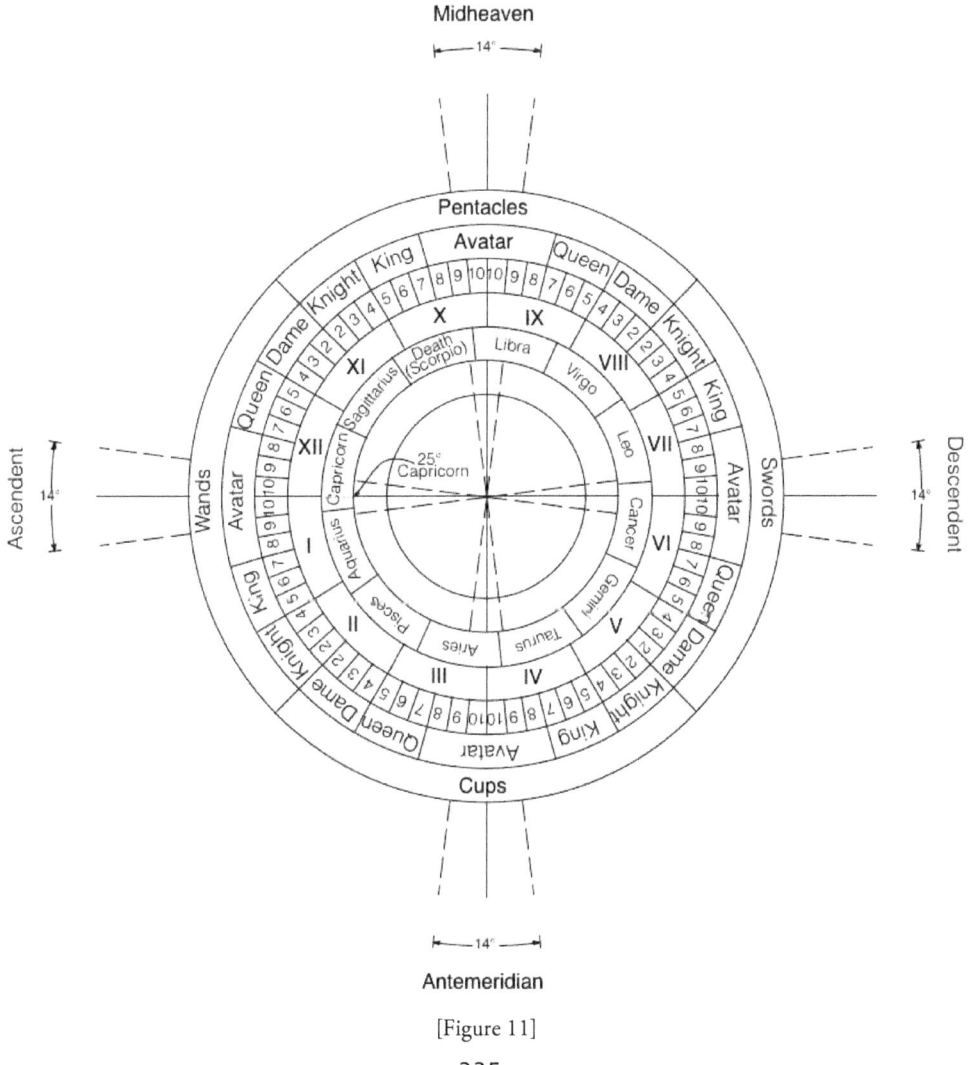

[Figure 11]

A PROPOSED SYSTEM

[3] The locations of the 10 planetary trumps are determined by pairing each trump from stack (c) with one drawn from stack (a). The results are given in [Table 7].

Card drawn from stack (a)	Card drawn from stack (c)
Inverted 7 of Wands	Judgment - ♂
Inverted Avatar of Wands	The Empress - ♀
Upright Avatar of Wands	The World - ♇
Inverted 9 of Swords	The High Priestess - ☽
King of Wands	The Tower - ♃
Inverted 4 of Pentacles	The Hanged Man - ☿
Upright 3 of Cups	The Sun - ☉
Inverted 4 of Swords	The Magician - ♅
Inverted Avatar of Pentacles	The Fool - ♆
Inverted Avatar of Cups	Fortune - ♄

[Table 7]

Each planetary trump is initially placed at the default 0° point of its pair drawn from stack (c) [Figure 12].

[4] The only significant planetary influences in a chart are the angular ones.[676] In the current spread, both Pluto and the Moon are angular – each in opposition to the other.

Angular Pluto connotes unexpected psychological shocks that can prove to be devastating. One may run the risk of having deep secrets exposed, experience unexpected confrontations with the police, being beaten, robbed, and even sudden death. The underlying effects of this configuration engender the impulse to flee from the circumstances at hand.[677]

The Moon's angularity and opposition to Pluto can be read to signify the urge to escape tyranny or authority, a fugitive state, feelings of oppression, etc.[678]

---

[676] See Fagan. *The Solunars Handbook*, page 97.
[677] See Fagan. *Primer of Sidereal Astrology*, pages 131-132.
[678] Fagan. *The Solunars Handbook*, page 103.

The corresponding reading for the default "0° to 0°" configuration is thus rather grim. However, four of these card drawings offer the possibility of shifting the planet in question into or out of angularity, thereby mitigating or enhancing benefic or malefic effects.

Since Pluto (The World) is matched with an upright Avatar of Wands, it is permissible to shift the planetary card 10° from its default position to occupy the denary position of the upright 8 of Wands, thus taking it off the angle and lessening its negative effects.

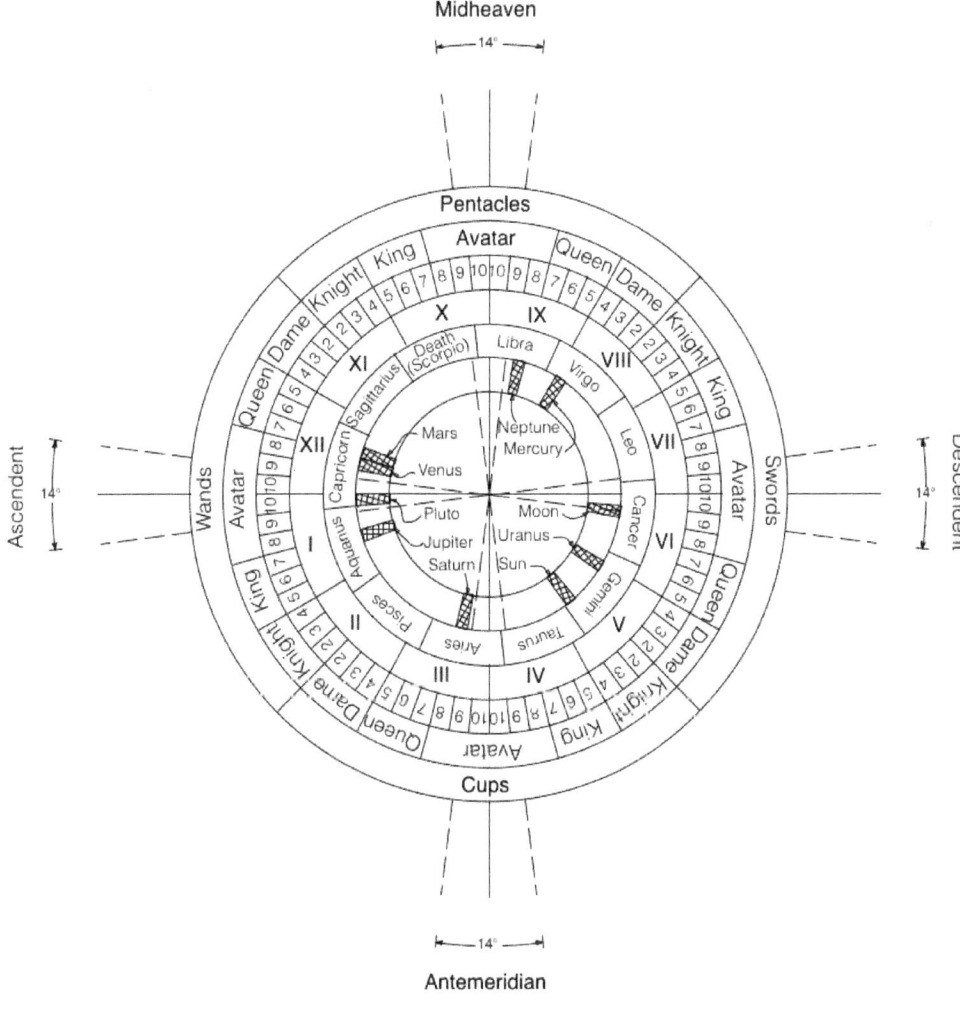

[Figure 12]

We are also able to shift the Venus card (The Empress) 10° to the position of the inverted 10 of Wands, thus making Venus angular. This configuration

A PROPOSED SYSTEM

brings good will, affection, favors, gifts, pleasantries, and stirs the Querent's affections into action. The resultant configuration of Venus to the Moon (each opposite the other) suggests popularity with the opposite sex.[679]

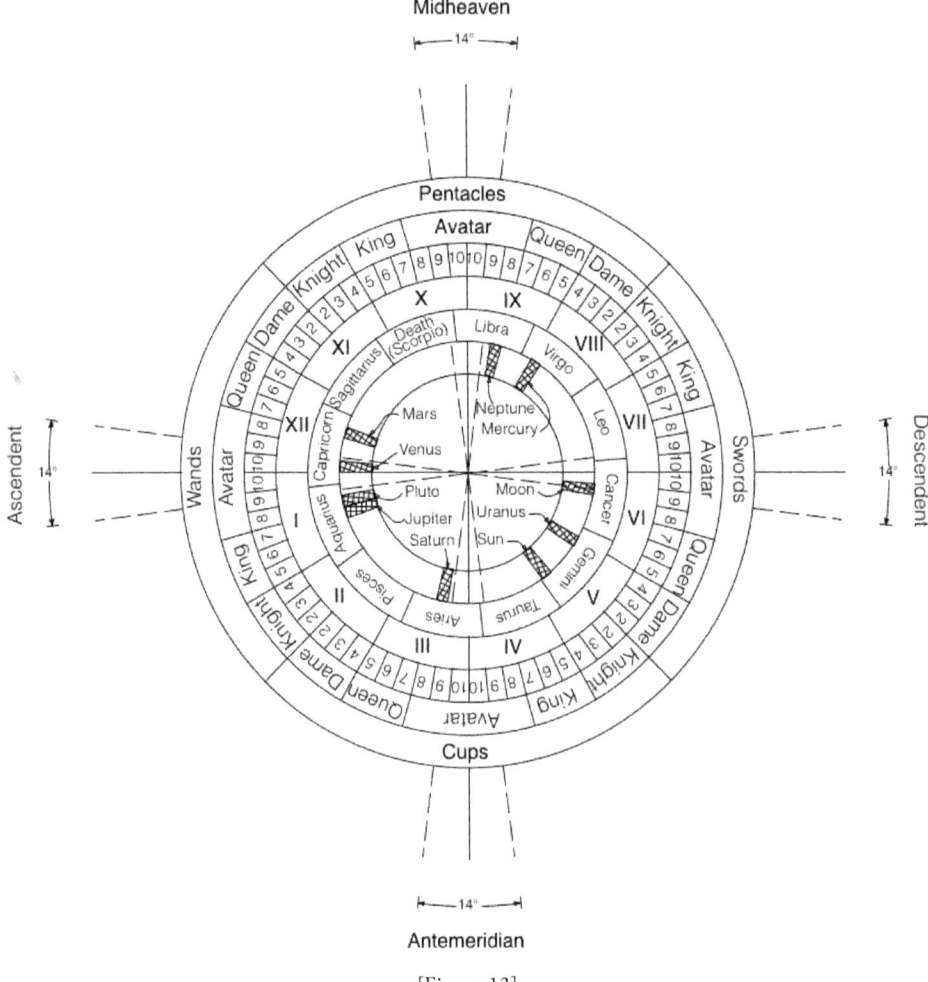

[Figure 13]

The Fool's pairing with the inverted Avatar of Pentacles offers the possibility of shifting Neptune from its default position of the inverted 8 of Pentacles to the inverted 10 of Pentacles, thus placing Neptune on the angle. Angular Neptune engenders a desire in the Querent to escape their responsibilities, becoming lost in their favorite distractions.[680] In the present

---

[679] Fagan. *Primer of Sidereal Astrology*, page 121.
[680] Ibid. Page 130.

case, the shift has caused Neptune to be in significant aspect to Venus (square), betokening the liability of the Querent to become ensnared in romantic hopes, which may lead to exploitation.[681] For this reason, the Avatar can do little to help the Querent, other than maintain the default position of the inverted 8 of Pentacles, thereby avoiding any negative effects of an angular Neptune.

The last possible adjustment from the default mode concerns the pairing of the inverted Avatar of Cups with the Fortune trump, to which Saturn is ascribed. In this instance, we have the possibility of shifting the position of the Saturnine trump from the inverted 8 of Cups to the inverted 10 of Cups, thereby making Saturn angular and configured with Venus. However, since this positioning suggests neglect by one's partner (up to the point of desertion), and in general, the angularity of Saturn presages malefic effects,[682] the Avatar once again chooses to shield the Querent from potential disaster by reversion to the default mode, which takes Saturn off the angle.

The final result of the operation is given in [Figure 13]. The Querent may face some imminent danger of slander or prosecution, possibly involving physical injury, but higher forces watch over them and without fail provide opportunities at each critical juncture that can turn such potentially negative situations into ones that promise affection and even success in romance.

---

[681] Ibid. Page 131.
[682] Ibid. Page 126.

Conclusion

OUR STUDY OF the Tarot has focused on the various myths and meanings encoded within its symbols, with particular emphasis being placed on their relation to the phenomenon of precession, and the Secret Doctrine of Aeonic succession. Much of the symbolism has been shown to derive from ancient Egyptian, Celtic, Hellenistic, and medieval customs and rituals, in some cases dating as far back as the Age of Taurus (c. 4200-2100 BCE), or even earlier.

The annual flooding of the Nile River, and its connection to the star Sirius was of paramount importance during this era, and it is demonstrated how much of the mythology and symbolism of later Ages and cultures are actually derivations from this fundamental mythos – in which the Mother Goddess was revered rather than vilified, as the case has subsequently become during the patriarchal Ages of Aries and Pisces. Thus, the Biblical account of the Fall of Humanity is interpreted as a record in astromythological form of the usurpation of the ancient stellar cosmology that gave preeminence to the Great Mother by the solar and patriarchal regime, wherein Jehovah became the central deity. Although it is undeniable that our current Age is still characterized by many aspects of a bellicose Jehovistic patriarchy, it is suggested that this is a paradigm losing its foothold, to be supplanted in the (relatively) near future with a more egalitarian cosmology.

The ancient Egyptians were quite aware of the phenomenon of precession well before it was "discovered" by the Greek astronomer Hipparchus, during the Hellenistic era. In fact, Plato recounts how the Egyptian priests of the 7th century BCE maintained that they had knowledge of multiple precessional cycles, and the concomitant periodic destruction of humanity through various global cataclysms – from which they were invariably saved, on account of their close proximity to the Nile. This cycle was recognized by ancient Hindu cosmologists, leading to their development of the concept of Yugas – wherein each Yuga represents a portion of the Great Year, with its own celestial characteristics that determine the relative level of spiritual advancement of humanity as a whole.

CELESTIAL ARCANA

The precessional cycle, and its various structural counterparts (i.e., the NEP, NCP, ecliptic, celestial equator, solstitial and equinoctial nodes, etc.) can really only begin to be fully grasped via reference to the celestial sphere, and it is likely that in past Ages the ingenious concepts that could unify the multitude of discrete celestial maps and observations into a harmonious whole were kept secret from the profane by the priestly caste. Indeed, the structural components of the celestial sphere are keys to unlocking much of the symbolism behind various esoteric constructs and symbols, such as the Kabbalistic Tree of Life, Yggdrasil, the Enneagram, the Great Seal, and the Porta Alchemica. These keys have been applied to the Major Arcana in a multifaceted approach that considers astromythology, equinoctial precession, and ancient modes of chronometry in the examination of trump attributions.

Previous tarot systems have been explored, drawing from various historical sources, such as Arthur Waite, Aleister Crowley, A.E. Thierens, Eliphas Levi, Paul Foster Case, and Antoine Court de Gebelin. The tradition of ascribing a Hebraic letter to each of the 22 trumps has also been addressed, and a new system proposed. This system interprets the threefold categorization of the Hebraic alphabet into the "3 mothers," "7 doubles," and "12 simples," according to the celestial mechanism of precession.

The work of Gerald Massey has been drawn from throughout this book; for though arcane, vast, and difficult his tomes may be, the spirit of true Gnosis has not failed to leave its indelible stamp therein. And this was duly noted by some of the most prominent Victorian occultists who shaped for our future time the doctrines of Aeonic succession and semiotics of tarot attributions. In fact, Massey was considered to be a self-initiate into the Secret Doctrine by H. P. Blavatsky, and Crowley listed him as one of the Chiefs of the Argentium Astrum. It would appear that many occultists coming after Massey utilized his extensive research into the ancient Egyptian mythos, weaving it into their own esoteric narratives, but without giving credit to the Chief Druid. Crowley, in particular, seems to have been familiar with the Masseian literature, and much that is obscure in his tarot doctrine has been clarified via reference to Massey's extensive writings on the astromythology of the ancient Egyptians.

We have discussed the Minor Arcana and its relationship to the Major Arcana, and it has been proposed that the two Arcana can be used as an astrological notation system in which any kind of chart (i.e., natal, progressed, etc.) may be constructed by utilizing specific combinations of the cards. It is postulated that much of the secrecy and taboo surrounding the Tarot originated from such a use during the medieval period, when prognosticative astrology was forbidden under threat of Inquisition. This hypothesis does not

CONCLUSION

discount the use of the Tarot for divination via the random drawing of cards, however, and a new 22-card spread has been provided for this purpose.

Analysis of the Major Arcana has indicated that its symbolism relates at one level with the realm of the fixed stars, the phenomenon of precession, and the Secret Doctrine relating to the character and implications of each successive Age throughout the vast cycle of the Great Year. Within these symbols, particular reference is made to the 12 zodiacal constellations, 7 circumpolar regions (or astronomes) of the Egyptian heptanomis,[683] as well as the 3 primary forces that govern the precessional phenomenon. These forces are represented by the fixed and "eternal" NEP, with its associated solar axial connotations tying it to the Tree of Life; the mutable axis of the NCP, suggestive of the Tree of Knowledge; and the magian reconciler between the two – that binding force which governs the dispensation of the unique qualities associated with each Age in its own particular time, analogous to the Holy Spirit of gnostic Christian doctrine.

These are matters pertaining to higher influences and cycles of time. The Minor Arcana can be seen as a framework within which such eternal and ineffable influences can be contextualized with respect to the limited field of view available to us in our relatively short lifespans; in essence, serving – at one level – as a symbolic system of mundane astrology. It is in this sense that the Tarot can be considered as a device for prognostication, and it is possible that the real methods employed for such application have been lost or deliberately occluded, passing down to us instead as a series of procedures based heavily on the completely random outcome of a series of cards drawn from a shuffled deck. This, however, does not preclude the possibility that the individual who has been gifted with true psychic abilities may in fact be able to use this or any other method as a framework within which to organize their noumenistic impressions.

This being said, the question arises as to whether the various systems of astrology offer any real basis for reliable prognostication. The methods which I have found to be the most promising were developed from the ancient Babylonian and Egyptian systems by sidereal astrologer Cyril Fagan. In his own words:

> I was a tropical advocate for over 20 years; so all the arguments are known to me. It was complete dissatisfaction with tropical astrology that led me to investigate the whole matter. My one regret is that I did not begin to be dissatisfied sooner. One can waste so many years, and that is precisely what I am trying to prevent our younger astrologers from doing.[684]

---

[683] And by extension, the corresponding planets which can be considered as "ruling" adjuncts to the particular Age or Aeon under consideration.

[684] Fagan. *The Solunars Handbook*. Page 130.

If – as it is currently being postulated – the Tarot has encoded at one of its fundamental levels of meaning a symbolic system of organization specifically intended for mundane astrological application, the question as to why a less cryptic form was not instead employed is answered by a consideration of the cultural milieu in which this arcane system originated. During the 13-14th centuries, the timeframe in which the Tarot as we know it is supposed to have arisen, the practice of astrology and its associated methods of self-study as a form of esoteric psychology was grounds for a certain and violent end to those unfortunate enough to be discovered practicing such "works of the devil."

According to Ouspensky:

> The letters of the Hebrew alphabet and the various allegories in the Cabala; the names of metals, acids and salts in alchemy; the names of planets and constellations in astrology; the names of good and evil spirits in magic – all these were but a conventional hidden language for psychological ideas. Open study of psychology, especially in its wider sense, was impossible. Torture and the stake awaited investigators. If we look still further into past Ages we shall see still more fear of all attempts to study man. How was it possible amidst all the darkness, ignorance and superstition of those times to speak and act openly? The open study of psychology is under suspicion even in our time, which is considered enlightened. The true essence of Hermetic sciences was therefore hidden beneath the symbols of Alchemy, Astrology and the Cabala.[685]

Utilization of a set of conventions similar to those previously elucidated would enable a group of people who shared a consensus of rules to depict any particular horoscope or discuss matters of the esoteric psychology of alchemy simply by pairing up a certain number of cards, such that to the outside observer it would appear that nothing more than a card game was transpiring. This is not to suggest that the Tarot was not used for gaming purposes during the medieval period, but it seems unlikely that this was their only application.

Those dangerous centuries comprising the Iron Age – and later, the Middle Ages – are associated with some of the darkest periods in the Hindu Yuga system.[686] If any credence can be given to this system, things can only get better... before they get much worse again, some 20,000 years in the future. However, if we consider these cycles to comprise a spiral through time rather than a series of superimposed circles, the possibility of an even larger cycle – orthogonal to the precessional one – presents itself, in which the existence of an ascending progression spanning millions of years may ultimately ameliorate the descending portion of the Great Year.

---

[685] Ouspensky. *A New Model of the Universe*.
[686] According to Yukteswar's Yuga system. See *The Holy Science*.

CONCLUSION

Due to the vast lengths of time inherent in such great cycles, in combination with the entropic effects of conflict, disaster, and time on the fidelity of history – not to mention the erroneous Aeonic terminology arising from the fixation of the vernal point in Aries – there could easily be 1,000-2,000 years of variance regarding the timing of Ages, and their associated characteristics. We tend to pride ourselves on our enlightened state, relative to that of our distant ancestors, but an objective assessment of current world affairs might give one pause to reconsider.

If we consider the Age of Virgo to be the central locus of the Golden Age, we should be almost free of the darkest influences of the Piscean nadir, as we begin to enter the Aquarian Age. If, as other systems appear to suggest, vernal point Leo marks the zenith of the Great Year, the worst is yet to come for this cycle.[687] Are we truly rising out of the darkest of Ages, or have we yet to experience the Great Winter of precession? Time will tell...

---

[687] However, note the concept of "esoteric inversion" as mentioned in the chapter entitled "The Domicile System and World Ages," which suggests a somewhat Dickensian interpretation, namely denoting the lopsided state of being for one group versus another, whether it be in terms of socioeconomic class, or spiritual advancement: "It was the best of times, it was the worst of times..."

About the Author

THE AUTHOR'S DIVERSE pursuits have led to a variety of occupations, including author, biologist, drafter, illustrator, painter, web developer, and tattoo artist. A unifying element throughout his work has been an interest in the confluence of symbolism and esthetics in nature, art, science, and technology, and how meaning is assigned to the various elements in a picture or pattern. This book is the result of independent research, which has been necessitated by the nature of the subject at hand; as in the words of Arthur Waite: "there is no canon of authority in the interpretation of Tarot symbolism." He holds a bachelor of Science in Biology from Indiana University, Bloomington, and currently resides in Louisville, Kentucky.

Bibliography

(various). 1867. *Dictionary of Greek and Roman Biography and Mythology.* Boston, Massachusetts: Little, Brown, and Company.
—. 1906. *The Jewish Encyclopedia.* Funk and Wagnalls.
Akerman, Susanna. n.d. *Levity.com.* Accessed 3 21, 2015. levity.com/alchemy/queen-christina.html.
Allen, Richard Hinckley. 1963. *Star Names - Their Lore and Meaning.* Mineola, New York: Dover.
al-Qāshānī, 'Abd al-Razzāq. 1991. *A Glossary of Sufi Technical Terms.* London: The Octagon Press Ltd.
Apuleius, Lucius. 1566. *The Golden Asse.* Translated by William Adlington.
Barlow, Nadine G. 2008. *Mars: An Introduction to its Interior, Surface, and Atmosphere.* Cambridge: Cambridge University Press.
Bennet, J. G. 1973. *Gurdjieff: Making a New World.* New York, New York: Harper & Row.
Bentley, John. 1825. *A Historical View of the Hindu Astronomy, from the Earliest Dawn of that Science in India, to the Present Time.* London: Smith, Elder, and Company.
Benton, Christopher P. n.d. *maqom.com.* Accessed 3 21, 2015. maqom.com/journal/paper14.pdf.
Berg, Elizabeth L., and Ruth Wan. 2009. *Cultures of the World: Senegal.* New York: Benchmark Books.
Blavatsky, Helena Petrovna. 1891. *The Agnostic Journal (1887 Letter to Gerald Massey)* 214.
—. 1887. *Isis Unveiled.* London.
—. 1888. *The Secret Doctrine.* London: The Theosophical Publishing Company, Limited.
Bromberg, Dr. Irv. 2015. *Moon and the Molad of the Hebrew Calendar.* 5 29. Accessed 5 29, 2015. sym454.org/hebrew/molad.htm.
Brusch-Bey, Henry. 1879. *A History of Egypt Under the Pharaohs.* Vol. 1. 2 vols. London: J. Murray.
Budge, E. A. Wallis. 1920. *An Egyptian Hieroglyphic Dictionary.* London: John Murray.
—. 1905. *The Book of Gates.* London: Kegan, Paul, Trench, Trubner and Company.
—. 1904. *The Gods of the Egyptians.* Dover edition, 1969. Vol. 2. Chicago: The Open Court Publishing Company.

Bunsen, Christian Karl Josias. 1859. *Egypt's Place in Universal History, an Historical Investigation in Five Books.* Translated by Charles H. Cottrell. Vol. 5. London: Longman, Brown, Green, Longmans, and Roberts.

Caquot, André, and Maurice Sznycer. 1980. *Ugaritic Religion.* Leiden: E. J. Brill.

Case, Paul Foster. 1920. *An Introduction to the Study of the Tarot.* New York, New York.

—. 1975. *The Great Seal of the United States: Its History, Symbolism and Message for the New Age.* Los Angeles, CA: Builders of the Adytum.

—. 1916-1917. *The Secret Doctrine of the Tarot.* The Word.

—. 1990. *The Tarot: A Key to the Wisdom of the Ages.* 1st Revised Edition. Los Angeles, CA: Builders of the Adytum, Ltd.

Castellanos, Juan de. late 16th century CE. *Elegías de Varones Ilustres de Indias.*

Chance, Jane. 1994. *Medieval Mythography: From Roman North Africa to the School of Chartres. A.D. 433-117.* Tampa, Florida: University Press of Florida.

Cole, John. 1824. *A Treatise on the Circular Zodiac of Tentyra, in Egypt.* London: Longman and Cole.

Collin, Rodney. 1993. *The Theory of Celestial Influence.* New York, New York: Penguin Arkana.

Cooper, J. C. 1978. *An Illustrated Encyclopedia of Traditional Symbols.* London: Thames and Hudson, Ltd.

Crowley, Aleister. 1909. *John St. John - The Record of the Magical Retirement of G. H. Frater, O∴M∴ (Liber DCCCLX).* London: Simpkin, Marshall, Hamilton, Kent.

—. 1929. *Liber LVIII.*

—. 1929. *Magick in Theory and Practice.*

—. 1991. *The A∴A∴.* Thame: Mandrake Press, Ltd.

—. 2000. *The Book of Thoth (Egyptian Tarot).* York Beach, Maine: Weiser.

—. 1929. *The Spirit of Solitude: An Autobiography. Subsequently re-Antichristened the Confessions of Aleister Crowley.* London: Mandrake Press.

Crowley, Aleister, and S. L. MacGregor Mathers. 1904. *The Lesser Key of Solomon.*

Culpeper, Nicholas. 1653. *Complete Herbal.*

Davis, Jr., George A. 1945. "The So-Called Royal Stars of Persia." *Popular Astronomy Magazine*, April: 149-159.

de Gebelin, Antoine Court. 1773-1782. *The Game of Tarots.* Vol. 8, in *Le Monde Primitif (The Primitive World),* by Antoine Court de Gebelin, translated by Donald Tyson, 365-410. Paris.

Der Gold und Rosenkreuzer (The Gold and Rosy Cross). 1785. *Geheime Figuren der Rosenkreuzer, aus dem 16ten und 17ten Jahrhundert (Secret Symbols of the 16th and 17th Century Rosicrucians).* Altona, Hamburg: Der Gold und Rosenkreuzer.

Drummond, William. 1866. *Oedipus Judaicus.* London: Reeves and Turner.

BIBLIOGRAPHY

Elliot, H. M. 1875. *The History of India as Told by its Own Historians - The Muhammadan Period. Posthumous Papers of H. M. Elliot.* Edited by J. Dawson. Vol. VI. London.

Evelyn-White, Hugh G. 1914. *The Homeric Hymns.*

Fagan, Cyril. 1971. *Astrological Origins.* St. Paul, Minnesota: Llewellyn Publications.

—. 1970. *The Solunars Handbook.* Tucson, Arizona: Clancy Publications, Inc.

—. 1950. *Zodiacs Old and New.* Los Angeles, California: Llewellyn Foundation for Astrological Research.

Fagan, Cyril, and Roy Firebrace. 1971. *Primer of Sidereal Astrology.* Tempe, Arizona: American Federation of Astrologers, Inc.

Faik-Nzuji, Clementine Madiya. 1996. *Tracing Memory: A glossary of Graphic Signs and Symbols in African Art and Culture.* Qubec: Canadian Museum of civilization.

Faraday, Winifred. 1902. *Divine Mythology of the North.* London: David Nutt.

Flamsteed, John. 1729. *Atlas Coelestis.* London: Flamsteed, Margaret; Hodgson, James.

Fox, Marvin. 1990. *Interpreting Maimonides.* Chicago, Illinois: The University of Chicago Press.

Freyle, Juan Rodriguez. 1636. *Conquista y Descubrimiento del Nuevo Reino de Granada.*

Gantz, Timothy. 1996. *Early Greek Myth: A Guide to Literary and Artistic Sources.* Vol. 1. Baltimore, Maryland: Johns Hopkins University Press.

Garstin, E. J. Langford. 1932. *The Secret Fire.* London: The Search Publishing Company, Ltd.

Ginsburgh, Yitzchak. 1990. *The Hebrew Letters.* Rechovot, Israel: Linda Pinsky Publications.

Graves, Robert. 2012. *The Greek Myths.* New York, New York: Penguin.

—. 2013. *The White Goddess.* Kindle edition. New York: Farrar, Straus and Giroux.

Gravrand, Henry. 1990. *La Civilisation Sereer.* Vol. 2. Dakar: Nouvelles Editions Africaines.

Grimm, Jacob. 1882. *Teutonic mythology.* Translated by James Steven Stallybrass. Vol. 1. London: George Bell and Sons.

Gurdjieff, Georges Ivanovich. 1950, 1993. *All and Everything: Beelzebub's Tales to His Grandson.* Aurora, Oregon: Two Rivers Press.

Hall, Manly Palmer. n.d. "The Manly P. Hall Archive." *The Manly P. Hall Archive.* Accessed 3 21, 2015. manlyphall.org/audio/astrotheology-2/astrotheology-part-2-of-5.

—. 1928. *The Secret Teachings of All Ages.* Paperback edition: Tarcher/Penguin 2003, New York. Originally published by: Philosophical Research Society. Los Angeles, CA.

Herodotus. 1890. "History." Translated by G. C. Macaulay. London: Macmillan.

n.d. *Holy Bible.*

Hyginus. 1482. "Astronomica." Translated by Mary Grant. Venice: Ratdolt, Erhard.
Kaplan, Aryeh. 1997. *Sefer Yetzirah - The Book of Creation in Theory and Practice*. Kindle. San Francisco, CA: Weiser.
—. 1979. *The Bahir: Illumination*. York Beach, Maine: Samuel Weiser, Inc.
Kaplan, Stuart. 1978. *The Encyclopedia of Tarot*. Vol. 1. New York, New York: U.S. Games Systems, Inc.
Kircher, Athanasius. 1665. *Arithmologia*. Bavaria: The Bavarian State Library.
Lambert, W. G. 1967. "Enmeduranki and Related Matters." *Journal of Cuneiform Studies* (21): 126-138.
Landgraf, Karl. 1824. *La Pierre Zodiacale du Temple de Denderah*. Copenhagen: Seidelin.
Le Page Renouf, Peter. 1892-1897. *The Egyptian Book of the Dead*. Translated by Peter Le Page Renouf.
Levi, Eliphas. 1892. *Magical Ritual of the Sanctum Regnum*. Translated by W. Wynn Westcott. London: George Redway.
—. 1896. *Transcendental Magic, Its Doctrine and Ritual (Dogma et Ritual de la Haute Magie)*. London: Rider and Company.
Lewis, G. C. 1860. *The Fables of Babrius*. Translated by John Davies. London: Lockwood and Company.
Lilly, William. 1852. *An Introduction to Astrology (Christian Astrology)*. London: G. Bell.
Lull, Raymond. 1305-1308. *Ars Generalis Ultima*.
Mackillop. 2004. *Oxford Dictionary of Celtic Mythology*. Oxford: Oxford University Press.
Maimonides. n.d. *Guide for the Perplexed*.
Mallet, Paul Henri. 1847. *Northern Antiquities*. Translated by Thomas Percy. London: H. G. Bohn.
Manhar, Nurho de. 1914. *The Zohar: Bereshith to Lekh Lekha*. Translated by Nurho de Manhar. New york, New york: Theosophical Publishing Company.
Massey, Gerald. 1881. *A Book of the Beginnings*. London.
—. 1907. *Ancient Egypt, the Light of the World*. London. masseiana.org.
—. 1888. "Gnostic and Historical Christianity (Lecture #4)."
—. 1887. "Luniolatry: Ancient and Modern (Lecture #7)."
—. 1900's. "The Hebrew and Other Creations Fundamentally Explained (Lecture #5)."
—. 1886. "The Historical Jesus and Mythical Christ (Lecture #1)." Springfield, Massachusetts: Star Publishing Company. masseiana.org/ml1.htm.
—. 1887. "The 'Logia of the Lord;' or, the Pre-Christian Sayings Ascribed to Jesus the Christ (Lecture #3)." London.
—. 1883. *The Natural Genesis*. London: Williams and Norgate.
—. 1887. "The Seven Souls of Man and Their Culmination in Christ (Lecture #9)."
Miller, Moshe. n.d. *Kabbalah Online*. chabad.org/kabbalah/article_cdo/aid/380410/jewish/The-Zohars-Mysterious-Origins.html.

BIBLIOGRAPHY

Moakley, Gertrude. 1954. "The Waite-Smith 'Tarot', A Footnote to the Waste Land." *Bulletin of the New York Public Library* 58: 471-475.

Molnar, Lawrence et al. 2017. "Prediction of a Red Nova Outburst in KIC 9832227." *The Astrophysical Journal.*

Molnar, Michael. 1998. "Symbol of the Sphere." *The Celator*, June: 1-2.

Ouspensky, Peter Demianovich. 1931. *A New Model of the Universe.* New York, New York: Knopf.

—. 1949. *In Search of the Miraculous.* New York, New York: Harcourt, Brace & World, Inc.

Ovid. 1717. *Metamorphoses.* Translated by et al John Dryden.

Papus. 1909. *Le Tarot Divinatoire.* Paris.

Paulos, J., and B. Hauck. 1986. "The Sirius Supercluster." *Astronomy and Astrophysics* (162): 54-61.

Pausanias. 2nd century CE. "Description of Greece." Translated by W. H. S. Jones. 1918.

Petticrew, Ian. n.d. *Gerald Massey.* Accessed 3 21, 2015. gerald-massey.org.uk/massey/index.html.

Pike, Albert. 1871. *Morals and Dogma of the Ancient and Accepted Scottish Rite of Freemasonry.* Charleston: The Ancient and Accepted Scottish Rite of Freemasonry - Supreme Council of the Southern Jurisdiction.

Plato. 1871. "Politicus (Statesman)." Translated by Benjamin Jowett. New York: C. Scribner's Sons.

—. 1871. *Timaeus.* Translated by Benjamin Jowett. New York: C. Scribner's Sons.

Pliny. 1601. *The Historie of the World. Commonly called, the Naturall Historie of C. Plinius Secundus.* Translated by Philemon Holland. London: Adam Flip.

Plutarch. 1908. *Plutarch's Morals: Theosophical Essays.* Translated by C. W. King. London: George Bell and Sons.

Pseudo-Apollodorus. 1st-2nd centuries CE. "Bibliotheca." Translated by Aldrich.

Ptolemy. 1822. *Tetrabiblos.* Translated by J. H. Ashmand. London: Davis and Dickson.

Purdon, Wayne. 2014. "Secret Mysteries of the Sun Revealed." *New Dawn Magazine*, February.

Rolleston, Frances. 1862. *Mazzaroth; or, the Constellations and Mizraim; or, Astronomy of Egypt.* London: Rivingtons, Waterloo Place.

Rosenberg, Diana K. n.d. "Fixed Stars." *Fixed Stars.* Edited by Christine Bennett. Accessed 2015. ittakes2.com.au/fixed_stars.htm.

Salmon, William. 1679. *The Soul of Astrology.* London.

Schonfield, Hugh J. 1984. *The Essene Odyssey.* Shaftesbury, Dorset: Element Books Ltd.

Schram, Peninnah. n.d. *torahaura.com.* Accessed 2015. torahaura.com/bible/learn_torah_with/ltw_5761/ltw_5761_tetzaveh/ltw_5761_tetzaveh.html.

Schweighardt, Theophilus. 1618. *Speculum Sophicum Rhodotauroticum (The Mirror of Wisdom of the Rosicrucians).*

Sitchin, Zecharia. 1976. *The 12th Planet.* New York, New York: Stein and Day.
Spencer, Edmund. 1596. *The Faerie Queene.*
Statius, Publius Papinius. 2003. *Thebaid.* Translated by D. R. Shackleton Bailey. Harvard, Massachusetts: Loeb Classical Library, Harvard College.
Steele, Robert, and Dorothy Singer. 1928. "The Emerald Table." *Proceedings of the Royal Society of Medicine,* January: 492.
Stenring, Knut. 2004. *Sepher Yetzirah.* Translated by Knut Stenring. Berwick, Maine: Nicolas-Hays.
Strong, James. 1890. *Strong's Concordance.*
Stuckenbruck, Loren T. 2003. "Revelation." *Eerdman's Commentary on the Bible.* Compiled by James D.G. Dunn and John William Rogerson. Grand Rapids, Michigan: Wm. B. Eerdman's Publishing, November 19.
Stumpf, Walter E., and Thomas H. Privette. 1991. "The Steroid Hormone of Sunlight - Solitrol (Vitamin D) As a Seasonal Regulator of Biological Activities and Photoperiodic Rhythms." *Journal of Steroid Biochemistry. Molecular Biology.* 39 (2): 283-289.
Tacitus. 1894. *The Agricola and Germania.* Translated by R. B. Townshend. London: Methuen and Company.
Taylor, Thomas. 1792. *The Hymns of Orpheus.* London: B. White and son.
Theosophical University Press. 1999. "Encyclopedic Theosophical Glosary." Pasadena, California: Theosophical University Press.
Thierens, A. E. 1930. *General Book of the Tarot.* Philadelphia: D. McKay Co.
Thomas, William, and Kate Pavitt. 1922. *The Book of Talismans, Amulets and Zodiacal Gems.* 2nd and Revised Edition. London: William Rider and Son, Limited.
Totten, Charles Adiel Lewis. 1897. *The Seal of History: Our Inheritance in the Great Seal of "Manasseh", the United States of America: Its History and Heraldry; and Its Signification Unto "The Great People" Thus Sealed.* Nabu Public Domain Reprint. New Haven: The Our Race Publishing Company.
U.S. Department of State, Bureau of Public Affairs. n.d. "The Great Seal of the United States." state.gov/documents/organization/27808.pdf.
Valentine, Basil. mid-17th century. *Chymical Wedding.*
various. n.d. *The Ancient Gods Speak: A Guide to Egyptian Religion.* Edited by Donald Redford. Oxford: Oxford University Press.
visual-arts-cork.com. n.d. *Art Encyclopedia.*
Waite, Arthur Edward. 1926. "Great Symbols of the Tarot." *The Occult Review.*
—. 1911. *The Pictorial Key to the Tarot.*
Webb, James. 1980. *The Harmonious Circle.* New York, New York: G. P. Putnam's Sons.
Wilhclm, Wagner. 1886. *Asgard and the Gods.* London: Swan Sonneschein, Le Bas & Lowrey.
Yukteswar. 1894. *The Holy Science.*

Index

A∴A∴
  sigil of
    and heptagram of asteroid cycle, *147, 148*
Aan, 220
Abraham, 13
abyss
  "Sun of the abyss" = Neptune = Full Moon ("nocturnal Sun"), 226
  Amenta during Age of Taurus = Pisces / Aquarius / Capricorn, 307
  and "bloody" Sun, Osiris Tesh-Tesh, 145
  and 40 years of Hebraic wilderness wandering, 300
  and Aquarius in Age of Taurus, 161–62
  and chalice in Levi's transmutational sigil, 76
  and circumpolar regions - Emperor is below, 111
  and crayfish of Moon trump, 265
  and Egyptian god Nun – Death trump, 195
  and Enneagram, "put-circle" of Ptah, 44–47
  and Fool, 299
  and *globus cruciger*, 71
  and Neptune, 75
  and Persephone, Virgo, Hades – Hermit trump, 150
  and Porta Alchemica, 74
  and Serer cosmological octohedron, 187
  and Sut, Cetus/leviathan – Ages of Gemini and Taurus, 128
  Aquarius/Pisces Australis, 255
  connected to *Nun* and *Mem*, 198
  of current Age = Scorpio / Sagittarius / Capricorn, 200
Aces
  and sidereal angles, 320–23
Adam, 81, 93, 99, 124, 142
  twin of Eve, 127
Adam Kadmon, 81
Adonis, 105
Aeddon. *See* Hu
Agni, 228–29
Air (element), 161
Akasa
  all-pervading etheric substance - cf. Archaeus, Astral Light, 210
Akhenaten, 138
alchemy, 73, 77, 104, 160, 343
Aldebaran
  and Schweighardt's Invisible College, 92
  fiducial star of Egypto-Babylonian zodiac, 231
Alef, 7, 83, 152, 277–79, 286–88, 296, 298, 301
Aleh
  Hebraic "hosts of heaven" – forms *Alhim/Elohim* when connected to *Mi* ("highest pole"), 149
Alhim, 149
All Souls Day
  All Hallows Eve (Halloween), 150
Allfather, 180–82
  Odin as NEP, 181
Alpha Leonis. *See* Regulus

Amenta, 72, 74, 76–77, 185–86, 193–95, 247, 265, 273–74, 281, 285, 300, 307
America
  American dollar and Great Seal, 62
  United States of
    and Manasseh, 13th Tribe of Israel, 63
Amphitrite, 225–27, 241
Am-smen, 45
analemma, 78, 129, 140, 293
Ancient of Days, 146, 149–50, 176, 298
angles
  of sidereal astrology, 304–7, 331
    and Aces, 313, 318
      effect of Moon and Mars being simultaneously on angles, 123
      effect of Saturn when on the angles, 67
Annu, 79
Antares
  and Schweighardt's Invisible College, 92
Antemeridian (astrological), *304, 305, 306*, 313
  and 22-card spread, 332
anthropomorph, 136–37, 169, 243, 245, 267
Anthropos
  "divine Anthropos" = crystallization of higher consciousness, 81
antimony, 103–6
  and Star Regulus, Porta Alchemica, 73
Anubis, 181, 218, 223, 266
Anunnaki, 147
Anup, 167, 220–22
Apap, 145, 194
ape, 96, 247
  "mount of ape" = Cepheus, 11, 24
  represents East/wind/Hapi in Egyptian scheme, 306
Aphrodite
  mirror of, 105
apocalypse, 40, 183, 282
  4 beasts of, 40
  from Greek "apocálypsis" = "uncovering", 282
Apollo, 78, 83, 271–72
Apt, 71, 145, 252
Aquarius
  "constellation of the Urn", 198
  Age of, 37
    and "Grip of the Lion's Paw"/Resurrection of Lazarus, 120–21
    and Crowley's Lust Atu, 143–46
    and Schweighardt's Invisible College, 90
    and the "Night of Brahma", 227–31
    and Winter of Great Year, 235, 247
    less accepting of misogyny, 126
  and abyss during Age of Taurus, 161, 198, 281
  "relictual" element in Crowley's Death Atu, 195
  and the Enneagram, 46
  and alchemical dissolution, 160–62
  and canines of Moon trump, 265
  and Crowley's "double inversion", 7
  and Crowley's "*kteis+fallos*" equation, 163
  and current vernal point, 111, 167, 269
  and Lent, 300
  and Enneagram, 40
  "centers", 325
  and Sirius Supercluster/Ursa Major Stream, 261
  and suit of Cups, Wands, 306
  and the "Two Truths" - solar sign of "expiration" during Age of Taurus, 185
  and the Enneagram, 43
  and the Great Seal, 64
  and the Tower of Belus, 245
  angel of Rider-Waite Fortune trump, 155
  angel of The World trump, 289
  associated with winter solstice during Age of Taurus, 31

INDEX

connection to the Great Mother, 255
hypothetically ruled by Mars/Jupiter in current Age, 34
nocturnal "water sign" of inundation during Age of Taurus, 307
and the Porta Alchemica, 74
nocturnal annunciator of summer Sun during Age of Taurus, 115, 134
and Fortune Trump, 155–56
one of 4 beasts of Apocalypse, 192
Saturn ascribed to in Ptolemaic (tropical) scheme, 29
Ara (constellation), 190, 274
Archaeus
   all-pervading etheric substance - cf. Akasa, Astral Light, 210
Argo Navis, 178
Argonauts, 178
   and Shemhamphorasch, zodiac, 177
Ariadne, 92, 96
   wreath or crown of immortality conferred to, 140
Arianrhod, 86–87
Aries
   Age of
      and "Aeon of Osiris", 143
      and "temptation of Lilith", 93
      and Anubis, 266
      and Chariot as another form of Emperor, 137
      and Deneb, 282
      and Egyptian god Sevekh, 112
      and Emperor, 110–12
      and Khepra, Khep, 268
      and Khunsu, Harpocrates, Horus, 273
      and Lamb of Revelation, 115, 192
      and locus of NCP, 202
         and Dragon of Medea, 213–15
         and Sevekh-Kronus, 224
      and luni-solar chronometry, 308

and Nebuchadnezzar, 291
and Ovid's *Metamorphoses*, Spenser's *The Faerie Queene*, 167–68
and Schweighardt's Invisible College, 91
and Sevekh-Kronus/Saturn eating his children, 247
and Spenser's *The Faerie Queene*, 172
and St. John's Revelation, 116
and tarrochi
   The Empress, 103
and Ta-urt's desertion of Sut for Horus, 127
and The Emperor, 83
and Tower of Belus, Nimrod, Rhampsinitus, 247–49
destruction of and Crowley's "*kteis+fallos*" equation, 163
Marriage of Bride and Lamb, 135–37
myth of Amphitrite provisionally dated to, 227
Ptolemaic domicile system derives from, 29–31
vernal point mistakenly fixed in Aries, 131
and Crowley's "double inversion", 7–9
and Enneagram, 43
constellation
   vernal point of Hypsomatic (Egypto-Babylonian) zodiac, 231
Ark
   Argo Navis
      and the Hanged Man, 178
   of Noah = "barque of Isis", 91
   of the World
      Corona Borealis, 92
Ark-stone (Druidic), 129
Asar, 235
   Orion, Osiris, 65
Ascendent (astrological), *304*, *305*, *306*, 314–15, 319
   and 22-card spread, 332
Asgard, 14

Asterios
    Minotaur son of Pasiphae, 122
Astraea
    and Ovid's *Metamorphoses*
        Justice trump, 167
    and Spenser's *The Faerie Queene*,
        168
    Virgo, 171
Astral Light
    all-pervading etheric substance - cf.
        Archaeus, Akasa, 210–12
astrology, 35, 38, 121, 193–94, 285,
        315, 324, 326, 331, 343
    an ancient form of astronomy, 31
    tropical, sidereal, 29
    various correlations with the
        Tarot, 342–43
astrotheology, 19
Aten, 138
Atlantis, 143
Atreus, 234
Atum, 85, 99, 142, 277, 282, 285
Autolycus
    and Shemhamphorasch, zodiac,
        177
Avatar
    Ace, 331–34
    and Iu as Wandering Jew, 298
    Hierophant/Initiator of each Age,
        120
Ayin, 152
Babel
    and Tower trump, 242
Babylonian
    astrology was sidereal, 304, 342
    zodiac was sidereal, 231
Bagge, Oluf Olufsen, 16–17
Bahir, 13, 19–23
Baphomet
    Holy Spirit, Great Mother, Ta-urt,
        Templaric Mete, 210–11
    Pan, 235
Barque of Isis, 256
beehive
    ancient symbol for Earth, and
        "put-circle" of Ptah, 44
Beit, 88, 152
Belus
    inventor of sidereal science,
        personification of Dog-Star,
        245
    Tower of, 245
Benetnasch
    star of Ursa Major used to site
        ancient Egyptian temples, 251
Bennu
    and Amphitrite, 227
    bird ("nictorax"), 236
    Canis Major, Sirius, 238
    soul of Ra, guide of gods in Tuat,
        Morning Star, 236
Bennu-Asar
    celestial region of Sirius and
        Orion, 65
    double-ship of Moon and Sun,
        235–37
bennu-bird
    phoenix, dove, Sirius, 65
    and Enneagram, 41
Ben-Pandira, Jehoshua
    supposed historical Jesus, 136
Berosus, 209, 267
    and "double ending" of Great
        Year, 238–39
Bes, 74
    and Porta Alchemica, 71
Bethlehem
    Star of, 261
Bifrost, 14, 207
    representative of Milky Way, 16
Binah, 39, 75–76
bird
    "mount of bird" = Cygnus, 11, 24
biune deity
    originally feminine, 128
    represented by Sun and Moon as
        masculine and feminine aspects
        of light, 32
    tzimtzum as feminine aspect of,
        267
black
    and Sut, 185
    banner of Death trump, Scorpio,
        183
    tongue of Agni, Kali Yuga, 229
Blavatsky, Helena Petrovna, 9, 210,
        226, 229–30, 235, 291

355

INDEX

and Egyptian origin theory of Tarot, 2
and Moon as vampiric feeder upon humanity, 221
blind (occult), 9, 13, 40, 153, 316
and Justice, Strength, 173
blue
and subconscious mind, Tower trump, 249
color of sky/heaven, 276
Boaz
pillar of, 87, 117
Bohme, Jacob
and "Rose of the Lily", 184
and *illuminati*, 184
Book of the Dead, 79, 150, 236
Bootes, 96, 127, 145, 205
and hermanubis of Rider-Waite Fortune trump, 156
and Minor Arcana
ancient Egyptian scheme of zootypical cardinal directions, 307
and Shemhamphorasch, zodiac, 177
bow
"bow in the sky"
and Milky Way, 205
and Sagittarius, 206
accoutrement of Neith, 200
and Galactic Equator, Iris, 202
ideograph
Gimel, Beit, 88
Brahma
light of, and Age of Virgo, Leo, 99
Night of
and Makara, 229–31
Throne of = Galactic Center, 37–38, 283
Britain, Ancient
and Druidic lingum/yoni stones, 129
and invention of writing, Gwydion, 201
Bronze Age
and Ovid's *Metamorphoses*, Spenser's *The Faerie Queene*, 167–69

Buddha
similar to Egyptian Ptah, divided the land into 9 parts after the Great Deluge, 45
Budge, E. A. Wallis, 186
Cabala. *See* Kabbalah
caduceus, 181, 202, 211
Calvary
3 crosses of and belt stars of Orion, 119
Cancer
a diurnal "water sign" of inundation
during Age of Aries, 131
during Age of Taurus, 46, 73, 76, 91, 131, 167, 206, 221, 226
Age of
"original" starting-point of Great Year, 241
and "summer" of Great Year, 34–37, 92–93, 126–27, 247
and Corona Borealis, 85–87
and Golden Age, 161
and Sut, hippopotamus astronome, 209
and "crib", "manger", 134
and "double ending" of Great Year, 238
and Amphitrite, 225
and Chariot, Clavis Corona, Corona Borealis, 310
and destruction of Tower of Belus, 245
and fall of Tower of Nimrod, 244, 247
and Gate of Humanity, 138
and heptanomis correspondence, 96
and hermanubis of Fortune trump, 157
and Moon trump, 267–69
and Ptolemaic domicile system, 29–30
and Schweighardt's Invisible College, 95
and Shamgar's ass' jaw, 174

and Sothic cycle during Age of
    Taurus, 32
and Spenser's *The Faerie Queene*,
    170
and symbolism of The Chariot,
    134–36
and Tower of Rhampsinitus, 248
and unicorn of "The Lion and the
    Unicorn", 96–98
ark of Khepra and "double ship",
    236
ruled by Moon in Age of Aries,
    309
tropical sign of summer solstice,
    130
Canis Major
  and Porta Alchemica, 75
  and ship of Bennu-Asar, 65, 235–
    36
Capricorn
  Age of
    and Winter of Great Year, 161,
      235, 247, 283
    and Yugas, 92
    and "double ending" of Great
      Year, 238
    and abyss during Age of Taurus,
      195
    and the Enneagram, 46
    and abyss of current Age, 200
    and abyss of Great Year, 268
    and Dagon, Makara, "blinded"
      transposition with Aquarius,
      229 30
    and Deneb ("Lord who cometh, in
      Cycnus"), 282
    and Diana of Ephesus, 134
    and heptanomis correspondence,
      97
    and Minor Arcana suits, 309
    and Pan's goat feet, 235
    and placement of angel's feet in
      Temperance trump, 207
    and Sola-Busca Devil trump, 208
    and Tower of Rhampsinitus, 248–
      49
    and Uranus, 318
    and Venus as "spark" of "double
      ship", 236
  ark of Khepra and "double ship",
    236
  as Triton: Mother = Amphitrite,
    Father = Poseidon, 225
  example in 22 card spread, 335
  nocturnal, 226
  nocturnal "water sign" of
    inundation during Age of
    Taurus, 307
  and the Porta Alchemica, 74
  Saturn ascribed to in Ptolemaic
    (tropical) scheme, 29
  tropical solstice point (winter), 268
Carina (constellation)
  and Hanged Man trump, 178–80
  and Porta Alchemica, 75
Case, Paul Foster
  and difficulties/advantages of
    connecting Tarot to Kabbalah,
    151–54
  and Egyptian origin theory of
    Tarot, 2
  and Fool trump, *Alef*, 278
  and Hanged Man trump, *Mem*,
    278
  and High Priestess trump, 172
  and lemniscate of Magician trump,
    82
  and Moon trump, *Kuf*, 267–68
  and Strength/Justice "blind",
    Lamed, 172–73
  and Sun trump, *Reish*, 277–79
Cassiopeia, 127, 255
  and *meskhen*, "birthplace", celestial
    source of inundation, 198–200,
    268–69
Castor, 127, 272
Cat's Eye Nebula, 76, 82, 90, 155
catastrophe
  changing of polestar seen as by
    ancients, 10, 144, 296
causal body, 81, 101
celestial axis
  and "tree of time", 228
  and *Ercole I d'Este* Star trump, 262
  and Magician trump, 78–79, 83
  and Sola-Busca Devil trump, 208
  and Tree of Knowledge, Lovers
    trump, 125

INDEX

in contradistinction to ecliptic axis, 79
intersection with ecliptic axis = "throne" of Blessed Holy One, 19
Talus anthropomorph of during Bronze Age, 169
celestial equator
  and *crux ansata*, 110
celestial influence
  of Galactic Center, 37
  of Sirius, 38
celestial macrocosm
  relation to human microcosm and Porta Alchemica, 77
celestial mountain
  Mount Meru, Alborz, Eden, 186
celestial pole, 107, 260, 298, 308
  ancient deity of connected to Neith, Uat, 200
  and "Ape of Thoth", 81
  and "girdle of Isis", 212
  and abyss – Emperor below, 111
  and circular Zodiac of Denderah, 213, 253
  and Great Sphinx, precession, 142
  and Lake of Sa, 213
  and Odin as "Spear Master", 181
  and Tef (ancient Genetrix), 296
  and Tower of Babel, 242
  and *uat* scepter, 200
  North=Horus, South=Sut, 294
  place of equipoise in stellar mythos, 167
  precession of and temptation of Lilith, 93
  shifts precessionally with vernal point, 24
celestial sphere
  3 divisions of and Neptune, Vishnu, 226
  and *globus cruciger*, 102
  and Hebraic Tree of Life, bronze vessel of Solomon, 19–24
  and Rosy Cross Lamen, 190–91
  and Schweighardt's Invisible College, 91–96
  and Serer cosmological octohedron, 187
  and the Enneagram, 39–44
  and the Porta Alchemica, 74–77
  and Visconti-Sforza World trump, 293–95
  and Yggdrasil, 14–18
  key to unlocking many esoteric symbols, 12
celestial water
  and Urnas/Uranus, 74
Celtic
  mythology
    and Arianrhod, Corona Borealis, 86
    and Kaer-Bediwyd, Corona Borealis, 92
centers
  Gurdjieffian "brains" or "chakras"
    and 4 Minor Arcana suits, 322
    and fixed zodiacal signs, 325–26
    emotional, 322
    instinctive, 323, 325
    intellectual, 322
    moving, 323, 325
    negative and positive halves of, 324
Cepheus, 58, 97, 127, 156, 208, 253, 255, 269, 280–81, 283, 290, 318
  "mount of ape", 24
  and Shemhamphorasch, zodiac, 177
  and sphinx of Rider-Waite trump, 161
  and Tower trump, 242–47
cerebellum, 267
Cetus, 128, 265, 268
chakra
  6th, 101
  7th, 101
  ajna, 64
Chariot (trump), *137*
  and Cups, 310
  Levi's version, *131*
  relation to High Priestess trump, 133–34
chatoteret, 134

Chet, 134, 138, 152
China, 45
   years anciently reckoned by eclipses in, 224
Chinese, 31, 45, 104
Chochmah, 39, 76
Christ, 78, 80, 117, 134, 136, 150, 261, 297
Christina
   queen of Sweden
      and alchemical laboratory, Porta Alchemica, 70
chronometry
   ancient Egyptian, and Sothic cycle, 135
   lunar, 134
   luni-solar, 244, 274
   shift from lunar to luni-solar, 135
   shift from lunar to solar, 141
   shift from stellar to luni-solar and Revelation, 112, 115
   shift from stellar to solar, 150
   stellar, 79, 158, 200, 261, 276, 298
chupah, 134
Cicero, 118, 238
circle of time
   of Enneagram, 40, 49
Clavis Corona
   "Key of the Crown" = central star of Corona Borealis, 85, 92, 98, 99, 165, 207, 310
   and Holy Grail, 99
Collin, Rodney
   and Enneagram, 40
   and Enneagram "essence types", 26–28
Columba, 41, 43, 75, 139
   and Enneagram "centers", 325
Coma Berenice, 104, 107
consciousness
   and Galactic Center, 118
   higher states of, 64, 77, 81, 100, 120, 126
   integration of subconscious and waking consciousness, 267
   intuitive "lunar" aspect of, 99
   lower states of
      and "Fall" of humanity, 92, 161
   normal state of, 250
   objective and subjective states of, 137
   of humanity
      and Great Year, 34, 93, 182
      and Sirius, 38, 66
Conver, Nicolas, 175
Cor Leonis, 247
Corona Borealis
   "keystone" of heptanomis, primary polestar, 24, 142, 149, 241
   and High Priestess, Age of Cancer, 85–87
   and Rider-Waite Chariot trump, 310
   and crown of Rider-Waite Death trump, 190
   and crown of Sola-Busca Nabuchodenasor trump, 292
   and floral crown of Rider-Waite Sun trump, 274
   and hermanubis of Rider-Waite Fortune trump, 155–57
   and hippopotamus, unicorn, 96
   and red hippopotamus, Whore of Babylon, 145
   and relation of Chariot to High Priestess, 133
   and Rider-Waite Justice trump, 164, 172
   and Schweighardt's Invisible College, *90*, 92
   and Sirius Supercluster/Ursa Major Stream, 261
   and Ta-urt's desertion of Sut for Horus, 127
   as primary "seat in the circle" ("*Bab* of *El*"), 146
   Kabbalistic "Crown of Crowns" = *Keter*, 276
cosmic egg
   "definitionless Cipher of the Mysteries", 16
   representative of celestial sphere, 16
Crater (constellation), 106, 146, 198–99
crayfish

INDEX

as representation of Pisces/Cetus in Moon trump, 265
crocodile, 96, 112, 198, 223, 229, 308
   represents West/earth/Sevekh in Egyptian scheme, 306
Crowley, Aleister
   and Egyptian origin theory of Tarot, 2
   and Emperor trump, vis à vis the Abyss, 111
   and Fortune trump, "*kteis+fallos*" equation, 163
   and Hanged Man trump
      Dalet, 316
      medieval interpretation of, 178–79
      Mem, 278
   and Lust trump, A∴A∴ sigil, 143–48
   and Moon trump, precession, 264–68
   and Piscean symbolism of Death trump, 195
   and Sun trump, *Reish*, 275–77
   and Temperance ("Art") trump, *Samech, Nun*, 204–5
   and the Hierophant trump, 120–23
crux ansata
   symbol of equinox, 110
crux immissa ("Latin cross"), 91
cube of space, 13, 19
Culhuacan, 221
Culpeper, Nicholas
   and *Complete Herbal*, Enneagram essence types, 26–28
Cups (suit), 305–7, 310, 325
Cybele, 84
Cycnus. *See* Cygnus
cyfriu
   Celtic "trinity" associated with Hu, 129
Cygnus, 91, 97, 156, 245, 290, 318
   and Judgment trump, 280–85
   and Shemhamphorasch, zodiac, 177
   and sphinx of Rider-Waite trump, 156, 161
   mount of bird, 24
   mythical suicide attempt of, 90
Cyid
   Serer "disembodied souls", 188
Dagon, 229
Daidalos, 122
Dalet, 151–54, 277, 279, 286, 318
Dame (court card), 331
Dan (Biblical), 317
David (Biblical), 63
   star of connected to Lyra, 150
Dayan Esiru
   Sumerian "Crown of Heaven", 91
de Gébelin, Antoine Court, 1
   and Tower of Rhampsinitus, 247–49
de Salzmann, Alexander, 40
Death
   valley of
      Amenta, 185, 195
Death (trump), 107, 183
   and Apocalypse, 191
   and Scorpio, 195
decan, 274, 333
   36 divisions of ecliptic, 19
decan system, 324
   "pips" hypothesized to have been used as, 313
declination, 36, 126, 148, 170, 283
   of ecliptic
      change in connected to myth of Phaeton, 217
   of Galactic Center in Northern Hemisphere at greatest angle during summer solstice in Age of Aquarius, 37
   of Sirius
      greatest in summer, 197
   seasonal effects of, and tropical astrology, 30
Delphinus (constellation)
   and King of Cups, 307
   and Legend of Amphitrite, 225–27
   known as "Job's Coffin"
      and Judgment trump, 280–84
Deluge, 217
denary, 318, 325–28, 331–35, 337

correspondences between Fagan's house system and Waite's denary keywords, 310
Denderah
  Temple complex, 251
  Zodiac of
    circular, 208, 213, 253, *254*
      Khunsu/Harpocrates, 273
    rectangular, 127
      Hapi/Aquarius, *256*
      Shu/Tefnut as Sagittarius, *128*
      Ta-urt, *258*
Deneb
  and Judgment trump
    NCP of Mars points to, 285
    the "Judge" or "Lord" in Cygnus, 280–84
Descendent (astrological), *304*, *305*, *306*, 313
  and 22-card spread, 332
Deucalion, 217
Devil (trump), 208, *210*, 235, 241, 249
  and Pentacles, 310
Diana (goddess)
  of Ephesus
    Dea Multimammae, 134, 204–5
diencephalon, 43, 77, 99
  "two-headed"
    and double-headed eagle, pineal gland, 64
Diodorus Siculus, 44
divination, 304, 313
  possible connection to astrological prognostication, 314
  random guessing versus astrological prognostication, 303–4
  Waite's ambivalence regarding, 302–3
Doepler, Carl Emil, 181
Dog-Star, 142, 222, 238, 245, 261, 263, 275
dolphin
  of King of Cups connected to Delphinus, 307
  of Poseidon = Delphinus, 225
domicile (astrology)

Ptolemaic system, 23, 29, 34, 97, 331
door in the stone
  and "Gate of Fair Entrance" – "open sesame", 185
double inversion
  part of Crowley's solution to the "planetary blind" involving Emperor, Star, Justice, and Strength trumps, *7*
dove
  and Enneagram, 40
    "centers", 325
    "phoenix cycle", 43
    Columba, Sirius, 41
  and Great Seal
    "phoenix cycle", 65
  and Pleiades, 122
  and Porta Alchemica, 75
Draco
  and "dragon of eclipse", "dragon of pole", 225
  and "temptation of Lilith", 93
  and crocodile zootype, 96
  and Emperor trump
    Sevekh, Great Beast, Great Harlot, 112–16
  and Empress trump, 107–8
  and Great Harlot, 11
  and Griepenkeri's Venus-Urania, Empress trump, *109*
  and Hanged Man trump
    Age of Aries, Odin's wolves, caduceus, 180–82
  and heptanomis, 24
  and Lovers trump, 125, 127
  and Magician trump
    ecliptic axis, 79
  and Minor Arcana
    ancient Egyptian scheme of zootypical cardinal directions, 307
  and Moon trump
    circumpolar Pool of Khep, 269
  and Porta Alchemica, 74, 76
  and Schweighardt's Invisible College, 91
  and Sirius Supercluster/Ursa Major Stream, 261

INDEX

and Star trump
    circular Zodiac of Denderah, 254–55
    ecliptic axis, 261
and Sun trump, 274
and Ta-urt, 205, 208
and World trump, 289–92

dragon
    "mount of dragon", 11, 24
    "wings of" = Ursa Minor, 181
    7-headed
        and Mount Meru, 31
        Revelation, 112
    and Porta Alchemica, 76
    and Sola-Busca World trump (Nabuchodenasor), 291
    flag of Aeddon (Hu), 277
    Hydra, Sevekh-Kronus, 145–46
    of drought, 58
    of eclipse, 223–25, 248
    of Medea, 211, 213–15
    of the deep, 128
    tail of, Thuban, 110

dragon-horse, 269
    Ta-urt, Draco, 114

Druids, 129
    and red dragon flag of Aeddon, 277

Durga
    consort of Shiva, 228
    Ursa Minor, 229

Durga Kali
    and Capricorn, 228

Dvipa
    Jambu-Dvipa (Meru), 31

eagle
    2-headed, Holy Roman Empire, 116
    and Enneagram, 40
    "centers", 325
    and Great Seal, 58–59, 64
        Porta Alchemica, 71
    of Zeus, Aquila, 244–47
    Scorpio
        beast of Apocalypse, 155, 191–93
        Galactic Center, 326
        tribe of Dan, 318

Earth
    and 8-year conjunction cycle with Sun, Venus, 147, 240–41
    and phoenix cycle
        position relative to Sirius, Sun, Venus, 65, 131
    and radiations from Galactic Center, 34–36
    and Thierens' attribution scheme, 2
    declination of
        and Phaeton, 148
    descent of soul to, 138
    horizon of
        and subterranean chamber of Great Pyramid, 177
    incarnation of souls upon
        and *Kuf, Reish, Zayin*, 267
    left by Astraea, 167, 172
    nearest black hole to in Orion Nebula, 203
    permeated by astral radiations, 144, 211
    purportedly visited by Anunnaki, 147
    Saturn furthest classical planet from, 29
    soul must escape from chains of, 211

Earth (element), 161
    and West/crocodile/Sevekh, 306

Easter Judgment
    observed as March Assizes in England, 282

Ecclesiastes, 21

ecliptic
    anciently divided into 10 constellations, 115
    and "planet of the crossing", 147
    and Charioteer's belt, 130
    and *crux ansata*, 110
    and Egyptian decan system, 19
    and Enneagram, 40, 43
    and Gate of Humanity, 107, 118–19, 138, 191
    and Gate of the Gods, 118–19, 191
    and *globus cruciger*, 102

and golden path of Moon trump, 265
and Midgard Serpent, 14
and ouroboros
   Devil trump, 213
   Magician trump, 78
and wreath of World trump, 289
guarded by 2 dogs
   Moon trump, 264
ecliptic axis
   and "tree of eternity", 228
   and 7-armed balance of heptanomis, 165
   and 8-pointed star of Star trump, 261
   and caduceus, 215
   and Enneagram, 42
   and *Ercole I d'Este* Star trump, 262
   and Magician trump, 78–79, 83
   and Mystic Rose, lily-lotus, 186
   and Schweighardt's Invisible College, 91
   and Sola-Busca Devil trump, 208
   and the Bahir, 21
   and Tree of Life
      central 4 sefirot, 17
   and Tree of Life, Lovers trump, 125
   and trunk of Yggdrasil, 14
   in contradistinction to celestial axis, 79
   intersection with celestial axis = "throne" of Blessed Holy One, 19
ecliptic pole
   and "girdle of Isis", 212
   and circular Zodiac of Denderah, 213, 253
   and *Keter*, 82
   and Porta Alchemica, 74
   and *Shin*, 76
   and the Bahir, 23
Eddas (Nordic), 14, 180
Eden
   Garden of
      celestial connotation, 125
   mountain of
      celestial connotation, 186

Egypt, 8, 19, 38, 71, 76, 85, 125, 128, 165, 184, 186, 195, 198, 206, 218, 236, 244, 264, 294, 300, 308
   north of equator by approx. 30°, 221
   years anciently reckoned by eclipses in, 224
Egyptian
   "put-circle" and Enneagram, 44–47
   Ali = Hebraic *Elohim*, 276
   astrological house system correspondence with Waite's denary keywords, *310*
   astrology and 2nd house, 193
   astrology based on pentadic divisioning, 324
   astrology was sidereal, 304, 342
   astronomical mythos assimilated by Christianity, 149
   Book of Hades, 185
   Book of the Dead, 193, 236
   circumpolar zootypes, 96
   deification of time/Kronus = Sevekh, 224
   dualistic "mount of equinox", 85
   eschatology was celestial, 167
   Hercules = Khunsu, 273
   hieroglyph for Sirius, 130, *131*
   hieroglyph *Nun*, *197*
   Judgment Hall, *Maati*, 185
   mysteries
      Jesus initiate of, 136
      Plato initiate of, 233–34
   origin of Hebraic *Kuf*, 268
   origin of Hebraic *Nun*, 195
   origin of Ptolemaic domicile system, 29
   origin theory of Tarot, 1–2
   Planisphere by Kircher, *259*
   rules of celestial orientation, 79–80
   scheme of zootypical cardinal directions, *307*
   Sothic and civil years, 33
   source of Amphitrite legend, 225
   stellar cult of Sut-Typhon and Revelation, 112
   symbolism of Great Seal, 58, 65

INDEX

system of astrological "term" distribution, 330
technique of harnessing Solar Force, 138
eirô. *See* Iris
Eleatic Stranger
and Plato's *Politicus*, 233–34
Electra, 202
Elohim, 149, 206, 276
El-Shaddai, 125
Emperor (trump), 83, 110–12, 116, 137
and Crowley's "double inversion", 7, 111
and Wands, 310
Minchiate Fiorentine, *102*
Tarot de Marseille, *103*
Empress (trump), 84, 102, 104, 108, 110, 173, 260
Minchiate Fiorentine, *102*
Tarot de Marseille, *103*
Enneagram
and "inner circulation", *48*
and "put-circle" of Ptah, *47*
and apocalyptic beasts, *41*
and celestial sphere, *42, 44*
and circumpolar precessional circle, 25
and complementary triplicities, *48*
and Great Seal, *59*, 62, 65
and Kircher's *Arithmologia*, *60*
and Lull's *Ars Generalis Ultima*, *59, 60*
and Tree of Life, *39*
essence types and Culpeper's *Complete Herbal*, 27
Equatoria, 200, 221, 244
equinox
"altar" of = Ara during Age of Taurus, 274
"mount" of both polar and ecliptical, 150
and *crux ansata*, 110
and *globus cruciger*, 102
and Great Sphinx, 142
and Greek cross, 86, 281
and Har-Makhu, 283

and Serer cosmological octohedron, 187
and Shu, Tefnut, 127
autumnal
and Death, Revelation, 193
and Persephone, 106, 150
and Schweighardt's Invisible College, Antares, 92
commonly associated with the West, 174
of the Great Year, 95
place of equipoise in solar mythos, 167
precessional shifting of
and *The Faerie Queene*, 171
vernal
and Child Horus, Khunsu, 273
and Easter, Lent, 300
and Hu, *Magnum Sublatum*, 277
and Schweighardt's Invisible College, Aldebaran, 92
and stellar mythos, judgment, 221
and triumph of Osiris, 136
commonly associated with the East, 174
of the Great Year, 282
today relictually associated with Aries, 111
Ergon, 101
and NEP, 101
Eridanus, 179–80
Eriginus
and Shemhamphorasch, zodiac, 177
eschatology
ancient Egyptian = celestial, 85, 126, 167, 204, 220
evolution of throughout precessional time, 206
Hebraic eschatology derived from celestial Egyptian eschatology, 269
Piscean
Death, Burial, Resurrection, 291

stellar feminine mode
anathemized, 244
transition from matriarchal to
patriarchal, 115
Esdras, Book of
apocryphal, 134–35
essence
and Enneagram types, 25
Culpeper's *Complete Herbal*, 27
and *innocentia inviolata*, 143
Euphemus
and Shemhamphorasch, zodiac, 177
Euphrates River, 44
Eurytus
and Shemhamphorasch, zodiac, 177
Eve, 94, 124, 150, 238
twin of Adam, 127
Exodus, Hebraic
and cult of Sut-Typhon, leprosy, 243
Eye of Providence, 58–60
and Great Seal, *59*
and Kircher's *Arithmologia*, *60*
faces (zodiacal), 326–31
72 proposed by Salmon in *Horae Mathematicae* (1679), 330
Chaldean Order of, 326
William Salmon's attribution scheme, *330*
Fagan, Cyril
and astrological "angles", 304
comparison of Fagan's houses with Waite's denaries, 311–12
house system = acronychal, 313
on "original zodiac", 231
on 2nd astrological house, 193
on ascription of elements to zodiacal signs, 308
on astrological "background angles", 320
on astrological effects of fixed stars, 68
on astrological effects of Moon/Mars configurations, 123
on astrological effects of Moon/Pluto configurations, 66
on astrological effects of Saturn, 66

on astrological rulership of sexual organs, 123
on dating of circular Zodiac of Denderah, 253
on dating of rectangular Zodiac of Denderah, 256
on Egyptian pentadic divisioning of ecliptic, 324
on erroneous nature of tropical astrology, 31, 342
on Seshat and "stretching of the cord" ceremony, 251
on sidereal location of current vernal point, 231
fallos
(Greek) similar to Hindu *lingam*, 163
false ego, 277–79, 287
Fire (element)
and Ages of Great Year, 161
and ecliptic axis, Porta Alchemica, Levi's pantacle, 75
and South/phoenix/Har, 306
flag-flower
and Iris, 200
Flamsteed, John, 271
Fomalhaut
and present vernal point, 92
and Schweighardt's Invisible College, 92
Fool (trump), *297*
and "0=1" formula, 7
and heptanomis
בגף (division), רתדר (frequency), 301
and *Mem*, 296
and ostrich feather, celestial pole, equinox, Sun trump, 278
and precession of polestar, 296
as Spirit prior to self-expression, 299
belt of symbolizes year, Great Year, 297
feather of
and *Maat*, 298
originally had no numerical value, 298
Visconti-Sforza

INDEX

headdress and Lent, precession, 300
Fortune (trump), 157, 163
  and Crowley's "*kteis+fallos*" equation, 247
  ascribed to Saturn in proposed system
    per Crowley's "*kteis+fallos*" equation, 163
  spun by Tower trump, 249
Founding Fathers, 63–64, 69
free trinity, 43
  of Enneagram, 40, 46
Freke, 181
French Revolution, 63
gabah
  Hebraic = "haughty", "high", "lofty", "taller", 21
Galactic Center, *203*
  and astral body, "centers", 326
  and crown of Rider-Waite Death trump, 207
  and Enneagram, 41, 43
  and Gate of the Gods, 107, 118, 203
  and Schweighardt's Invisible College, *95*
  and standard of Rider-Waite Sun trump, 274
  and Yugas, Great Year, 35–38, *36*, 66, 126, 283
Galactic Equator
  and chalices of Rider-Waite Temperance trump, 199
  and Gate of Humanity, 118–19, 139, 191
  and Gate of the Gods, 118–19, 191
  intersection with ecliptic, 107, 119
  and Ennegram, 43
  primeval "bow in the sky", 202
Gate of Fair Entrance, 185
Gate of Fair Exit, 185
Geirtýr
  "Spear God" (appellation of Odin), 181
Geirvaldr
  "Spear Master" (appellation of Odin), 181

Gemini
  a diurnal "water sign" of inundation
    during Age of Aries, 131
  Age of
    and expulsion from paradise, 94, 98
    and Fall of humanity, 95, 126
    and mercurial serpent of Enki/Ea, 99
    and Mercury/Odin, 176
    and *meskhen*, 200, 268
    and Ta-urt's betrayal of Sut for Horus, 127
    and Yugas, 92
    Autumnal Equinox of Great Year, 291
  and Ptolemaic domicile system, 29
  and Sun trump, *272*
  conflation with Hercules, 272
  constellation
    and Gate of Humanity, 118, 139
    and monks of Rider-Waite Hierophant trump, 119
    and rectangular Zodiac of Denderah, 256, *258*
    and Sut/Horus, Shu/Tefnut, 127
    summer solstice in 146 AD, 32
Genetrix, 46, 127–28, 134, 145, 149, 157, 244
Gere, 181
gestation, 300
  9 months of, and "put-circle" of Ptah, 45
Gimel, 88, 152
Gjallarbru, 16
globus cruciger, 103–4, 110, 116, 280
  and pomegranate, 105
  and Porta Alchemica, 71
  as equinoctial symbol, *103*
  symbol of worldwide supremacy, 45
gnostic
  Christ, 261
  Holy Spirit, 342
  ogdoad, 80

Golden Age
   and Berosus' accounts of Great Conflagration, Great Flood, 267
   and crown of Sola-Busca Nabuchodenasor trump, 292
   and Galactic Center; Ages of Virgo, Leo, Cancer, 34–38
   and Great Sphinx, 83, 142
   and head of Draco, Rider-Waite World trump, 290
   and Lyra, 150
   and nimbus of Rider-Waite Hanged Man trump, 181
   and Plato's *Politicus*, 234
   bounded by ascending/descending Mercurial Ages, 98
   end of, and Corona Borealis, 92
Golden Gate (Gate of the Gods), 107, 118
Grail, Holy, 146, 310
Gravrand, Henry
   and Serer diagram of celestial octohedron, 187
Great Beast, 112–16, 128
   and changing of polestar, 144
Great Flood, 267
   occurred circa 8,000 BCE per Berosus (Age of Cancer), 209
Great Mother, 46, 71, 112, 125–28, 205, 208, 212, 252, 255, 268, 289, 298, 300
Great Seal
   and Enneagram, *59*, 62
   design committee est. July 4, 1776, 58
   eagle of actually phoenix, 64
   early version depicted Moses; heliacal rising of Sirius, 58
   of Rosicrucian design, 62
   stars of, *71*
Great Work, 64, 99, 101, 161
Great Year, 83, 94–96, 108, 126, 142, 159, 172, 181, 235, 238, 247, 263, 265–66, 268, 276, 282, 284–86, 289, 291, 342
   "double ending" of, 238
   and Fool, Iu, Wandering Jew, 297–98
   and inversion of Venusian pentagram, Devil trump, 240–41
   and phoenix cycle, 238
   and Ptolemaic domicile system; Hindu Yuga system, 34–38
   divided into 4 World Ages, 161
Greece, 244
Greek cross, 91, 281
   and equinox, 86
green
   associated with "upper", spring, summer, 276
   color of Emperor's *globus cruciger* = spring, 103
   color of renewal, 276
Griepenkerl, Christian, 108
Gurdjieff, George Ivanovich
   and Enneagram, 39–41
   and Great Seal, 59, 62, 65
   and Enneagram "essence types", 26–28
Gwydion, 201
Hades, 80, 106, 150, 193
   Judge of the dead
      and Mors, Judgment trump, 285
Hall, Manly Palmer, 16, 19, 62, 64, 73, 120, 177, 185, 271, 282
Hangagod
   and Age of Virgo, development of proto-writing, 176–77
Hanged Man (trump), 279, 286, 316
   and Frolich's "Sacrifice of Odin", *176*
   as circumpolar precessional symbol, *182*
   attributed to Pisces by Thierens, 2
   connected to Sun trump, 278
   Conver's version, *175*
   medieval interpretation of, *179*
Hapi, 197–98, *197*, 255, 261
   represents East/wind/ape/ in Egyptian scheme, 306
Har
   represents South/fire/phoenix in Egyptian scheme, 306
Har-Makhu

INDEX

god of double equinox, horizon, 283
god of double horizon, 245
Harpocrates, 273–74
Harris, Marguerite Frieda
   illustrator of Crowley's "Thoth" tarot deck, 195, 331
Hathor, 44, 122, 135, 198, 256
   and Khunsu/Harpocrates, Hercules, 273–74
   Temple of, 128
Haunch
   constellation of
      meskhen, Cassiopeia, 198
Hebraic
   "horse of Nimrod" = Pegasus, *243*
   Ancient of Days
      analogous to Odin, 176
      superseded Egyptian Genetrix, 149
   association of Adam with equinox, 142
   Davidian lineage, 63
   inscription on Porta Alchemica, 75
   mystical texts, 13, 17
      Book of Esdras, 134
   Tree of Life (*Etz Chaim*), *14*
   word for Orion = *Kesil* ("fool"), 296
Hebraic alphabet
   and Crowley's "double inversion", 8
   and Major Arcana ascriptions – suggested to be fallacious, 2
   as "base 4" numerical system, 152
   proposed correspondence with Major Arcana, *317*
Hebraic calendar
   beginning of month and year marked by New Moon, 256, 264
Hei, 4, 7, 111, 277
   and Tetragrammaton, 152
Heine, Friedrich Wilhelm, 14
heliacal
   rising of Sirius, 131
      and "double ending" of Great Year, 238
   and "phoenix cycle", 65, 120, 131, 236, 241
   and Amphitrite, 225
   and astrological considerations of Founding Fathers, 68
   and marriage of Bride and Lamb, 135
   and slaying of lion by Samson, 173
   and Sothic cycle, 32
Helios, 217
Hell, 168, 193
   descent into underworld – connected to Virgo, 150
heptanomis, 10–12, *11*, 23–26, *86*, 94, *97*, 101, 107, 110, 114–16, 126, 142, 146, 149, 165, 207, 208, 251, 270, 284, 296, 318, 342
   and Fool, *Maat*, 298
   and Magician, Empress, High Priestess, 83
Hercules
   and Porta Alchemica, 76
   and Shemhamphorasch, zodiac, 177
   conflation with Gemini, 271–74
   constellation, 260
      "mount of Man", 115
      acronychal regulus of inundation during Ages of Aries, 244
   and Atlantis, Great Sphinx, heptanomis, 143
   and heptanomis, 24, 85, 98, 142, 270
      Age of Leo, Galactic Center, 283
   and Schweighardt's Invisible College, *90*
   and Schweighardt's *Mirror of Wisdom*, 95
   Egyptian equivalent = Horus, 83
   Labors of = celestial allegory, 141
   stellar hero before solar one, 141
   twin to Iphicles, 270
hermaphrodite
   Egyptian Hapi depicted as, 255

Hermes Trismegistus
  Emerald Tablet of, 98, 100–101
Hermetic Order of the Golden Dawn
  and "blinded" trumps, 8
  transposition of Strength and
    Justice trumps, 5
Hermit (trump), 150–52, 154, 279
Herodotus, 234, 247
  History
    and precession, 232–33
Heru-ra-ha, 273
Hetep, Mount, 275
  and Ara, horizon, circumpolar
    paradise, 274
Hierophant (trump), 107, 117, *119*,
    120, 279
  referenced in Rider-Waite Death
    trump, 191
High Priestess (trump), 301
  and Isis, 86
  and Secret Church, Schweighardt's
    Invisible College, 88
  pillars of distinct from pillars of
    Hierophant trump, 117
  relation to Chariot trump, 133–34
  relation to Justice trump, 172
Hindu
  Great Year = 24,000 years, 24
  lingam and yoni; Chariot trump,
    129
  Yuga system, 35, 38
    and Golden Age, 92
hippopotamus, 96–98, 134, 208, 253,
    307
  "mount of hippopotamus" =
    Corona Borealis, 11
  North/water/Typhon in Egyptian
    scheme, 306
Holy Roman Empire, 114–16
Holy Spirit, 78, 342
  Baphomet, 211
Hoor-pa-kraat. *See* Harpocrates
horoscope
  "angles" of, 304
  constructable using Major and
    Minor Arcana, 318–23
  sample reading, 332–39
  of American Declaration of
    Independence, *68*

  of American Declaration of War,
    67
horseshoe
  and "put-circle" of Ptah,
    Enneagram, omega (Ω), 44–47
Horus
  "birthplace" of was season of
    inundation, 194
  Aeon of, 121, 144, 163
  and 8-pointed star, 261
  and lily-lotus, 197
  and Porta Alchemica, 71
  and rectangular Zodiac of
    Denderah, *258*
  and Sun trump, 271, 273–76
  and Visconti-Sforza World trump,
    294
  as Khunsu/Harpocrates, 273
  battles with Sut during eclipse, 223
  Greek equivalent = Hercules, 83
  lion of "Lion and Unicorn", 96
  overcomes drought in Egyptian
    mythos, 145
  served as "door in the stone", 185
  suckled by Hathor, Neith, 198
  Ta-urt deserted Sut for, 126
  twin of Sut, 127–28
  white crown of, 276
  white god, 185
house (astrological/zodiacal), 2, 29
  "esoteric inversion", 121
  2nd house = "*Haidou pyle*", "door of
    Hades", Scorpio, Mars, 193
  2nd house connected to
    Mors/Thanatos, 285
  and "faces", 329
  and Aces, 313
  and pairing of denaries and court
    cards in "random" card reading
    method, 333
  Astraea as 6th house of tropical
    zodiac per Edmund Spenser,
    171
  Fagan's system acronychal with
    respect to Ptolemaic system,
    313
  houses and proposed attribution
    system, 320

INDEX

may need to shift according to precession, 31–34
of Hathor/Hera = "hat-hor" ("house of Horus"), 273
of lion-faced god Tum connected to Ages of Leo and Taurus, 85
possible to maintain fixed system, if taken as relative to the Great Year, 34–36
proposed correlation between Waite's denary descriptions and 12 houses, 310–15
Ptolemaic house system apparently derived from Age of Taurus, 31
Ptolemy's rationale for house system questionable, 29
ruling planet of house = "significator", 315
significance of in astrological prognostication according to William Lilly, 315
Thierens starts from 1st zodiacal house in his Tarot attribution scheme, 3

Hu
  Celtic sun god of horizon, Aeddon, Hua, Hva, 129, 276
Hua. *See* Hu
Hva. *See* Hu
Hydra
  constellation
    acronychal counterpart to Cetus/leviathan, 128
    and circular Denderah Zodiac, *146*
    and Great Seal "dragon of drought", Pharaoh, 58
Hyginus, 272
Ice Age
  Hanged Man, proto-writing, Lyra, 176
Ichthus, 108
Ida, 82
  and Moon, 100
ideograms
  ancient development of connected to origin of Tarot, 1

Ihuh, 125
illuminati, 184
immortality, 81, 96, 101, 104, 211, 248
  elixir of and *Peil lingam/*sword, 163
Imperial Globe. *See globus cruciger*
India
  years anciently reckoned by eclipses in, 224
Indra, 248
informing triangle
  of Enneagram, 39–41, 43, 49, 59
  and Great Seal, 59
innocentia inviolata
  and "spiritual essence", 143
inundation
  a possible basis for story of Rhampsinitus' Tower, 249
  and "abyss of source", 161
  and "birthplace" of Horus, 194
  and "bow in the sky", Sagittarius, Temperance trump, 205–7
  and "phoenix cycle", 65, 241
  and "put-circle" of Ptah, Enneagram, 45
  and "put-circle" of Ptah, Porta Alchemica, 73, 76
  and Amphitrite, 225–27
  and bennu-bird, nictorax, phoenix, 236, 238
  and Cassiopeia/Cepheus, 255
  and Crater (constellation), 146
  and Diana of Ephesus, 134
  and equipoise between flood/drought, 167, 174
  and fall of Belus' Tower, 245
  and fall of Nimrod's Tower, 244
  and Full Moon as nocturnal harbinger of, 219
  and Horus, 127, 185
  and lily-lotus, *Nun*, Temperance trump, 197–201
  and lion zootype, 85
  and Pool of Khep, Moon trump, 268
  and Shamgar, Samson, 174
  of Nile

and red waters, "Tesh-Tesh",
    145
inversion
    occult
        nocturnal "esoteric" sign
            counterpart to diurnal
            "exoteric" sign, 36–38
        of *globus cruciger* and sigil of
            Venus, 103
        of Minor Arcana cards, 320
        of Strength/Justice per Waite,
            Crowley, Case, 7–9
        of Venusian "pentagram of
            conjunction" via precession,
            241
Invisible College, *89*
    and Fortune trump, 159
Iphicles
    twin to Hercules, 270
Iris
    and "bridge" of Milky Way, 204
    and flag-flower, rainbow, 201
    possible derivation from Greek *eirô*
        ("I join"), 202
Iron Age
    and feet of Hanged Man trump,
        181
    and Ovid's *Metamorphoses*,
        Spenser's *The Faerie Queene*,
        Talus, Astraea, 167–72
    Winter of Great Year, 290
Ishtar
    "Lady of the Mountain" – goddess
        of NCP, 107
    and 8-pointed star, 261
Isis
    Aeon of, 144
    and dove, Porta Alchemica, 75
    and High Priestess, 87
    and Persephone, 106
    and Scarlet Woman, Sirius, 120
    and Ship of the Bennu-Asar, 65
    as Dea Multimammae, 205
    as Full Moon, 136, *219*
    as Sirius, Moon, that which is
        "above ground and visible",
        218–20
    barque of
        and Ark of Noah, 91

symbolized by Pasiphae's
    hollow cow, 123
conceived by smelling lily-lotus,
    184
girdle of, 211
    and precessional binding of
        NCP/NEP, 212
    and *sa* of Ta-urt, *213*
Isis-Serkh
    scorpion goddess (Scorpio), 194
Israel
    12 Tribes of, 19
    13th Tribe of
        and Manasseh, United States,
            63
    and shift from stellar to solar
        worship, 125
    generic term signifying humanity,
        19
Iu
    "original" Wandering Jew, 133
Iu-sa. *See* Iu
Jachin
    pillar of, 87, 117
jackal
    "mount of jackal" = Ursa Minor,
        11, 24, 96
    and Anubis, Polaris, 181
    as figure of celestial pole, 264
    of circular Denderah Zodiac, 213,
        253
Jambu-Dvipa, 186
Jason (Argonaut)
    and Age of Aries, Golden Fleece,
        Medea, 214
    and Porta Alchemica, 76
Jehovah, 82, 90, 125, 244
Jerusalem, 78, 134
JHVH
    tetragrammaton, 158, 159
Joseph (Biblical), 64
judgment
    and Amenta, 285
    and Anubis, 220–21
    and Dan, eagle, Scorpio, 317
    hall of
        and *maat*, 165, 282
    of Great Harlot, 113
    seal of

and *Tav*, 287
seat of
   and celestial pole, 167
Judgment (trump), 172, 282, 285
   Rider-Waite version, *282*
   transposition with World trump, 317
   Visconti-Sforza version, *280*
Jupiter
   and Culpeper's *Complete Herbal*, 27
   and Enneagram
      feminine "essence type", 25
   and heptanomis correspondence, 97
   and horoscope of Declaration of Independence, 66
   and Neptune, Sagittarius, 318
   and Piscean Age
      and Crowley's "*kteis+fallos*" equation, 163
   and Porta Alchenica, 72
   and Tower of Belus, 247
   ascription to Tower trump, 247
   nocturnal ruler of
      Pisces/Sagittarius
   and heptanomis, *26*
Justice
   and *maat*, 165
   and Manifest Destiny, 63
Justice (trump)
   and Libra
      inversion with Leo/Strength, 5–8
   and Swords, 310
   inversion with Strength, 140, 153, 173
   relation to High Priestess, 172
Kabbalah
   true meaning of hidden to avoid persecution, 343
Kabiri, 79
Kaer-Bediwyd
   "Ark of the World" = Corona Borealis, 92
Kali, 228
   as one of the 7 tongues of Agni/Shiva = Ursa Minor, 229

Kaplan, Aryeh, 7, 21
Kepler, Johannes, 91
Keridwen
   and yoni stone, 129
Kesil
   Hebraic Orion, "fool", 296
Keter, 39, 76, 276
   and Cat's Eye Nebula, 82
Ketos (constellation). *See* Cetus
Khamsin winds
   and Tyhpon/Scorpio, 194
Kheft, 146, 201
Khep, 268
Khepra, 268
   ark of and "double ship", 236
Khunsu, 136
   and Harpocrates, Hercules, Scorpio, Hera, 273–74
King (court card)
   and intellectual parts of centers, 326
   of Cups, 307
   of Pentacles, 305
   of Wands, 305
   replaced by Knight in Harris-Crowley Tarot, 326
Knight (court card), 331
   replaces King in Harris-Crowley Tarot, 326
Krater. *See* Crater (constellation)
kteis
   (Greek) similar to Hindu *yoni*, 163
Kuf, 152, 163, 266–68, 277
kufa (boat)
   and "put-circle" of Ptah, 44
Kumaras
   Hindu gods exempt from attachment
      and Capricorn, Dagon, Makara, Aquarius, 229
Lamb
   of Revelation, 114–16, 135, 192
Lamed, 152, 173–74
latitude
   and "seasons" of Great Year, 35
   effect on celestial observations, 221–23
lemniscate, 78

and analemma, Visconti-Sforza
World trump, *294*
of Magician trump, 79–83
of Rider-Waite Magician trump,
79
of Rider-Waite Strength trump,
140
of Solar Force and de Gébelin's
Chariot trump, *137*
Leo
a diurnal "water sign" of
inundation
during Age of Aries, 131
during Age of Taurus, 76, 91,
120, 128, 131, 167, 206, 226
and Porta Alchemica, *74*
Age of
and "summer" of Great Year,
34–37, 92, 247
purported origin of double-
lion imagery, 85
and Galactic Center, 126, 283
and Golden Age, 161
and Great Sphinx, 142–43
and Hercules (constellation), 83
and Strength trump, 141–44
impending Age per occult
"inversion formula", 37, 120
and "double ending" of Great
Year, 238
and Enneagram, 40, 43
"centers", 325
"put-circle" of Ptah, 46
and heptanomis correspondence,
96
and King/Queen of Wands, 305
and Sothic cycle during Age of
Taurus, 32
and Strength trump, 140
completes Scorpio/Death, 193
inversion with Libra/Justice, 5–
8
and Sun trump, 270, 274
and the "Two Truths" - solar sign
of "inspiration" during Age of
Taurus, 185
constellation
and circular Denderah Zodiac,
146
and Empress trump, 107
and Great Seal, 58
and Regulus, Empress trump,
104
antiscion of Crater, and Locri
Persephone pinax, 105
true house of Sun during Age
of Taurus, 31–32, 46, 174
lion of Rider-Waite Fortune
trump, 155
lion of The World trump, 289
one of 4 beasts of Apocalypse, 192
ruled by Sun in Ptolemaic
domicile system, 29
Leo Minor
and Queen of Wands, 305
Lernaean Hydra, 271
Levi, Eliphas (Alphonse Louis
Constant), 2, 83, 130, 326
and "Astral Light", Akasa, Holy
Ghost, Baphomet, Goat of
Mendes, 210–11
and *Urim, Thummim*, 88
pantacle of
and Great Seal, Enneagram,
60–63, *62*
sigil of, *75*
and celestial sphere, Porta
Alchemica, *76*
Libra
Age of
and Griepenkeri's Venus-
Urania, Empress trump, 108
and Revelation, 115
anciently midpoint of flood
season, not year *in toto*, 174
and "Aeon of Osiris", 144
and circular Denderah Zodiac
Khunsu/Harpocrates, 273
and Justice trump
Corona Borealis, *maat*, 165
inversion with Leo/Strength, 5–
8
Strength/Justice inversion, 164
and Justice trump, Swords, 310
and justice, *maat*, 167
and Schweighardt's Invisible
College, *95*, 99

and Spenser's *The Faerie Queene*,
Astraea, 172
constellation
   anciently joined with Scorpio,
204
   point of equipoise between flood
and drought
      during Age of Taurus, 167
Light of the World
   and Hermit trump, 149
      Christ, Ra, Sun, 150
   Taht-Aan
      lunar representative of Sun, 220
Lilith, 93, 96, 99
Lilly, William
   and "Ptolemaic" attribution
scheme
      Christian Astrology, 309
   and querent, significator, 314
lily
   conflation with Mystic Rose, 184
   conjunction of Sun with Sirius,
185
   lily-lotus, 184–87, 196–98
      scent of impregnates Isis, 185
   rose of the lily, 184
lingum, 129
logos
   Brahmanic, 37–38
      emanations from Galactic
Center, 126
Lotus
   8-leaved
      Hindu Golden City of Gods,
45
   of Earth
      "seed-cup" of = cone of
precessional motion, 186
      Mount Meru, *187*
   of Immensity, 31
Lull, Raymond, 59
Lyra, 24, 97, 150, 271
   and Griepenkeri's Venus-Urania,
Empress trump, *109*
   and Hanged Man
      Ice Age, proto-writing, 176–78
   and Revelation, 115

and Shemhamphorasch, zodiac,
177
and subterranean chamber of
Great Pyramid, *178*
lyre
   and Gemini, 271
   and Lyra, 108
maat, 165–67, 172, 220
Maat (Maut), 282
   Egyptian vulture goddess, 298
Maati, 185
Macara. See Makara
Magician (trump), 140, 297
   and Apollo, Hercules, 83
   and celestial axis, ecliptic axis, *79*
   and gnostic Christ, ogdoad, 80
   attributed to *Alef*, 278
Magnum Sublatum
   "Great Upraising", vernal equinox,
277
Maimonides, 8–9
Major Arcana, 13, 111, 153, 318, 342
   and Minor Arcana
      may have originally been
intended as completely
separate decks, 302
   not occult, but mystical per Waite,
vii
   proposed correspondence with
Hebraic alphabet (according to
Crowley), *8*
   proposed correspondence with
Hebraic alphabet (according to
precessional model), *317*
Makara
   anagram of Kumara, 229
   and Night of Brahma, 229–31
Malchut, 39
malefic
   astrological effects
      ameliorated by Avatar in 22-
card spread, 334
   and Saturn, 29, 339
      Declaration of
Independence, 66
Manasseh
   and "13th Tribe" of Israel, United
States, Aquarius, 63–64

mankind
   "mount of mankind" = Hercules,
      11, 24
Marduk, 245
Mars
   and Culpeper's *Complete Herbal*,
      27
   and Enneagram
      masculine "essence type", 25
   and heptanomis correspondence,
      97
   and horoscope of Declaration of
      Independence, 66
   and Marduk, Har-Makhu, 245
   and Pluto, Scorpio, 318
   and Porta Alchenica, 72
   and Scorpio
      Pasiphae legend, 123
      Tartarus, 193
   diurnal ruler of Aries/Scorpio
      and heptanomis, *26*
   relationship to Judgment trump,
      285–86
Massey, Gerald
   and double representation of *maat*
      (circumpolar and equinoctial),
      167
   and Enneagram, put-circle, abyss,
      44–46
   and lily-lotus, Mystic Rose,
      Jambu-Dvipa, Rose-Apple
      Tree, 186
   and precessional shifting of
      planetary domiciles, 31–33
   theory of heptanomis, rejection of
      occult secrecy, 9–12
Mathers, Samuel Liddell MacGregor
   and Hanged Man trump
      medieval interpretation of, 178
   changed planetary attributions of
      Sefer Yetzirah, 5
Medea, 211, 213
Medieval Era, 271, 343
medulla, 99, 267
Mem, 63, 152, 198, 278–79, 287, 296,
   298–301
Mercury
   and Anubis, Hermes, Hermanubis,
      223
   and Culpeper's *Complete Herbal*,
      27
   and Enneagram
      masculine "essence type", 25
   and Hanged Man trump, 316
   and heptanomis correspondence,
      97
   and horoscope of Declaration of
      Independence, 67
   and Odin, Hanged Man trump,
      176–77
   and Porta Alchenica, 72
   and Shemhamphorasch, zodiac,
      177
   and Tarot origin, 1
   diurnal ruler of Gemini/Virgo
      and heptanomis, *26*
   unites ascending and descending
      portions of Great Year, 98
Meru, 31, 98
   and ecliptic axis, spinal cord, 77
   and Shiva, 228
   symbol of precession, *33*, *187*
meskhen, 200, 268
   constellated as Cassiopeia, 198
Mi
   Hebraic "highest pole" – forms
      Alhim/Elohim when connected
      to *Aleh* ("hosts of heaven"), 149
Midgard
   and Tree of Life, *18*, 19
   Serpent of
      and ecliptic, 14
Midheaven (astrological), *304*, *305*,
   306, 313, 319
   and 22-card spread, 332
Milky Way, 16, 74, 198–99, *203*, 205,
   207, 245, 283
   "bridge" of, *203*
Mimir, 175
Mina
   sign of Pisces
      and Night of Brahma, 230–31,
      230
Minchiate, 102, 110
Minor Arcana, 1, 302, 305, 310, 318,
   320, 325, 331, 342
   and Major Arcana

INDEX

may have originally been intended as completely separate decks, 302
and standard playing deck of 52 cards, 23
pips (denaries), 319, 324
Minos
  and Talus, 168
  figure of vernal Sun during Age of Taurus, 122–23
Moakley, Gertrude, 300
Monoceros, 96
Moon
  ½ of ancient biune deity (Sun + Moon), 32
  and Chariot trump, 130, 134
  and Corona Borealis, Kaer-Bediwyd, 92
  and Culpeper's *Complete Herbal*, 27
  and eclipse, Typhon, 223–24
  and Emerald Tablet, 99–100
  and Enneagram feminine "essence type", 25
  and High Priestess trump, 86–90
  and horoscope of American Declaration of War, 66
  and horoscope of Declaration of Independence, 67
  and Lilly's "face" system, 329
  and Porta Alchemica, 71–72
  and Schweighardt's Invisible College, 99
  and Strength trump, 145
  and Thierens' attribution scheme, 2
  and Tower of Rhampsinitus, 248–49
  angular and opposite to Pluto engenders fugitive urge to escape oppression, 336
  associated with Taht (Thoth), 220–21
  conjunction with Sun metaphor for balancing rational and intuitive minds, 266
  Full
    and Khunsu/Harpocrates, 273–74
    and Neptune, Poseidon, 123, 226
    and Osiris, 46
    and Pasiphae, 122
    and Virgin Birth, 136
    as Isis, 218
    passenger with Sun in "double ship", 236
  New
    and rectangular Denderah Zodiac, Hebraic calendar, 256
  nocturnal ruler of Cancer, 29
  and heptanomis, *26*
  purported vampiric feeder upon humanity, 221
  ruler of Cancer in Ptolemaic system, 309
Moon (trump), 264–66
Morganwy, Myfyr, 129
Morning Star, 235–36
Mors (Greek Thanatos)
  minister to Pluto/Hades associated with Mars, 285
Moses, 211, 214, 269
  depicted on early version of Great Seal, 58
Most Ancient Order of Druids
  Massey Chosen Chief of, 10
mount of the equinox, 85, 91, 96, 207, 274
mum
  Hebraic "blemish" - connected to *Mem*, 287
Mystic Rose, 184, 186–87, 190, 276
nadir
  of Great Year
    and "The Lion and the Unicorn", 96–98
    and Fortune trump, 161
    and Hanged Man trump, 181
    and Jupiter, Tower of Belus, 247
    and Moon trump, 265, 268
    and Plato's *Politicus*, 235

point of equipoise between involution and evolution, 291
of winter Sun
and Death trump, 195
of winter Sun in Age of Taurus = Aquarius, 46
NCP (North Celestial Pole)
and "stretching of the cord" ceremony, *253*
and Avatars throughout Great Year, 108
and Chariot trump, 133
and Corona Borealis *Lapis Exilli*, 99
and Culhuacan, 221
and Dragon of Medea, *214*
and Dragon/Serpent of Sevekh, *215*
and Empress trump, 107
and Enneagram, 41, *42*, *44*
and Fool trump
and Iu, 298
and Hanged Man, Odin, 181–82
and heptanomis, 10, *97*, 142
and High Priestess trump, 87, 93
and Iris, 202
and Magician trump, 83
and Minor Arcana, 307–8
and Moon trump, 266
and Moses/Shu-Anhar, 269
and Mount Meru, *187*
and Neith, Uat, 200
and parergon, 101
and Rider-Waite Fortune trump, *156*, *160*, 162–63
and Rider-Waite Justice trump, 165
and Serer cosmological octohedron, 188
and Sun trump, 274–76
Mystic Rose, "lotus of immensity", 276
and Tarot of Marseille Sun trump, *271*
and Ta-urt's desertion of Sut, 127
and Temperance trump, 200
and Tree of Knowledge, 342
and Tree of Life, 21, 23

and Visconti-Sforza Judgment trump, 280
and Visconti-Sforza Sun trump, *270*
and World trump, 295
and Yggdrasil, *16*, *17*
and Zohar, *Aleh*, *Mi*, 149
associated with Sevekh during Age of Aries, 112
feminine counterpart to vernal ecliptic locus, 111
of Mars
and Judgment trump, 285
Nebuchadnezzar, 291
king of Babylon during Age of Aries, 291
Nefertiti, 138
Neith, 198
goddess of the bow, 200
goddess of the North, 200
Nemean Lion, 141, 271
NEP (North Ecliptic Pole)
and "Blessed Holy One", 93
and "mooring post" of Ta-urt, 210
and "Sun behind the Sun", 263
and Dragon/Serpent of Sevekh, *216*
and Empress trump, 107–8
and Enneagram, *42*, *44*
and Ergon, 101
and Fool Trump
and primordial Mother, 299
and Ptah, 298
and Hanged Man, Odin, 181
and *Keter*, 276
and Mount Meru, *187*
and Porta Alchemica, 76
and Rhea/"Mother of the Mountain", 85
and Rider-Waite Fortune trump, *156*, 160
and Rider-Waite High Priestess trump, 87
and Rider-Waite World trump, *290*
and Serer cosmological octohedron, 188
and Sirius, 275
and Sola-Busca World trump, *292*

INDEX

and Sun trump, 278
  Mystic Rose, "lotus of immensity", 276
and Tarot of Marseille Sun trump, *271*
and Tree of Life, 21, 342
and Visconti-Sforza Judgment trump, 280
and Visconti-Sforza Sun trump, *270*
and Visconti-Sforza World trump, *294*
and World trump, 295
and Yggdrasil, *16*, *17*
and Zohar, *Aleh*, *Mi*, 149
Nephte
  and Sagittarius
    and "bow in the cloud", 206
Neptune
  and "nocturnal Sun", 121–23, 226
  and Jupiter, Sagittarius, 318
  and Porta Alchemica, 72–75
  and Poseidon, Amphitrite, 225–26
  and Thierens' attribution scheme, 2
NGP (North Galactic Pole), 104, 107, 200–202, 260
  and Yggdrasil, *16*, *17*
Nibiru, 147–48
nictorax
  "bennu-bird" = Sirius, 236
Nile River, 46, 141, 145, 167, 197–98, 206, 217, 226, 238, 273, 294, 308
Nimrod, 242, 244–47
  and Tower trump, 242
  fall of Tower astronomical, precessional, 247
Nine Stone Rig
  and "put-circle" of Ptah, 45
Niobe, 217
Nirvana, 45, 77
Noah
  Ark of = "barque of Isis", 91
Northern Crown (constellation), 96, 274
Northern Hemisphere

and Ptolemaic domicile system, 36–38
Nun, 152, 196–200, 204, 206, 286, 317
  Egyptian "abyss", 195
  Egyptian hieroglyph, *197*
Oannes
  Triton, Capricorn, 225
occult
  "blind", 9–10, 40, 153
  French "occult revival", 1
  relationship between Makara/Kumara, Capricorn/Aquarius, 229–31
  use of Great Seal and American dollar as pantacles, 62
Odin
  and Hanged Man trump, 175–77, 179–82
  as representative of NEP
    various forms correlate to precessional stations of NCP, 108
ogdoad, 80
omega (Ω)
  and "put-circle" of Ptah Enneagram, 46
  horseshoe, 44
Ophiuchus, 91, 194, 274
  and Sirius Supercluster/Ursa Major Stream, 261
Orion
  and Chariot as another form of Emperor, 137
  Asar, Osiris
    and ship of Bennu-Asar, 65
  constellation
    "arm" of = current locus of Sun at summer solstice, 32
    and Gate of Humanity, Hierophant trump, 117–20
    and rectangular Zodiac of Denderah, *258*
  Hebraic word for = *Kesil* ("fool"), 296
Orion Arm
  of Milky Way, *203*
Orion Nebula

gave birth to Sirius and Sun, 203
Orpheus
  and Cygnus symbolized in Judgment trump, 281
  and Shemhamphorasch, zodiac, 177
Osiris
  "coffin" of
    analog to Job's Coffin (constellation), 281
    "swallowed" by Sut during eclipse, 223
  Aeon of, 143–44
  and bennu, phoenix, 238
  and Chariot as another form of Emperor, 137
  and Fool trump, 297
    Green Man, 301
  and Full Moon, 46
  and Hierophant trump, 120
  and ship of Bennu-Asar, 235
  and valley of Amenta
    Death trump, 193–95
  as Lord of the Waters, 296
  as nocturnal Sun, 72, *219*
    "Osiris Tesh-Tesh", 145
    and Hierophant trump, 122–23
  as representative of humanity, 285
  Asar, Orion, 65
  representative of polestar and Sun, 43
  tomb of
    and Great Pyramid, 177
ostrich, 278
ouroboros, 19, 213
  and Enneagram, 40
Ouspensky, Peter Demianovich
  and Egyptian origin theory of Tarot, 2
  and Enneagram, 40
  and Enneagram "essence types", 26–28
  and hidden psychological meaning of astrology, alchemy, Kabbalah, 343
  and Moon as vampiric feeder upon humanity, 221
Ovid, 167
Page (court card), 331

replaced by Princess in Harris-Crowley Tarot, 326
Pan, 235
Pantacle, 61
Papus (Gérard Anaclet Vincent Encausse)
  and astrological/zodiacal trump attributions, 2
  and Egyptian origin theory of Tarot, 2
  and Temperance trump, 196–99, 204
Parergon
  and NCP, 101
  and Schweighardt's Invisible College, pineal gland, 99–101
Pasiphae
  figure of vernal Full Moon during Age of Taurus, 122–23
  hollow cow of symbolic of barque of Isis, 123
Pedjeshes
  Egyptian "stretching of the cord" ceremony, 200, 251, *253*
Pei, 152, 163
Pelydr, 129
  "beam" formation of Druidic "Seven-stone" and "Ark-stone" to represent solstitial and equinoctial Sun, 129
Pentacles (suit), 305, 325, 335
pentagram
  and Devil trump
    and Venus, phoenix cycle, Sirius, Great Year, 240–41
  and Levi's Chariot trump
    and Sirius, inundation, phoenix cycle, 130–32
  and Rosicrucian Lamen, celestial sphere, *190*
  and Serer cosmological octohedron, *189*
pentagram (inverted)
  and Mystic Rose, Death trump, 190
Persephone, 84, 105–6, 150
Perseus Arm
  of Milky Way, 204
Phaeton, 148, 217

INDEX

Philistines, 173–74
phoenix
  "bennu bird", Sirius, 238
  and bennu, nictorax, Sirius, 236
  and Enneagram
    "phoenix cycle", 43
  and Great Seal, 64–65
  and Porta Alchemica, 75
  bennu-bird, dove, Sirius, 65
    and Enneagram, 41
  represents South/fire/Har in Egyptian scheme, 306
Phoenix (constellation)
  and "abyss of source", SCP during Taurean Age, *307*
phoenix cycle, 105, 120
  and Chariot trump, 131
  and Hierophant trump, 120
  and Sirius, Venus, Great Year, Devil trump, 237–41
Phoroneus
  Hellenistic "1st man", 217
pineal gland
  "seat of the soul"
    and Enneagram, 43
  and diencephalon, double-headed eagle, 64
  and higher states of consciousness, 100, 137
  and Schweighardt's Invisible College, 99
Pingala, 82
  and Sun, 100
Pisces
  Age of
    and "Aeon of Isis", 144
    and Anubis, 265
    and Astraea
      Ovid's *Metamorphoses*, 167
      Spenser's *The Faerie Queene*, 168
    and birth of Christ (Sun in Pisces) from Virgin (Full Moon in Virgo), 136
    and Chariot as another form of Emperor, 137
    and Nordic Eddas, 181
    and tarrochi

    The Empress, 103
    and Winter of Great Year, 235
    approx. 360 years remaining in, 231
  and abyss during Age of Taurus
    "relictual" element in Crowley's Death Atu, 195
  and the Enneagram, 46
  and Crowley's "double inversion", 6
  and Crowley's "*kteis+fallos*" equation, 163
  and Death trump, 190
  and heptanomis correspondence, 97
  and Moon trump, 265–66, 268–69
  and Night of Brahma, 230
  and Ptolemaic domicile system, 29–30
  and Schweighardt's Invisible College, 90
  attributed to the Hanged Man, according to Thierens, 2
  hypothetically ruled by Venus/Mars in current Age, 34
  nocturnal "water sign" of inundation during Age of Taurus, 226, 307
  and the Porta Alchemica, 74
Pisces Australis (constellation)
  "abyss of source", 255
Pius, Antoninus, 32
planetary sphere, 29, 82
Plato
  Politicus
    and precession, 233–35
  Timaeus, 148
    and Phaeton, recurring cometary event, 216–17
Pleiades, 122, 202, 256
  and dove, 122
  fiducial stars of Egypto-Babylonian zodiac, 231
Pliny, 215–16, 238
Plutarch, 126, 226
  On Isis and Osiris
    and Typhon as eclipse, 218–20
Pluto

and horoscope of American
Declaration of War, 66
and horoscope of Declaration of
Independence, 66
and Mars, Scorpio, 318
angular and opposite to Moon
engenders fugitive urge to
escape oppression, 336
Judge of the dead
and Mors, Judgment trump,
285
on angle connotes unexpected and
potentially devastating shock,
336
Polaris, 93, 181, 290
and caduceus, 214
and dragon of Medea, 213
Pollux, 127, 272
pomegranate, 84, 105–6
Porta Alchemica, *70*, 73–74, 77
Poseidon
and Amphitrite, 225
and Shemhamphorasch, zodiac,
177
as nocturnal Sun, 121–23
Prakriti, 99, 126, 137, 249, 267
and Purusha
mediation via Solar Force, *137*
precession, 11, 23, 29, 31, 34, 43, 78,
85, 92, 115, 133, 144, 161,
172, 180–81, 192, 194, 205,
232, 238, 243, 265, 275, 285,
298, 301, 342
Ptolemaic (tropical) domicile
system fails to account for, 29
pretense
of occultists according to Massey,
10
Princess (court card)
replaces Page in Harris-Crowley
Tarot, 326
Providence, 287
Pseudo-Apollodorus, 122
Ptah
"put-circle" of
and Enneagram, 44–47
and Porta Alchemica, 72
as NEP, 298
figure of the celestial pole, 301

Ptolemy, 67
and concept of astrological "faces",
"terms", 329–30
perpetuated classical planetary
domicile system, 29
Ptolemaic attribution scheme, 309
Ptolemaic domicile system, 23, 29–
31, 83, 97, 313, 331
Ptolemaic Great Year = 25,920
years, 24
Puanta, 71
Puppis (constellation)
and Hanged Man trump, 178
Puranas, 31, 229
Purusha, 99, 126, 137, 249
and Prakriti
mediation via Solar Force, *137*
put-circle, 44–47, 71
and Enneagram, *47*
and Enneagram, "the nine", and
ankh, 45
pyramid
of Great Seal, 62
similar to "informing triangle"
of Enneagram, 58
Pyramid, Great
and Lyra
Hanged Man trump, 177
Pyrrha, 217
Qoph (Kuf). See *Kuf*
Queen (court card), 70, 326, 331
of Pentacles, 305
of wands, 305
querent
and divination, 303
connection between tarot readings
and astrological
prognostication, 315
Ra, 194, 236
"Light of the World", 150
rabbinic law, 13, 40
Rachel (Biblical)
and Dan, judgment, 317
Ra-Hoor-Kuit, 273
Râhu, 248
Rannut, 252
red
and conscious mind, Tower
trump, 249

INDEX

and Sut, 185
associated with "lower", autumn, winter, 276
color of Empress' *globus cruciger* = autumn, 103
color of lower Sun, nether region, feminine source, 276
color of Nile at beginning of flood season
red tides, 145
dragon flag of Aeddon
standard of Rider-Waite Sun trump, 277
vernal equinox, 277
hippopotamus of Ta-urt, Apt, 145
setting Sun, underworld, 145
whore of Babylon, 145
Regulus
Alpha Leonis
The Empress, 104–6
and Porta Alchemica, alchemy, 73
and Schweighardt's Invisible College, 92
fiducial star of Egypto-Babylonian zodiac, 231
Reish, 152, 267, 277, 279
Remigius
of Auxerre, 285
Rerit
circumpolar genetrix = multi-breasted sow; Dea Multimammae; Diana of Ephesus, 133–34, 205
Revelation, Book of
and Cetus/leviathan, 128
and Great Beast, Sevekh, Aries, 112
and heptanomis, 10
and Scorpio, Death trump, 191
reworked from Egyptian *Book of the Dead*, 193
Rhampsinitus
and Tower trump, 247–49
Rhea, 84–85, *84*
Rolleston, Frances, 282, 318
Roog
Serer supreme deity = Universal Source, 188

Rose
Mystic, 183
of Horizon, 183
of Sharon, 183
of the Lily, 184
Rose-Apple Tree
symbol of precession, 186
Rosicrucian, 81, 88, 183, 190
ru
and *globus cruciger*, 45
Ruach, 278
Ruach Elohim
and Porta Alchemica, 75–76
Sa
Lake of = celestial pole, 213
Sabean
celestial orientation, 80
chronometry, 83, 111, 141
Sagitta (constellation), 207
Sagittarius
Age of
and Judgment trump, 283
and Yugas, 92
anciently depicted with scorpion's tail, 204
and "bow in the sky", Temperance trump, 205
and abyss of current Age, 200
and Enneagram, 41, 43
and Gate of the Gods, 118
and Neptune, 318
and placement of angel's feet in Temperance trump, 207
and rectangular Denderah Zodiac, 256
and *Samech*, 204
and Schweighardt's Invisible College, *95*
and Thaumas, Electra, 202
constellation
and Shu/Tefnut, 127
Sagittarius Arm
of Milky Way, *203*
St. John, 114
St. Peter, 117
Samech, 152, 204
Samson, 141–42, 173
and heliacal rising of Sirius, 173

Sarmoung Brotherhood
  purported esoteric Sufi order, 25
Saturn
  and Culpeper's *Complete Herbal*, 27
  and Enneagram
    masculine "essence type", 25
  and filial cannibalism, Tower of Belus, 247
  and heptanomis correspondence, 97
  and horoscope of Declaration of Independence, 67–68
  and Porta Alchenica, 72
  and Uranus, Capricorn, 318
  ascribed to Fortune in proposed system
    per Crowley's "*kteis+fallos*" equation, 163
  diurnal ruler of Aquarius/Capricorn
    and heptanomis, 26
  ruler of Capricorn in Ptolemaic (tropical) scheme, 29, 31
Scarlet Woman, 120, 146
  and changing of polestar, 144
Schweighardt, Theophilus (Daniel Mögling), 88, 91, 94, 99, 157
Scorpio
  Age of
    and Judgment trump, 283
    an air sign during Age of Taurus, 307
    ancient sign of inundation
      and Golden Age of Leo, 194
    and abyss during Age of Aquarius, 195
    and abyss of current Age, 200
    and Death trump, 190
    and eagle, Tribe of Dan, 318
    and Enneagram, 40, 43
      "centers", 325
      "put-circle" of Ptah, 46
    and Gate of the Gods, 107, 118
    and heptanomis correspondence, 97
    and Khunsu/Harpocrates, Hercules, 273–74
  and placement of angel's feet in Temperance trump, 207
  and Pluto, 318
  and Schweighardt's Invisible College, *95*, 99
  attributed to Temperance trump by Papus, 196
  constellation
    anciently joined with Libra, 204
  eagle
    beast of Apocalypse, 192
    eagle of Rider-Waite Fortune trump, 155
    eagle of The World trump, 289
    Moon debilitated in
      and Pasiphae legend, 122–23
    one of 4 beasts of Apocalypse, 192
Sebek-Horus
  associated with Ursa Minor, 127
Secret Chiefs, 7
Secret Church, 88
Sefer Yetzirah, 4–5, 13, 23, 173, 204
  Short, Long, and Gra variations, 7
sefirot
  and Tree of Life, 17
    Enneagram, *39*
Serer
  cosmological octahedron
    and pentagram, *189*
  cosmological octohedron, *188*, *189*
Serk (Serk-t)
  Egyptian goddess associated with Scorpio, 122
Serpenes (constellation)
  and Sirius Supercluster/Ursa Major Stream, 261
serpent
  "lion-serpent" of Crowley's Lust trump = Leo/Hydra, 145
  7-headed
    and Hydra, Ursa Minor, 145
  and celestial axis
    Lovers trump, 124
  and Scorpio, 122
  and standard of Dan = symbol of Scorpio, 318
  and temptation of Lilith
    precessional interpretation, 93, 96, 99

INDEX

as Sut pursuing Ta-urt after her
desertion, 126
of Rider-Waite Fortune trump
and Typhon, Ursa Minor, 156
Seshat, 251, *252*
and Pedjeshes, 200, 224
Seshet. *See* Seshat
Set, Seth. *See* Sut
Sevekh, 112, 145, 224, 247, 308
represents West/earth/crocodile in
Egyptian scheme, 306
Sevekh (Sebek/Sobek)
Egyptian crocodile-headed god
associated with Draco, Aries,
112
Sevekh-Kronus, 146
corresponds to Ursa Minor, 114
seven
"Seven above Twelve"
and Sefer Yetzirah, heptanomis,
23
astronomes of heptanomis, 11
gems of Chinese Empress' sacred
cap (possibly connected to
heptanomis), 45
Seven-stone (Druidic), 129
Shamgar
connected Kabbalistically with
Lamed, 173
Shekinah, 93
shemhamphorasch
72 spirits contained within
Solomon's bronze vessel, 22
Shin, 75–76, 152, 277–78, 286, 295–96, 301, 317
Shiva, 77, 226, 228–29
Shu, 58, 127, 206, 294
twin of Tefnut, 127–28
Shu-Anhar, 269
sidereal
astrology
"angles", *304*, *305*, *306*
in contradistinction to tropical
astrology, 29–31
Egypto-Babylonian zodiac, 231
science
invented by Belus, 245
significator

connection between tarot readings
and astrological
prognostication, 315
Silver Gate (Gate of Humanity), 107,
118, 138, 191, 203
and Hierophant trump, 117–19
Sirius
and 8-pointed star, 261
and Enneagram, *42*, *44*
"centers", 325
"phoenix cycle", 43
and Gate of Humanity, 139
and Great Seal, 58, 65–66
and Gurdjieff's "burying the dog
deeper", 40
and horoscope of Declaration of
Independence, 68
and lily-lotus, 185
and marriage of Bride and Lamb,
135
and NEP, 275
and pentagram, *131*
and phoenix cycle, 131
and Porta Alchemica, 75
and Scarlet Woman, 120
and Sirius Supercluster/Ursa
Major Stream, 261
and Sothic cycle, 32
and stellar cult of Sut-Typhon, 112
birthed from Orion Nebula, 203
celestial influence of, 38, 115
heliacal rising of
and Amphitrite, 226, *228*
and Isis, 218
and phoenix cycle, 235–41
and Samson, 173
personified by Belus, 245
Sitchin, Zechariah
and Anunnaki, Nibiru, asteroids,
A∴A∴ sigil, 147–48
Siva. *See* Shiva
skepticism
and subjective nature of
divination, 304
Smith, Pamela Colman
illustrator of "Rider-Waite" Tarot,
vii
Solar Force

384

and mediation of Prakriti,
  Purusha, *137*
Solinus, 238
Solomon, 71, 87
  bronze vessel of, *22*
  temple of was celestial, 136
Solon, 217
solstice
  sign of
    shifts according to precession,
      32
  summer
    anciently occurred when Sun in
      Scorpio, 194
    and Diana of Ephesus, 134
    and Galactic Center, precession,
      35, 37, 126
    and rectangular Denderah
      Zodiac, *258*
    and Regulus
      Age of Taurus, 92
    and Solomon's Temple, 88
    and Sun conjunct Sirius
      949 AD, Monastic
        Christianity, 65
    and triumph of Sun, Chariot
      trump, 129
    and Zeus as regulus, 244
    height of Egyptian flood season,
      120
    Sun in Leo during Age of
      Taurus, 31, 46, 174, 185, 194
  tree of
    and Sola-Busca Ipeo trump, *209*
  winter
    and abyss, 195
      Sut, 282
    and Galactic Center, precession,
      37, 283
    and Solomon's Temple, 88
sophistry
  of tropical astrology per Fagan, 31
Sothic
  cycle, *33*
    and Akhenaten, Nefertiti, 138
    and Amphitrite, 241
    and marriage of Bride and
      Lamb, Jehoshua Ben-
      Pandira, 135–36

year
  and Khepra the "closer",
    "clasper", 268
Sothis. *See* Sirius
soul
  and Enneagram
    dove, 326
  coaxed by Spirit of God to
    incarnate
    and *Kuf*, 267
  conscious of itself
    and Fool trump
      Mem, 299
  descent of to Earth
    and *Chet*, 138
  devoured by Ammut demon if
    found wanting, 221
  entering body
    and Fool trump
      closed *Mem*, 299
  enters world through Gate of
    Humanity
    Taurus, Orion, Sirius, 107
  eternally returning
    and Fool trump, 297
  refinement of
    and *Tav*, 287
Sourya, 231
Southern Crown (constellation), 274
Southern Hemisphere
  and Ptolemaic domicile system,
    36–38
space in six directions
  Hebraic = שש קצוות ("six
    corners"), 31
Sphinx, Great
  and Atlantis, 143
  and Golden Age, 83, 142
Spica, 200, 224, 251
  fiducial star of Egypto-Babylonian
    zodiac, 231
spinal column
  and moving/instinctive centers, 82,
    323
  and Sushumna Nadi, Kundalini,
    Azoth, 100
Spiritus Sanctus, 92
Sravana

INDEX

23rd lunar mansion and Vishnu, 226
Star (trump), 260–62
   and Crowley's "double inversion", 7, 111
Star of Zion, 61, 73
Star Regulus
   and antimony, Porta Alchemica, 73
Strength (trump)
   and *innocentia inviolata*, 143
   completion of Death trump per Crowley, 193
   inversion with Justice, 140, 153, 173
   Rider-Waite version, *140*
   Visconti-Sforza version, *141*
Sturluson, Snorri, 180
sub rosa
   activities of the Rosicrucians vis-à-vis the Great Seal, 62
subconscious
   atavistic impulses
      and *Kuf*, Moon trump, 267
   mind, emotions
      and Prakriti, 249–50
   possibly involved in tarot divination, 304
summer
   of Great Year
      and Cancer, Leo, 34
   ruled by Horus, 127
Sun
   "once rose where it now sets"
      precessional explanation, 234
   ½ of ancient biune deity (Sun + Moon), 32
   and 8-pointed star, 261
   and Culpeper's *Complete Herbal*, 27
   and Emerald Tablet, 99–100
   and Hermit trump, 150
   and horoscope of American Declaration of War, 66
   and horoscope of Declaration of Independence, 66
   and Kaer-Bediwyd, 92
   and Lilly's "face" system, 329
   and lily, 184
   and Mystic Rose, 183
   and Porta Alchemica, 72
   and Schweighardt's Invisible College, 95, 99
   and Shemhamphorasch, zodiac, 177
   and Thierens' attribution scheme, 2
   as astrological "lens", 35–36, 126
   as Osiris, 43
   birthed from Orion Nebula, 203
   conjunction with Moon
      metaphor for balancing rational and intuitive minds, 266
   diurnal ruler of Leo, 29
   of the abyss
      Neptune, Poseidon, Amphitrite, 227
   ruler of Leo during Age of Taurus, 31, 174, 185
   triumph of at summer solstice
      and Chariot trump, 129
supernova
   hypothetical origin of Nibiru, 148
   references in Schweighardt's Invisible College, 90
Supreme Spirit, 299
Sushumna, 82
   and spinal column, 100
Sut
   and Great Beast, 145
   and Sun trump, 271
   and Visconti-Sforza World trump, 294
   battles with Horus during eclipse, 224
   betrays Osiris, 281
   black/red god, 185
   desertion by Ta-urt, 127
   male hippopotamus, 208
   red crown of, 276
   star of = Polaris, 290
   swallows Osiris during eclipse, 223
   twin of Horus, 127–28
   unicorn of "Lion and Unicorn", 96
Sut-Anup

and Ursa Minor, Sut-An, Satan, 157
Sut-Typhon
  cult of
    and Draco, Ursa Minor, 112
    and leprosy, 243
swan
  Cygnus, 90, 177, 245, 281
Swords (suit), 306, 325
taboo-time
  connected with New Moon and menstruation in anciet Egypt, 264
Taht. *See* Thoth
Taht-Aan
  "Light of the World"
    as lunar representative of Sun, 220
Talus
  and Minos, 168
  and Spenser's *The Faerie Queene*, 168
  and Spenser's *The Faerie Queene*, Taurus, 172
Tao-Teh-King, 299
Tarot
  spread
    22-card, 334
Tarot
  "Ta-Rosh" = "pictures of Mercury", 1
  and 18th century French occult revival, 1
  and Hebraic alphabetic sequence, 7
  Antoine Court de Gébelin's origin theory, 1
  celestial rhythms, and, 1
  current form originated in Medieval Ages, 318
  Egyptian origin theory, 1
  foremost consideration is determining the "meaning behind meaning" per Waite, 196
  no authoritative symbolic canon of, vii
  of Marseille (*Tarot de Marseille*), 102, 270
  origin said to be connected to humanity's development of writing, 1
  origin theories unproven, 1
  possible Indian origin of, 1
  prehistoric humanity, and, 1
  Rider-Waite, 274
  similarity to ancient Mayan codices, 1
  Sola-Busca, 208, 291, 295
  spread
    "gypsy", 331
    10-card/Celtic cross, 331
    22-card, 331
    7th-card, 331
    horseshoe, 331
    royal, 331
  suit/direction/sign correspondence, 308
  Thoth's original paintings of the gods = "Ta-Rosh," or "pictures of Mercury", 1
  Yetzirac zodiacal attribution sequence, 5
tarrochi
  created circa 1300-1400 AD, 103
Tartarus, 193
Tatius, Achilles, 204
tau, 45, 76, 310
  and *globus cruciger*, 45
Ta-urt, 93, 126, 145, 205, 252–58, 261, 274
  and Baphomet, Holy Spirit, Great Mother, Templaric Mete, 210–13
  and Revelation, Whore of Babylon, 112–14
Taurus
  Age of
    and "Aeon of Osiris", 144
    and "temptation of Lilith", 93
    and Anubis, 266
    and Chariot as another form of Emperor, 137
    and locus of NCP and Sevekh-Kronus, 224
    and *meskhen*, 200
    and Phoenix constellation, 308

and Ptolemaic domicile system, 29
and special significance of Spica, 251
Autumnal Equinox of Great Year, 291
origin of Ptolemaic domicile system, 31
Scorpio considered an air sign during, 307
Sun in Leo during summer solstice, 185
and Enneagram, 40, 43
"centers", 325
"put-circle" of Ptah, 46
and heptanomis correspondence, 97
and Minos, 122
and Spenser's *The Faerie Queene*, Talus, 172
and suit of Pentacles, 305
bull of Rider-Waite Fortune trump, 155
bull of The World trump, 289
constellation
    and Gate of Humanity, 118, 139
    one of 4 beasts of Apocalypse, 192
    rose at sunset in October during Taurean Age – time of mating cattle and plowing, 308
Tav, 152, 286, 317
    said to embody secret of reincarnation, 287
Tawaret. *See* Ta-urt
tefilin, 295
Tefnut
    twin of Shu, 127–28
Temperance (trump), 196–97, 204
    relationship to Judgment trump, 285
Tet, 152
Tetragrammaton
    and Fortune trump, 155, 157, *159*
Thanatos (Roman Mors)
    minister to Pluto/Hades
    associated with Mars, 285
Thaumas, 202

Thierens, Adolph Ernestus
    and Egyptian origin theory of Tarot, attribution scheme, 2–5
Thoth
    and Moon, 220
    and origin of Tarot, 1
    ibis-headed scribe, 220
Thuban
    and "dragon of eclipse", "dragon of pole", 225
    and caduceus, 214
    and dragon of Medea, 213
    and Empress trump, 107
    and Gere, Freke, 181
    and Pedjeshes, 251
Thummim, 88
Tiamat, 274
    a form of Great Mother, 252
    interpreted by Sitchin as planet, 147
Tiferet, 19
Tigris River, 44
tortoise, 97
    "mount of tortoise" = Lyra, 11, 24
Totten, Charles Adiel Lewis, 63
Tower (trump)
    and Belus, 245
    and Crowley's "*kteis+fallos*" equation, 163
    as force that turns Wheel of Fortune, 249
Tree of Life, *14*
    and 9 Nordic Worlds, *18*
    and Enneagram, *39*, 44
    and Hebraic "cube of space", *20*
    and Lovers trump, 124
    and NEP, 342
    Etz Chaim, 14
    functionaries of
        and celestial sphere, 17
Triton
    as Capricorn; Mother = Amphitrite; Father = Poseidon, 225
tropical
    astrology
        erroneous per Fagan, 31

in contradistinction to sidereal
astrology, 29
zodiac
and Astraea as 6th house, 171
Tuat, 186, 236
Tum, 85, 285
Two Truths, 184
and "double ending" of Great
Year, 240
fire/water, summer/winter, 238
types
Crowley's notion of, 326
Enneagram essence types
masculine and feminine, 25
planetary, 27, *28*
Typhon
and Rider-Waite Fortune trump,
157
and Scorpio, Osiris, 122, 193
as eclipse, 218–25
North/water/hippopotamus in
Egyptian scheme, 306
typhoon
and Typhon/Sut (Set, Seth), 194
Tzadik, 4, 6–7, 111, 152, 277
Uat
goddess of North
and Neith, Iris, Seshat, Kheft,
200–201
underworld
and abyss during Age of Taurus,
46
and barque of Isis, cow of
Pasiphae, 123
and color red, abyss, Osiris Tesh-
Tesh, 145
and Gate of Fair Entrance, 185
and Gate of Fair Exit, 185
and Hebraic wandering in
wilderness, 300
and Judgment trump
Atum, Deneb, Cygnus, 282
and Persephone, Empress trump,
106
and Porta Alchemica
Levi's transmutational sigil, 76
Milky Way, 74
Neptune, abyss, 75

and Serer cosmological
octohedron, 187
Nordic, 16
unicorn
and Sut, hippopotamus,
Monoceros, Age of Cancer, 96–
98
uninitiated
supposedly being protected by
occult "blind", 9
Universal Medicine
union of Mystic Rose and Cross,
183
Uranus, 72, 74
and Saturn, Capricorn, 318
and Thierens' attribution scheme,
2
Urim, 88
urn
symbol of inundation and abyss –
connected to Crater and
Aquarius in Age of Taurus, 198
Urnas
ancient Egyptian Uranus = Milky
Way, 74
Ursa Major
and Age of Aries
Great Mother's desertion of
Sut, 127
and Age of Taurus
Great Mother, 46
Ishtar, 107
Rhea, 85
and *Alhim, Elohim*, 149
and crocodile zootype, 96
and hermanubis of Rider-Waite
Fortune trump, 156
and Lilith, Ta-urt, 93
and Seshat, 200
Pedjeshes, 251
and stellar cult of Sut-Typhon, 112
and Ta-urt, whore of Babylon, 145
Ursa Major Stream
Sirius Supercluster
and Star trump, 261
Ursa Minor
anciently known as "wings of the
dragon", 181
and "mount of jackal", 24, 96

INDEX

and Age of Aries
   Sevekh, 112
and circular Denderah Zodiac
   jackal, NCP, 213
and dragon of Medea, 214
and Durga, 229
and Revelation, 112–16
   7-headed dragon, 112
   Sevekh-Kronus, 114
and serpent of Rider-Waite
   Fortune trump, 156
and Sevekh, Great Beast, 145
and Sevekh-Kronus, 224

Vatican
   coat of arms, 117

Vav, 134, 279
   and Tetragrammaton, 152

Vega
   and Golden Age, 66
   and Hanged Man trump, Great Pyramid, 177
   and Hermit trump, 151
   and Tower of Belus, 245

Vela (constellation)
   and Hanged Man trump, 178

Venus
   and 8-year conjunction cycle with Earth, Sun, 147
   and Amphitrite, 226–28
   and Culpeper's *Complete Herbal*, 27
   and Empress trump
      globus cruciger, 102
   and Enneagram
      "phoenix cycle", 43
      feminine "essence type", 25
   and Great Seal
      phoenix cycle, 65
   and heptanomis correspondence, 97
   and horoscope of Declaration of Independence, 66
   and phoenix cycle, 235–41
   and Porta Alchenica, 72
   nocturnal ruler of Taurus/Libra and heptanomis, *26*
   spark of the ship of Bennu-Asar, 236

Venus-Urania
   and Empress trump, 108

vernal point
   and Enneagram, 43
   shifts precessionally with polestar, 24

vernal point (sidereal)
   currently at approx. 5° Pisces, 30, 231

vernal point (tropical)
   fixed in Aries, 30

Virgin, 134, 136, 150, 171

Virgo
   a diurnal "water sign" of inundation
      during Age of Gemini, 128
      during Age of Taurus, 76, 91, 131, 206, 226
         and Porta Alchemica, 73
   Age of
      and Galactic Center, 126
      and Golden Age, 34, 37, 66
      and Mercury/Odin, 176
      and Revelation, 115
   and "Aeon of Isis", 144
   and Crowley's "double inversion", 6
   and Empress trump, *104*, 107
      Locri Persephone Pinax, *106*
   and heptanomis correspondence, 97
   and Hermit trump
      Halloween, "crust of Hades", Persephone, 150–51
   and Isis, 106
   and Sothic cycle during Age of Taurus, 32
   and Spenser's *The Faerie Queene*, Astraea, 171–72
   and The Empress, 103
   Full Moon in and Virgin Birth, 136

Vishnu, 1, 45, 226
   as representative of NEP
      various Avatars correlate to precessional stations of NCP, 108

Waite, Arthur Edward

and Egyptian origin theory of
    Tarot, 2
and Fool trump
    "0=1" formula, 8
and Hanged Man trump, *Mem*,
    278
and Magician as form of Apollo,
    83
comparison of Fagan's houses with
    Waite's denaries, 311–12
no canon of authority regarding
    Tarot symbolism, vii
Wandering Jew, 133, 298
Wands (suit), 305, 319, 325, 337
Water (element), 161
    North/hippopotamus/Typhon,
    306
Webb, James, 39
white
    "city of the white wall" =
        circumpolar region, 276
    and Horus, 185
    color of "upper", 276
Wind (element)
    and East/ape/Hapi, 306
winter
    of the Great Year, 235, 283
    ruled by Sut, 127
Yggdrasil, 14, *15*, 16, 175
Yi King, 31
yoni, 129, 137, 163
Yud, 152, 154, 279
    and Tetragrammaton, 152
Yugas
    and Great Year, Galactic Center,
    36, *93*
Zayin, 126, 134, 151–54, 267
zenith, 29, 46, 308
Zeus, 193, 244
Zeus Belus, 245
zodiac
    and Enneagram, 40, 43
    and Porta Alchemica, 71
    and put-circle of Ptah, 45
    Crowley's "double loop", 7
    of Denderah
        rectangular, 127
Zohar, 13, 149

www.ingramcontent.com/pod-product-compliance
Lightning Source LLC
Chambersburg PA
CBHW070409100426
42812CB00005B/1683